Implementation of Enzymatic Processes and Lactic Bacteria in the Food Industries

SCIENCES

Agronomy and Food Science,
Field Directors – Jack Legrand and Gilles Trystram

Food Bioprocesses, Subject Head – Mohamed Ghoul

Implementation of Enzymatic Processes and Lactic Bacteria in the Food Industries

Coordinated by
Mohamed Ghoul

iSTE

WILEY

First published 2024 in Great Britain and the United States by ISTE Ltd and John Wiley & Sons, Inc.

ISTE Ltd
27-37 St George's Road
London SW19 4EU
UK

www.iste.co.uk

John Wiley & Sons, Inc.
111 River Street
Hoboken, NJ 07030
USA

www.wiley.com

Library of Congress Control Number: 2023952359

British Library Cataloguing-in-Publication Data
A CIP record for this book is available from the British Library
ISBN 978-1-78945-137-5

ERC code:
LS9 Applied Life Sciences, Biotechnology, and Molecular and Biosystems Engineering
LS9_5 Food sciences (including food technology, food safety, nutrition)

Contents

Preface

Mohamed GHOUL
LRGP, CNRS, Université de Lorraine, Nancy, France

To meet growing consumer demand for healthy, natural, additive-free products with low environmental impact, plant-sourced raw materials have emerged as a relevant alternative to derivatives from the petrochemical industry. However, most of these raw materials are complex and difficult to digest as they are. They therefore require numerous physicochemical and mechanical processing, as well as the implementation of numerous processing reactions.

The use of enzymes and micro-organisms as catalysts to hydrolyze, condense and/or fractionate these raw materials is well adapted. Enzymatic processes enable processes to be implemented under mild conditions and targeted molecules to be produced. Similarly, given the selectivity and specificity of these catalysts, separation and/or purification operations downstream of the production stages are simpler and less costly.

This book describes the latest developments in the use of biocatalysts in the food industry.

Part 1 is dedicated to a presentation of current enzyme processes in the food industry (general characteristics of enzymes, classifications and operating modes of the main enzymes used in the food industry, enzyme production for food applications), ways of improving enzyme processes (enzyme engineering, process engineering) and the main enzyme processes in the food industry (bakery and confectionery, malting and brewing, starch and its derivatives, milk, fats, meat and seafood products, flavors and additives, fruit juices and wines).

Part 2 begins by describing a family of micro-organisms that are widely used in food processing and in the formulation of healthy foods. These are lactic acid bacteria and bifidobacteria (taxonomic concepts and definition of "probiotic", probiotic market and health claims, production of probiotics and metabiotics, industrial applications, development and innovation prospects, impact of lactobacilli on digestive health, etc.). Secondly, we describe the processes used to encapsulate lactic acid bacteria (atomization, extrusion, emulsification, coating), the most commonly used encapsulation matrices and the process by which bacteria adhere to these matrices. Finally, we present the best-known fermented foods and the new approaches under development in this field.

January 2024

Introduction

Mohamed GHOUL
LRGP, CNRS, Université de Lorraine, Nancy, France

The raw materials used in the various production processes for foodstuffs and non-food active ingredients have a wide variety of structures (cellulose, hemicellulose, lignin, proteins, phenolic compounds, fats, etc.). This structural complexity means that some of these materials cannot be directly assimilated by the consumer, and do not have the desired biological activities and functionalities (nutrition, health, nutraceuticals, etc.).

Digestion and transfer of the active ingredients of these raw materials through cell walls, and/or their functionalization, therefore require one or more hydrolysis and rearrangement stages, which involve several types of reactions (synthesis, hydrolysis) that are catalyzed in vitro or in vivo by complex enzyme systems. These enzyme systems can either be made up of enzymes or enzyme cocktails extracted from plant, bacterial or animal matrices through the application of appropriate unit operations, or through the use of whole cells (bacteria, yeast, fungi, etc.).

To ensure that these catalytic systems (extracted enzymes or whole cells) operate under optimum conditions, a number of factors (pH, temperature, concentration, redox potential, oxygen level, etc.) need to be carefully adjusted. Similarly, in order to preserve and maintain the activity of certain catalytic systems over a long period of time, it is necessary to carry out a protection step, such as encapsulation or immobilization. These protective operations are carried out by applying the appropriate matrices, enabling adequate transfer of substrates between the macro- and micro-environment of the catalytic system, thus maintaining the required catalytic activities and functionalities.

Various sources and classes of enzymes and bacteria exist naturally, and are already used effectively in many food and non-food industries. However, to enhance the catalytic activities of these enzymes and micro-organisms, a great deal of work is regularly devoted to discovering new bacteria and enzymes, as well as improving their catalytic activities. This work either uses classical selection, cloning and mutation tools, or tools involving rational design through the application of various molecular modeling approaches.

Part 1 of this book is devoted to a presentation of (i) the different classes of industrial enzymes; (ii) enzyme production techniques; (iii) current and future avenues for enzyme improvement; (iv) techniques for intensifying enzymatic processes; and (v) a description of the main industries using industrial enzymes.

Part 2 of this book is dedicated to the description of a group of bacteria that are widely used in food processing, namely, lactic acid bacteria. A general presentation of these bacteria is given, followed by a more detailed discussion of the bifido-bacteria subgroup.

In order to maintain and prolong their activity, these bacteria can be encapsulated using a variety of techniques. The encapsulation techniques best suited to lactic acid bacteria that are used on an industrial scale are described, as well as bacteria–matrix interactions and their role in retaining bacteria in microparticles, during storage and gastrointestinal transit. Lastly, the most common industrial applications involving lactic acid bacteria are presented.

PART 1

Enzymatic Processes in the Food Industry

Introduction to Part 1

Isabelle CHEVALOT
LRGP, CNRS, Université de Lorraine, Nancy, France

Enzymes are biological catalysts that regulate the metabolic pathways of all living systems. They are globular proteins that act as catalysts, accelerating the speed of a reaction by lowering the energy of activation. Unlike chemical catalysts, enzymes are of natural origin, operate under mild conditions of temperature and pressure, exhibit high specificity and are biodegradable (Subin et al. 2015). Due to these main characteristics, enzymes have commercial applications in a variety of industries. They have been used for millennia in food processing, such as bread making, brewing, cheese and wine making. However, the large-scale introduction of enzymes into manufacturing processes in the 1970s paved the way for the development of numerous food processing and additive productions (Mishra et al. 2017). The industrial enzyme market was estimated at 5.9 billion USD in 2020, and is expected to reach 8.7 billion USD by 2026, with a growth rate of 6.5% in terms of value[1]. Growing environmental concerns and advances in enzyme engineering and green chemistry have boosted their industrial use. Technological innovations that reduce the consumption of chemical products are contributing to market growth. Industrial enzymes are widely used in the food and beverage industry, but also in pharmaceuticals, textiles, animal feed, biofuels, cosmetics, detergents and paper. Overall, human and animal nutrition applications account for between 55% and 60% of the global enzyme market, with an estimated annual growth between 6% and 8%. Half of the enzyme market (around 50%) concerns the category of enzymes catalyzing carbohydrate hydrolysis (mainly amylases and cellulases/hemicellulases). Proteolytic enzymes account for 25–30% of the global

1 MarketsandMarkets.com.

Implementation of Enzymatic Processes and Lactic Bacteria in the Food Industries,
coordinated by Mohamed GHOUL.

enzyme market, while lipases represent 10% of the global enzyme market (Guerrand 2017).

In the food industry, enzymes are mainly used as additives in foods, or as processing aids in the production of foods and beverages, to improve texture, flavor and quality, as well as to extend shelf life. Enzymes also have applications as analytical tools in the food industry. However, enzyme activity and stability are influenced by operating conditions, such as pH, temperature, substrate concentration and the presence of metal ions. Enzymes require pH and temperature conditions to maintain optimum activity and stability, which may not entirely correspond to industrial process conditions. Furthermore, the reactions and substrates of interest to industry may differ from the enzyme's natural reactions and substrates. In order to overcome these difficulties and offer high-performance catalysts on a large scale, several strategies are being considered, based either on the process itself or on the biocatalyst.

Recent developments in enzyme biotechnology, nanobiotechnology, metagenomics and proteomics have contributed to the design and development of tailor-made enzymes, with new properties and catalytic activities. There have also been numerous innovations in enzyme immobilization media. Enzymes can be used in free or immobilized form, for example, bound to an inert support, so as to enable repeated and/or continuous use of the enzyme, and also to increase their stability. Implementing a process based on immobilized enzymes on an industrial scale requires careful assessment; indeed, in addition to potential technical problems, such as loss of activity during immobilization and mass transfer limitations, it is necessary to consider the economic impact of using immobilized enzymes, including the cost of immobilization, the cost of the immobilization media and the chemical compounds used for the immobilization procedure (Sheldon and van Pelt 2013). More recently, processes have been developed, involving microwave and/or ultrasonic technologies, to increase catalytic efficiency. This intensification of biocatalytic processes is also conceived through the design of new micro-structured reactors.

Recent advances in biocatalysts and processes have played an important role in the discovery and development of new enzymes, as well as in the implementation of new processes for food and beverage processing, and co-product valuation.

Today, the integration of enzymes in food processing is well known, and research dedicated to improving enzymes and optimizing the processes that use them is still in progress. The aim of Part 1 of this book is to provide an up-to-date review of the use of enzymes in food processes, their production, enzyme engineering methods for their improvement and biocatalytic process intensification technologies.

1

General Characteristics of Enzymes

Isabelle CHEVALOT[1], Mohamed GHOUL[1] and Seraphim PAPANIKOLAOU[2]

[1] *LRGP, CNRS, Université de Lorraine, Nancy, France*

[2] *LFMB, Agricultural University of Athens, Greece*

Enzymes are proteins that speed up chemical reactions. This acceleration process is called catalysis, and enzymes act as catalysts for chemical reactions. In enzymatic reactions, the molecules present at the start of the reaction are called “substrates”. Enzymes transform the substrates into different molecules called “products”. As with all catalysts, enzymes are not consumed by the reactions they catalyze and do not alter the equilibrium of these reactions. However, enzymes differ from most other catalysts by being much more specific. Numerous works detail the structure, functions, mechanisms, kinetics and technologies of enzyme catalysts (Pelmont 1997; Buchholz et al. 2005; Cornish-Bowden et al. 2005; Illanes 2008).

1.1. Notion of catalysis

Two factors are specifically involved in the rate laws for a chemical transformation: the pre-exponential factor (A) and the activation energy (Ea). The classical expression for the rate constant is:

$$k = Ae^{(-Ea/RT)} \qquad [1.1]$$

Implementation of Enzymatic Processes and Lactic Bacteria in the Food Industries, coordinated by Mohamed GHOUL.

This constant is only valid for a given system. From a set of reagents, through an appropriate reaction pathway, the system evolves toward the formation of products. *Ea* corresponds to the height of the potential barrier (Figure 1.1).

Unlike thermodynamics, which does not consider the path followed (the principle of initial and final states), chemical kinetics is largely dependent on it. By modifying this path, the potential barrier is altered. If *Ea* increases, the reaction-rate constant decreases and vice versa.

The principle of enzymatic catalysis is related to a reduction in the energy required for the reaction (Figure 1.1). The activation energy corresponds to the amount of energy absorbed by the substrate molecules required to move to an unstable transition or activated state, in which bonds are more fragile and easier to break. The transition state is located at the top of the energy barrier.

The enzyme reduces the activation energy by creating an environment in which the transition state is stabilized.

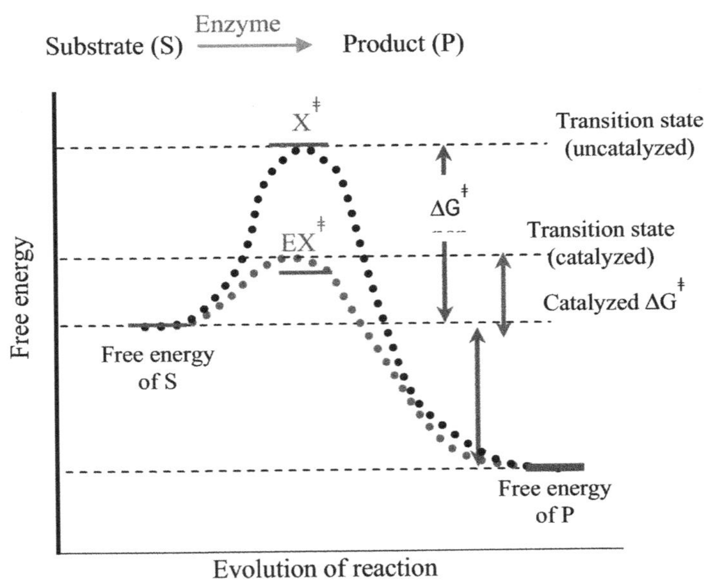

Figure 1.1. *Schematic representation of the role of enzymes as biological catalysts, which act by lowering the activation energy of reactions, enabling the formation of reaction intermediates with lower activation energy. For a color version of this figure, see www.iste.co.uk/ghoul/implementation.zip*

1.2. Notion of specificity

One of the most remarkable properties of enzymes is their specificity. Some enzymes have a very strict specificity, only catalyzing one particular reaction. Other enzymes will be specific to a particular type of chemical bond or functional group.

This specificity is directly related to the structure of enzymatic proteins. Indeed, native proteins fold into a unique functional tertiary conformation that confers their biological activity, which may be that of a catalyst. This overall structure of the macromolecule enables a particular region to adopt a spatial structure, recognized by the protein's specific ligand. In the case of enzymes, this particular region is known as the active site. Within the active site, a distinction is made between amino acids that constitute the binding site (these amino acids have no chemical functions involved in the reaction) and amino acids that constitute the catalytic site. The active site is made up of a small number of amino acids which, in most cases, are not contiguous in the polypeptide chain. These amino acids are characterized by side chains, whose chemical nature (ionizable groups) and structure (steric hindrance) are specifically adapted to substrate recognition. Enzyme specificity is determined by the complementary form, charge, hydrophilic/hydrophobic nature of the substrates and their three-dimensional structure.

Three types of specificity can be distinguished as follows:

– *Chemo-specificity*: it includes enzymes that are specific to certain functional groups, such as the amino, phosphate or methyl groups.

– *Regio-specificity*: it includes enzymes that are specific to a functional group, depending on its position within the structure of the molecule.

– *Stereo-specificity*: it includes stereospecific enzymes that only act on a steric or optical isomer and not on their homologous isomers.

1.3. Nomenclature

The International Commission on Enzymes, founded in 1955, has established an enzyme nomenclature system based on the type of reaction each enzyme catalyzes. Under this system, enzymes and coenzymes, their units of activity, standard assay methods and the symbols used to designate reaction kinetics are all grouped together in a single system. All enzymes are assigned a four-digit number according to the class, subclass and sub-subclass in which they have been classified. Each enzyme has been assigned a code number, consisting of four digits separated by dots. The first digit indicates the main class to which the enzymes belong:

– *Oxidoreductases* catalyze redox reactions in which hydrogen or oxygen atoms, or electrons are transferred between molecules. This extended class includes dehydrogenases (hydride transfer), oxidases (electron transfer to molecular oxygen), oxygenases (oxygen transfer from molecular oxygen) and peroxidases (electron transfer to peroxide). The second digit of the code indicates the reducing equivalent donor involved in the reaction. For example, glucose oxidase (EC 1.1.3.4 or β-D-glucose: oxygen 1-oxydoreductase), laccase (EC 1.14.18.1) and lipoxygenase (EC 1.13.11.12) are major enzymes of this class used in the food industry.

– *Transferases* catalyze the transfer of an atom, or group of atoms, between two molecules, apart from certain enzymes included in other groups (such as oxidoreductases and hydrolases). The International Commission recommends that the names of transferases end with X-transferase, where X is the transferring group. For example, glucanotransferase (EC 2.4.1.19) is used for the modification of starch into cyclodextrins. The second digit describes the transferred group.

– *Hydrolases* include enzymes catalyzing the hydrolytic cleavage of bonds, such as C – O, C – N, C – C, and a few others. They are classified according to the type of hydrolyzed bond. They are currently the most common class of enzymes used in enzyme technologies. In the food industry, the majority of enzymes classified as such include α-amylases (EC 3.2.1.1), β-amylases (EC 3.2.1.2), lactases (EC 3.2.1.23), lipases (EC 3.1.1.3) and proteases, including aminopeptidase (EC 3.4.11), trypsin (EC 3.4.21.4), subtilisin (EC 3.4.21.62), papain (EC 3.4.22.2), ficin (EC 3.4.22.3), pepsin (EC 3.4.23.1) and chymosin (EC 3.4.23.4).

– *Lyases* are involved in the non-hydrolytic elimination of certain groups by breaking various chemical bonds. This elimination often generates double bonds. The second digit in the classification indicates the broken bond. These include aldolases, decarboxylases and dehydratases. For example, acetolactate decarboxylase (EC 4.1.1.5) is used in the beer industry.

– *Isomerases* are enzymes that can catalyze various molecular isomerization reactions. These enzymes catalyze geometric or structural changes within a molecule. Depending on the type of isomerism involved, they may also be referred to as racemases, epimerases, cis-trans isomerases, isomerases, tautomerases, mutases or cycloisomerases, such as phosphoglucose isomerase (EC 5.3.1.9).

– *Ligases*, also known as synthetases, catalyze the synthesis of new bonds between two molecules. They link molecules by covalent bonds in biosynthetic reactions. These reactions require energy through hydrolysis of an ATP diphosphate bond or a similar triphosphate: this property justifies the difficulty of their application on an industrial scale. The second digit indicates the type of bond synthesized.

The second digit designates the enzyme sub-class, defined according to its mechanism of action. The third digit designates the nature of the molecule acting as acceptor in the case of electron transfer. The fourth digit is a sequence number within the group and sub-group. When an enzyme ends in 99, it is incompletely characterized.

Table 1.1 shows the six classes, the type of reaction catalyzed by the enzyme and an example for each class.

No.	Classes	Reaction	Examples
1	Oxidoreductases	Oxidation-reduction reactions	Glucose oxidase (EC 1.1.3.4)
2	Transferases	Functional group transfer	Hexokinase (EC 2.7.1.2)
3	Hydrolases	Hydrolysis reactions	Carboxypeptidase A (EC 3.4.17.1)
4	Lyases	Addition of double bonds	Pyruvate decarboxylase (EC 4.1.1.1)
5	Isomerases	Isomerization reactions	Malate isomerase (EC 5.2.1.1)
6	Ligases	Bond formation (C–C, C–S, C–O and C–N) with ATP cleavage	Pyruvate carboxylase (EC 6.4.1.1)

Table 1.1. *Enzyme nomenclature*

1.4. Mechanism of enzyme catalysis

Understanding the mechanism of enzyme catalysis is essentially based on the identification of the amino acids making up the enzyme structure, and their functional groups directly involved in catalysis. Methods for the determination of these active groups are based on the determination of the enzyme's primary structure, as well as on the analysis of changes in the enzyme resulting from variations in various environmental factors, such as pH or temperature, the use of reagents specific to certain functional groups or structural analogs of the substrate.

The speed of an enzymatic reaction depends on the efficiency of the collisions between the substrate and enzyme molecules that will form the transition state. This efficiency depends on the orientation of the molecules in relation to one another and on a minimum energy required, which is the activation energy, as described above.

There are two main types of catalytic mechanisms:

– General acid-base catalysis: this is the most common mechanism in enzyme catalysis. The reaction is accelerated by the transfer of a proton. For example, this proton may come from the imidazole group on the side chain of histidine.

– Nucleophilic (or covalent) catalysis: this mainly concerns enzymatic reactions involving several substrates, as in the case of enzymes catalyzing a two-substrate "ping-pong" mechanism. Part of the first substrate is transferred to the enzyme via the formation of a covalent bond; this part of the substrate is then transferred to the second substrate.

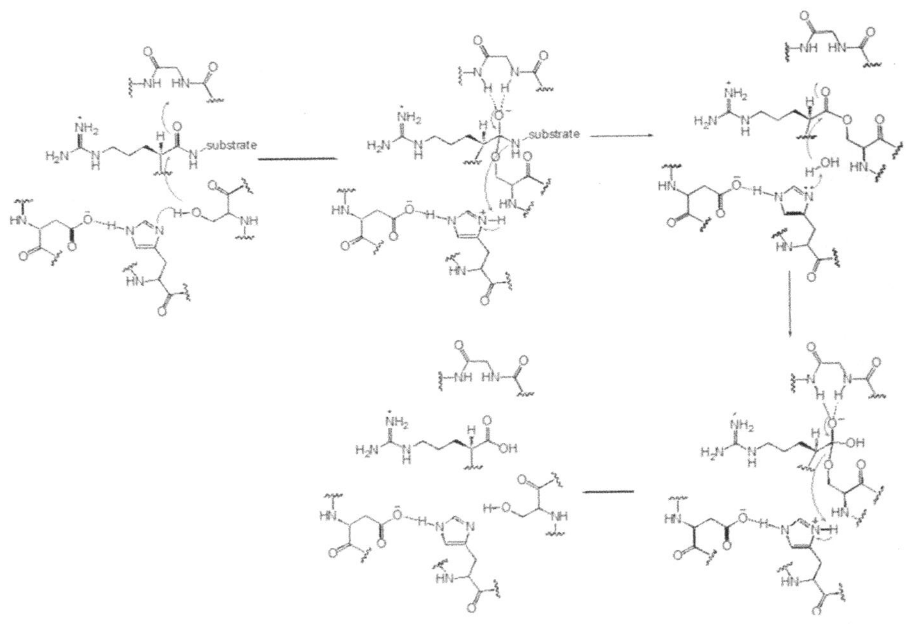

Figure 1.2. *Mechanism of serine protease enzymes. For a color version of this figure, see www.iste.co.uk/ghoul/implementation.zip*

This is the case for serine proteases, such as chymotrypsin, which share a cleavage mechanism based on a catalytic triad comprising serine, aspartic acid and histidine. This triad promotes polarization of the OH group on serine. The polypeptide is inserted into the serine protease in such a way that the carbonyl group is close to serine. The serine OH group attacks the carbonyl group, and the histidine nitrogen accepts the serine OH. An intermediate enzyme-substrate association is formed, and the hydrolysis product is released (Figure 1.2).

1.5. Single-substrate enzyme kinetics

The study of enzyme kinetics aims to identify and describe the mechanisms of biochemical reactions catalyzed by enzymes (enzymatic reaction), by studying their speed, that is to say, their evolution over time. Starting with isolated enzymes and moving toward organized, integrated metabolic systems, enzyme kinetics is used to quantitatively describe the catalytic properties of enzymes, as well as the mechanisms involved in regulating them. Enzymatic kinetics is an essential element in understanding how enzymes work and helps define the conditions for implementing biocatalytic processes in the food industry. The speed of an enzymatic reaction (V) is expressed by:

– the amount of substrate lost per unit of time with:

$$V = -\frac{d(S)}{dt} \qquad [1.2]$$

– or the quantity of product formed per unit of time with:

$$V = \frac{d(P)}{dt} \qquad [1.3]$$

A theory to explain the catalytic action of enzymes was proposed by Arrhenius at the end of the 19th century. He proposed that the substrate and enzyme form an intermediate transition state known as the ES complex. This can be represented schematically, as shown below:

$$E + S \underset{k_{-1}}{\overset{k_1}{\rightleftharpoons}} ES \xrightarrow{k_2} E + P$$

The existence of this intermediate ES complex has been demonstrated experimentally, and the reaction takes place with the formation of the product(s); the enzyme then returns to its original form, once the reaction is complete.

The determination of enzyme kinetics is a fundamental means of describing, predicting and calculating how enzymes can bind to substrates, in order to transform them into products, as well as determining the efficiency of catalysis.

For this purpose, at the beginning of the last century, a quantitative theory of enzyme kinetics was proposed, which led to the following equations being written, based on the assumption of a pseudostationary state such that:

Speed of appearance of ES = speed of disappearance of ES:

$$\frac{d(ES)}{dt} = 0 \qquad [1.4]$$

$$k_1(S)(E) = (k_{-1} + k_2)(ES) \qquad [1.5]$$

$$(E_T) = (E) + (ES) \qquad [1.6]$$

$$k_1(S)[(E_T) - (ES)] = (k_{-1} + k_2)(ES) \qquad [1.7]$$

$$k_2(S)(E_T) = k_2(ES)\left[(S) + \frac{k_{-1}+k_2}{k_1}\right] \qquad [1.8]$$

$$k_2(ES) = \frac{k_2(E_T)(S)}{\left[(S)+\frac{k_{-1}+k_2}{k_1}\right]} \qquad [1.9]$$

$$V_0 = k_2(ES) \qquad [1.10]$$

$$V_m = k_2(E_T) \qquad [1.11]$$

$$K_m = \frac{k_{-1}+k_2}{k_1} \qquad [1.12]$$

$$V_0 = \frac{V_{max}.S}{K_m+S} \qquad [1.13]$$

K_m is also known as the Michaelis–Menten constant. It is the apparent dissociation constant of the enzyme-substrate complex. K_m is the substrate concentration $[S_0]$ required to reach half the maximum velocity (in other words, when $v_0 = 1/2\ V_{max}$). In cases where k_1 and k_{-1} are much larger than k_2 (in other words, k_1, $k_{-1} >> k_2$), the Michaelis–Menten equilibrium hypothesis is valid and $K_m \cong K_s$ is used to describe the enzyme's affinity for the substrate.

The catalytic constant k_2 or k_{cat} is also known as the turnover number. It can be obtained from the expression $V_{max} = k_2\ [E_T]$. Turnover indicates the maximum number of substrate molecules that can be converted into product per enzyme molecule, per unit of time. Turnover varies from 1 to 10^4 per second for most enzymes.

The term k_{cat}/K_m is used to designate the catalytic efficiency of an enzyme and is used to classify them. A high value of k_{cat}/K_m (close to that of k_1) indicates that the substrate is closely bound to the enzyme, and the frequency of collisions between the enzyme and the substrate is the limiting factor.

The Michaelis–Menten graph representing v_0 as a function of [S] does not allow us to accurately determine K_m and V_{max}, since v_0 only reaches V_{max} at infinite substrate concentrations. To overcome this problem, Linweaver and Burk (1934) proposed a linearization of the Michaelis–Menten equation using a double inverse representation (Figure 1.3):

$$\frac{1}{v_0} = \frac{K_m}{V_{max}} \frac{1}{[S]} + \frac{1}{V_{max}} \qquad [1.14]$$

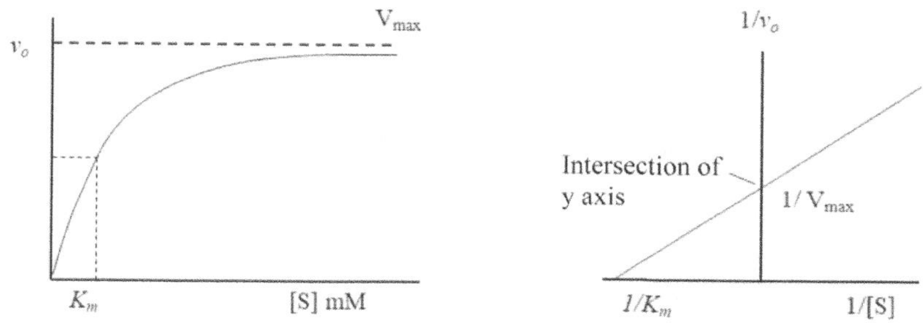

Figure 1.3. *Graphical representation of the Michaelis–Menten equation and its linearization using the Lineweaver–Burk method*

1.6. Effects of the environment on activity

Several factors can affect the speed at which enzymatic reactions take place, such as the substrate/product/cofactor concentration (which will influence kinetics and inhibition by excess substrate/product), the presence of effectors (inhibitors, activators) acting through various mechanisms and the environment (pH, temperature, solvent, oxidizing/reducing environment, etc.). Enzyme stability is an important factor to consider in the food industry.

1.6.1. *Effect of pH on enzyme activity*

All enzymes have a specific pH value or pH range for optimum activity. A change in pH can affect enzyme activity, with extremely low or high pH values potentially leading to changes in enzyme structure. The pH value at which an enzyme shows its maximum activity is called the optimum pH. This is an important criterion for their use in the food industry. The relationship between enzyme activity and pH is generally described by a bell-shaped curve. The rate of reaction is highest at the optimum pH and decreases on either side of this optimum pH. This effect may be linked to irreversible enzyme degradation due to extreme pH levels, or to an effect on the ionization state of the enzyme or substrate.

1.6.2. *Effect of temperature on enzyme activity*

Temperature is certainly the most important parameter to control in food processing. As with pH, each enzyme has its own optimum temperature range, and enzyme performance generally improves with increasing temperature, which is linked to molecular agitation. The number of collisions increases, thus raising the energy required to increase the reaction speed. Above the optimum temperature, enzymes can be denatured, moving from an active to an inactive form. In addition to the optimum temperature, denaturation kinetics over time are also crucial parameters. Depending on the process temperature, a wide range of enzymes is available for various purposes in the food industry.

Denaturation kinetics follow a first-order law. Thus:

$$\frac{d(E)}{dt} = -k_d(E) \quad [1.15]$$

$$(E) = (E_0)e^{-k_d t} \quad [1.16]$$

$$Ln(E) = Ln(E_0) - k_d t \quad [1.17]$$

E is the enzyme activity at time t, E_0 is the enzyme activity at time t_0 and K_d is the denaturation constant.

In practice, residual enzyme activity is monitored after different heating times. The graphical representation $Ln\ E/E_0 = f(t)$ is used to determine the slope $-k_d$.

1.6.3. *Effect of inhibitors on enzyme activity*

Enzyme activity can be inhibited by various substances, such as small molecules or ions that can bind to the enzyme. Inhibitors are substances that reduce the speed of enzyme-catalyzed reactions. They can either act on the substrate or combine directly with the enzyme. There are two types of inhibitors that directly interact with the enzyme: reversible inhibitors, which bind to enzymes reversibly and can be removed by dialysis to restore enzyme action, and irreversible inhibitors, which cannot be removed by dialysis. In the case of reversible inhibition, dissociation of the enzyme-inhibitor (EI) complex is much faster than in the case of irreversible inhibition, as there are no covalent interactions between the inhibitor and the enzyme.

When the inhibitor binds to the enzyme, because of its structural similarity to the substrate, and competes for the same binding site as the substrate, this is known as competitive inhibition. The inhibitor prevents the substrate from binding to the active site. A competitive inhibitor competes with the substrate for the binding site, resulting in a decrease in enzymatic activity, due to the reduced binding between the substrate and the enzyme. Competitive inhibition can be overcome by increasing the substrate concentration. Competitive inhibition not only depends on the concentration of the inhibitor and the substrate, but also on their relative affinity for the binding site. Thus, the degree of inhibition is greater when the inhibitor concentration is higher than that of the substrate. Conversely, when substrate concentration is high, the inhibitor competes with the substrate for unavailable binding sites, thus reducing the degree of inhibition. When substrate concentrations are much higher, inhibitor molecules are displaced by substrate molecules, leading to non-significant inhibition. As a result, V_{max} remains unchanged. However, the apparent K_M clearly increases as a result of inhibition and is referred to as K'_M:

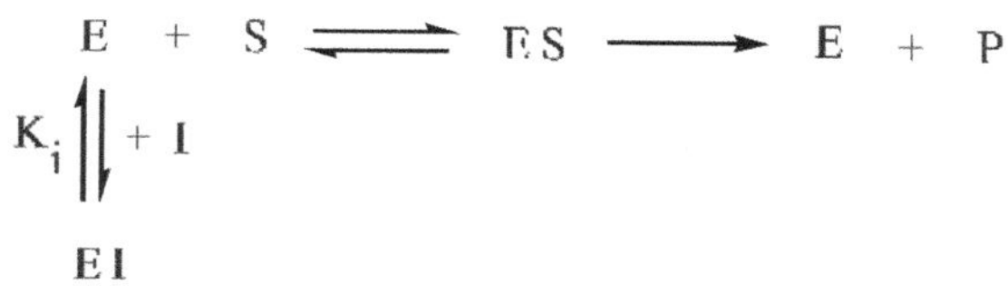

In the presence of a competitive inhibitor I, the Michaelis–Menten equation becomes:

$$V_0 = \frac{V_{max}.S}{K_M(1+\frac{I}{K_I})+S} \quad [1.18]$$

For example, the hydrolysis of pectin by a polygalacturonase (PG) has been described, demonstrating the presence of strong competitive inhibition of the product, with an inhibition constant K_I of 3.13 g/L (Belafi-Bako et al. 2007).

In non-competitive inhibition, there is no competition for the substrate binding site; the inhibitor can bind to the enzyme regardless of substrate binding; this means that the binding sites corresponding to the substrate and inhibitor molecules are different. Non-competitive inhibitors are substances which, when added to the enzyme, can modify the enzyme so that it cannot bind to the substrate. Non-competitive inhibitors can bind to the enzyme at the same time as the substrate. A complex is formed, consisting of the enzyme, the substrate and the inhibitor (ESI). Both EI and ESI complexes are non-productive. As the inhibitor cannot be detached from the enzyme by a higher concentration of substrate (as is the case of competitive inhibition), the apparent V_{max} changes. However, as the substrate can still bind to the enzyme, the K_m remains the same:

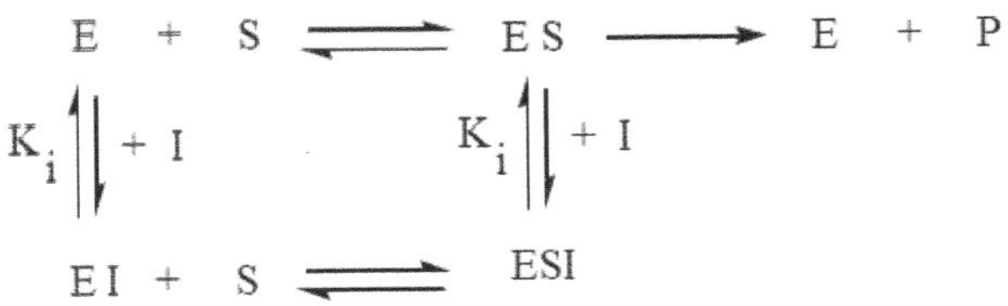

In the presence of a non-competitive inhibitor I, the Michaelis–Menten equation becomes:

$$V_0 = \frac{\frac{V_{max}}{(1+\frac{I}{K_I})}.S}{K_M+S} \quad [1.19]$$

For example, a plant-based flavonoid called luteolin has been shown to exert an effective inhibitory effect on peroxidase activity in a non-competitive manner. The interaction between luteolin and peroxidase induces the formation of a complex due to hydrogen bonds and hydrophobic interactions (Li et al. 2021).

While competitive inhibitors bind to free enzymes and form the EI complex, uncompetitive inhibitors only bind to the enzyme–substrate complex. Here, the inhibitor does not compete for the substrate binding site, but binds to a completely different site. Both V_{max} and K_m are modified:

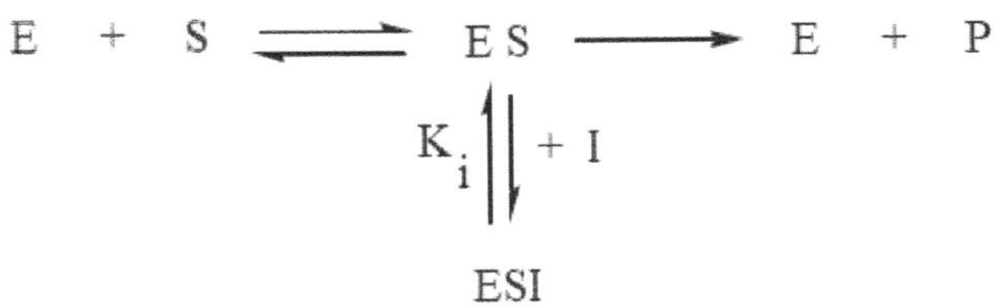

In the presence of an uncompetitive inhibitor I, the Michaelis–Menten equation becomes:

$$V_0 = \frac{\frac{V_{max}}{(1+\frac{I}{K_I})}.S}{\frac{K_M}{(1+\frac{I}{K_I})}+S} \qquad [1.20]$$

This is the case, for example, with the enzymatic browning of apples, which is linked to polyphenol oxidases, but can be limited by oligochitosans, which are inhibitors of these enzymes. It is thus shown that the type of inhibition is uncompetitive, and that these oligochitosans could therefore be considered as new types of anti-browning agent (Guo et al. 2018).

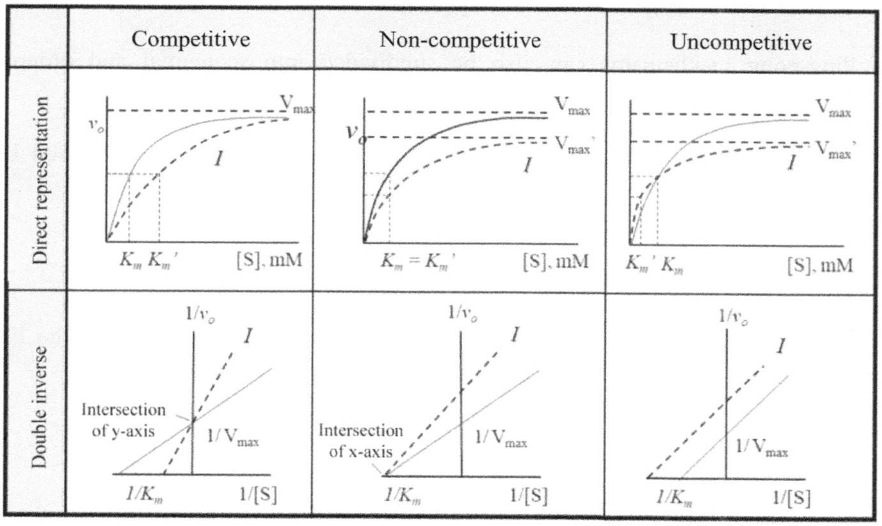

Figure 1.4. *Impact of different types of inhibitors on enzyme kinetics. Direct and double inverse representation to determine the catalytic parameters K_m and V_{max} for competitive, non-competitive and uncompetitive inhibition*

Figure 1.4 summarizes the direct and double inverse representations of enzyme rates in the presence of competitive, non-competitive and uncompetitive inhibitors.

1.7. Multi-substrate enzyme kinetics

When more than one substrate is involved, the reaction mechanism can be divided into two main categories: sequential and ping-pong.

In sequential mechanisms, all substrates combine with the enzyme before the reaction takes place. Sequential mechanisms can be divided into two subgroups (ordered or random), depending on whether or not there is a predetermined sequence of bonds between the substrates and the enzyme.

In ping-pong mechanisms, the product is formed before all of the substrates have bound to the enzyme, which means that the enzyme can exist as two catalytically active species, each recognizing a substrate and transforming it into a product, with a conformational change into the other species.

Ping-pong mechanisms can also be subdivided into sequential and ordered mechanisms, but only apply to reactions with more than two substrates.

Mechanisms can be represented according to the nomenclature proposed by Cleland (1963).

1.7.1. *Ordered sequential mechanism*

In this case, enzyme E has no binding site for substrate B and forms the EA complex alone with A, before being able to bind B to give the EAB complex.

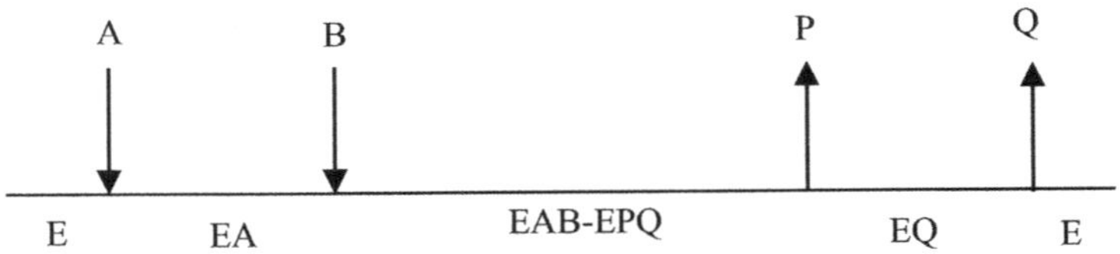

1.7.2. *Random sequential mechanism*

In this case, the enzyme can initially bind either of the substrates, so that the two secondary enzyme complexes, EA and EB, can be formed, and then bind the other substrate to generate the reactive ternary complex EAB.

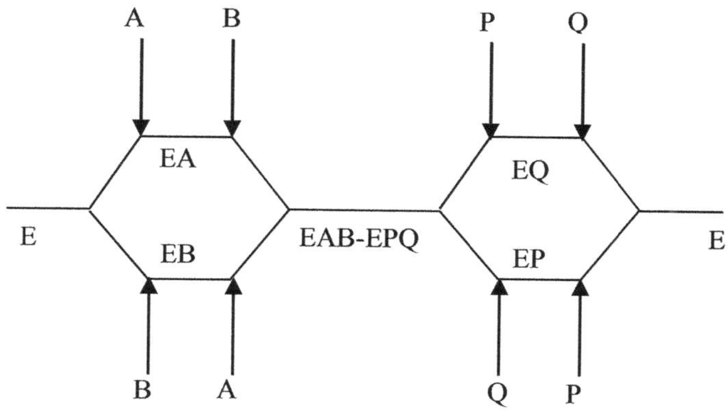

1.7.3. *Ping-pong mechanism*

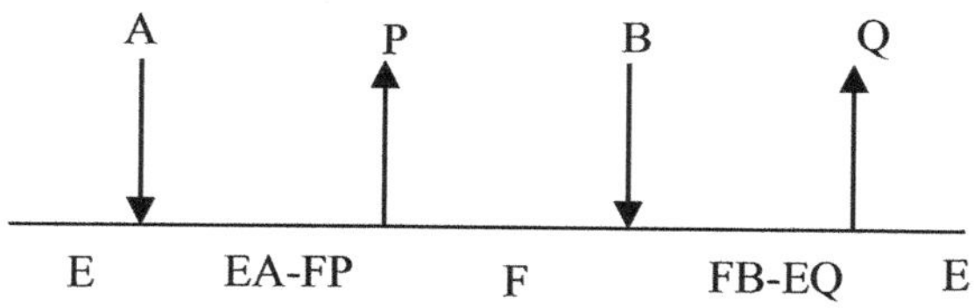

2

Classification of Enzymes Used in the Food Industry

Latifa CHEBIL and Mohamed GHOUL
LRGP, CNRS, Université de Lorraine, Nancy, France

Enzymes are complex molecules produced by various microorganisms (α-amylase, β-amylase, glucose isomerase, pullulanase, cellulase, invertase, lipase, lactase, catalase, etc.), animals (trypsin, chymotrypsin, pepsin, chymosin, catalase, amylase, lipase, etc.) and plants (α-amylase, β-amylase, β-glucanase, bromelain, papain, etc.). Both in vivo and in vitro, they catalyze many reactions that are necessary for the life of biological systems. These catalytic properties have been exploited by man for centuries in order to transform raw materials into products characterized by sought-after features for nutritional, health, well-being and environmental purposes (Mantovani et al. 2005; Raveendran et al. 2018).

According to sources, between 3,500 and 5,000 enzymes have been identified and listed. To facilitate this classification, the International Union of Biochemistry and Molecular Biology (founded in 1955) has proposed a simplified nomenclature, assigning a code to each enzyme. This code is made up of the letters "CE", followed by four numbers separated by dots, reflecting the type of activity and the position of the enzymes' action on the various substrates. This approach has made it possible to group enzymes into six major classes, according to their activity: oxidoreductases (CE1), transferases (CE2), hydrolases (CE3), lyases (CE4), isomerases (CE5) and ligases (CE6) (Kuddus 2018).

Implementation of Enzymatic Processes and Lactic Bacteria in the Food Industries, coordinated by Mohamed GHOUL.

2.1. Oxidoreductases (CE1)

This enzyme class catalyzes oxidoreduction reactions. The main sub-classes are glucose oxidases, peroxidases, polyphenol oxidases, lipoxygenases, catalases, dehydrogenases, oxygenases and hydroxylases. These enzymes are used in the food industry to modify the texture, color, nutritional value and shelf life of products. Table 2.1 illustrates the use of this class of enzymes in some food applications (Semenova 2006; Chandrasekaran 2016).

Product	Function	Sub-class
Milk	Removes hydrogen peroxide	Catalase (CE 1.11.1.6)
Breads and pastries	Strengthens cohesion and whitens dough	Glucose oxidase (CE 1.1.3.4), Lipoxygenases (CE 1.13.11)
Wine and beer	Enhances aroma and improves must clarification	Laccases (CE 1.10.3.2)
Wine and beer	Enhances color	Polyphenol oxidases (CE 1.10.3.1)
Fruit juices	Removes oxygen	Glucose oxidase (CE 1.1.3.4)
Fruit juices	Enhances color and stabilizes juice	Laccases (CE 1.10.3.2)

Table 2.1. *Main uses of oxidoreductases in the food industry*

The best-known reactions that are catalyzed by these enzymes are the enzymatic browning of fruit and vegetables, the oxidation of fatty acids and the formation of lactones.

2.2. Transferases (CE2)

This class of enzymes transfers radicals or atomic groups between a donor substrate and an acceptor substrate. Depending on the functional group transferred, a distinction is made between methyltransferases, glycosyltransferases and phosphotransferases. These enzymes are used in the meat tenderization process, in the gelation and polymerization of milk proteins, and in the synthesis of oligosaccharides and cyclodextrins. Table 2.2 lists some of the uses of transferases in food applications (Semenova 2006; Chandrasekaran 2016).

Product	Function	Sub-class
Milk	Gelation, polymerization	Transglutaminases (CE 2.3.2.13)
Meat	Modification of viscoelastic properties	Transglutaminases (CE 2.3.2.13)
Syrup	Synthesis of fructose oligomers	Fructosyltransferase (CE 2.4.1.9)
Starch	Cyclodextrin production	Glycosyltransferase (CE 2.4.1.19)

Table 2.2. *Some transferase applications in the food industry*

2.3. Hydrolases (CE3)

Product	Function	Sub-class
Milk	Milk coagulation-cheese production	Proteases (chymosin (CE 3.4.23.4), papain (CE 3.4.22.2))
Milk	Protein hydrolysis	Proteases (CE 3.4.21.1)
Milk	Lactose hydrolysis	Lactases (CE 3.2.1.108)
Milk	Cheese maturation	Aminopeptidases (CE 3.4.21.4), trypsin (CE 3.4.11)
Product	Function	Sub-class
Milk	Flavor development	Lipases (triacylglycerol lipases (CE 3.1.1.3))
Bread and pastries	Improvement of yeast fermentability, color and volume by sucrose hydrolysis	α-Amylases (CE 3.2.1.1) and β-amylases (CE 3.2.1.2)
Bread and pastries	Gluten hydrolysis and improvement of texture and rheology	Proteases (CE 3.4.21.112)
Pastries and Bakery	Improvement of pasta and bread quality	Cellulases (CE 3.2.1.4), endo-1,4-β-xylanase (EC 3.2.1.8)
Malt, beer and wine	Polysaccharide hydrolysis	α-Amylases (CE 3.2.1.1) and β-amylases (CE 3.2.1.2), α-glucosidases (CE 3.2.1.20), cellulases (CE 3.2.1.4), invertases (CE 3.2.1.26)
Malt, beer and wine	Improvement of filtration and clarification	Pectinases (CE 3.2.1.15), hemicellulases (CE 3.1.1.73), cellulases (CE 3.2.1.4)
Juices and fruits	Improvement of extraction and clarification	Pectinases (CE 3.2.1.15)
Juices and fruits	Improvement of liquefaction	Cellulases (CE 3.2.1.4)
Syrup	Polysaccharide hydrolysis	α-Amylases (CE 3.2.1.1) and β-amylases (CE 3.2.1.2), β-glucosidases (CE 3.2.1.21), cellulases (CE 3.2.1.4)

Table 2.3. *Main hydrolase applications in the food industry*

This enzyme class hydrolyzes a substrate in the presence of water to form two or more products. The best-known subclasses are α-amylases, β-amylases, glucoamylases, lactases, cellulases, proteases, invertases, pectinases, phosphatases and esterases. These enzymes are widely used for the hydrolysis of starch, liquefaction, as well as the clarification of fruit juices and wine. These enzymes can be added at various stages of the raw material transformation process, or produced and activated in situ, like in the barley malting process. Table 2.3 shows some applications involving these enzymes (Chandrasekaran 2016; Raveendran et al. 2018).

2.4. Lyases (CE4)

These enzymes catalyze the removal of a group of atoms from a substrate by cleaving C–C, C–O, C–S, C–N type bonds. This mechanism is different from the hydrolytic mechanism characterizing the hydrolases presented above. The best-known sub-classes are decarboxylases, aldolases, dehydratases and some pectinases. For example, acetolactate decarboxylase is used in the beer maturation process. Aldolases are used in the removal of certain aromas, or in the formation of aromas in beer and other beverages. Table 2.4 summarizes the use of some enzymes in this class in food processing (Mantovani et al. 2005).

Product	Function	Sub-class
Beer	Beer maturation	Acetolactate decarboxylases (CE 4.1.1.5)
Alginate	Oligomer production	Alginate lyases (CE 4.2.2.3)
Alginate	Oligomer production	Guluronate lyases (CE 4.2.2.3)
Fruit juices	Clarification	Pectin lyases (CE 4.2.2.10)

Table 2.4. *Some examples of industrial applications for lyases*

2.5. Isomerases (CE5)

This class includes enzymes that catalyze the rearrangement of groups of atoms in a molecule by isomerization. The major sub-classes are isomerases (glucose isomerases, xylose isomerases), epimerases and racemases. The isomerization of glucose to produce fructose is a solid example of the use of this enzyme class in the food industry. Table 2.5 gives some examples of the application of this enzyme class in the functioning of food products (Chandrasekaran 2016).

Product	Function	Sub-class
Glucose	Fructose production	Glucose isomerases (CE 5.3.1.5)
D-Xylose	D-Xylulose: conversion of hemicellulose into ethanol	Xylose isomerases (CE 5.3.1.5)

Table 2.5. *Industrial applications of isomerases*

2.6. Ligases (CE6)

Ligases, also known as synthases, are a class of enzymes that catalyze the creation of covalent bonds between two molecules. There are many different types of ligases depending on the type of covalent bond they catalyze (carbon–carbon bond, carbon–nitrogen bond, carbon–sulfur bond, etc.). They generally use the energy of an ATP (or GTP) molecule to catalyze the reaction. Examples of ligases include aminoacyl tRNA synthetases (which bind amino acids to corresponding tRNAs) or DNA ligases (which join two pieces of DNA together) (Chandrasekaran 2016).

3

Mode of Action of the Main Enzymes Used in the Food Industry

Catherine HUMEAU[1], Mohamed GHOUL[1] and Seraphim PAPANIKOLAOU[2]

[1] *LRGP, CNRS, Université de Lorraine, Nancy, France*

[2] *LFMB, Agricultural University of Athens, Greece*

The most commonly used substrates for the enzymatic production of food items and/or functionalized derivatives are those of the "starch", "lignocellulose", "pectin", "lipid" and "protein" type. The enzymes involved in transforming these substrates are mainly hydrolases (amylases, cellulases/hemicellulases, polygalacturonases/pectinesterases, lipases and proteases), oxidoreductases (laccases) and lyases (pectin lyases). The modes of action of these enzymes will be described according to the substrate used.

3.1. Enzymes involved in the starch degradation reaction

Starch is one of the most abundant renewable raw materials in nature, found in edible commodities like corn, potato, cassava, wheat, etc. It is a branched polysaccharide, composed of glucose monomers in which α-glycosidic [α-(1,4)] bonds predominate; its structure also includes branches, characterized by the presence of α-(1,6) bonds. Fractions that only contain α-D-glucose polymerized with α-(1,4) bonds constitute the amylose isoform, and branched fractions [α-(1,4) and α-(1,6)] constitute amylopectin (Balatsouras 2006; Aggelis 2007; Galiotou-Panayotou

Implementation of Enzymatic Processes and Lactic Bacteria in the Food Industries, coordinated by Mohamed GHOUL.

2007). The starch processing industry has been the driving force behind the development of hydrolase production processes, as well as the development of genetic engineering and molecular modeling tools to produce stable, competitively priced enzymes. Today, these enzymes are available in purified forms or in cocktails that are adapted to the nature of the raw materials to be processed.

These enzymes are grouped together under the generic term "amylases" or glycoside hydrolases (GH). They are responsible for the degradation/decomposition of starch. Depending on their mode of action, they are classified into three sub-groups:

– Endoamylases, which hydrolyze the α-1,4 bonds of amylose and amylopectin forming the foundation of starch and glycogen (Bhutada et al. 2017; Filippousi et al. 2019) to release oligosaccharides and dextrins (Aggelis 2007, 2017). This group mainly includes α-amylase (EC 3.2.1.1).

– Exoamylases, which include β-amylase (EC 3.2.1.2), α-glucosidase (EC 3.2.1.20) and glucoamylase or amyloglucosidase (EC 3.2.1.3). Their action leads to the release of low molecular weight sugars such as glucose, maltose and oligosaccharides (Galiotou-Panayotou 2007; Saini et al. 2017).

– Enzymes that catalyze the hydrolysis of amylopectin's α-1,6 bonds. Pullulanase (EC 3.2.1.41) and isoamylase (EC 3.2.1.68) belong to this group (Saini et al. 2017).

The action of α-amylase on the starch structure is "random". It takes place at the α-1,4 internal bonds of amylose to release linear oligosaccharides, and at the α-1,4 and α-1,6 bonds to form dextrins (Balatsouras 2006; Aggelis 2007). In contrast, β-amylase, which is an "exo-enzyme", breaks the outer α-1,4 bonds of the nonreducing end of starch, releasing maltose. Thus, the action of β-amylase on amylose only leads to the formation of maltose, whereas its effect on amylopectin mainly leads to the formation of maltose and dextrins containing three α-D-glucose molecules, with α-1,6 branching (Balatsouras 2006).

Enzymes known as glucoamylases (or amyloglycosidases) catalyze the breaking of the outer α-1,4 and α-1,6 bonds (and sometimes also the α-1,3 bond) (Balatsouras 2006), releasing α-D-glucose as the end product of the enzymatic reaction (Aggelis 2007; Saini et al. 2017). Thus, the combined action of the α-amylase and glucoamylase enzymes on starch will result in the production of α-D-glucose (Galiotou-Panayotou 2007).

Pullulanases are so-called debranching enzymes that hydrolyze the α-1,6 bonds of amylopectin, releasing dextrins that only contain α-1,6 bonds (Saini et al. 2017).

The main substrate for pullulanases is the pullulan biopolymer, which is a polysaccharide synthesized by the *Aureobasidium pullulans* fungus grown on a medium containing glucose or other sugars (such as cellobiose). This polysaccharide is composed of polymerized maltotriose units, linked by α-1,6 bonds (Aggelis 2017; Saini et al. 2017); its synthesis is favored through nitrogen substrate limitation (Galiotou-Panayotou et al. 1998; Aggelis 2017).

3.2. Enzymes involved in the degradation of lignocellulosic substrates

Lignocellulosic biomass is made up of macromolecules such as cellulose, hemicelluloses and lignin. This source of raw material is renewable and abundant (Payne et al. 2015). It is a potential substitute for derivatives from petroleum refining and processing (Koutinas et al. 2014; Aggelis 2017; Obeng et al. 2017). The use of this biomass as a carbon source in various bioprocesses is in full swing, enabling the synthesis of value-added products such as green biofuels (Koutinas et al. 2014; Sarris and Papanikolaou 2016; Valdés et al. 2020; Narisetty et al. 2022a). Given the growing demand for energy from renewable resources and concerns about climate change, it is likely that the use of lignocellulosic biomass will become the main source for energy production, in the short to medium term (Papanikolaou and Aggelis 2011; Payne et al. 2015). However, the use of this raw material requires an industrial-scale production of enzymes that are capable of degrading these materials. These include hydrolytic enzymes (cellulases/ hemicellulases) and oxidative enzymes (laccases) (Binod et al. 2011; Payne et al. 2015; Obeng et al. 2017). The efficiency of lignocellulosic material degradation is strongly influenced by the pre-treatments that are applied prior to the enzymatic reaction (Koutinas et al. 2014; Sarris and Papanikolaou 2016; Valdés et al. 2020; Narisetty et al. 2022b). These pre-treatment stages either rely on physical techniques or on the use of fungi (*Pleurotus* sp., *Lentinula edodes*, *Ganoderma* sp., etc.). These fungi release enzymes such as laccases, lignin peroxidases and hemicellulases (Diamantopoulou and Philippoussis 2001; Philippoussis et al. 2001, 2011; Philippoussis 2009; Economou et al. 2017, 2020; Melanouri et al. 2022a, 2022b).

Cellulases are enzymes that break down cellulose, either into a hydrophilic, water-insoluble polymer composed of repeating units of β-D-glucose linked by β-1,4 glycosidic bonds or into shorter chain polysaccharides such as cellodextrin, cellobiose and β-D-glucose. Depending on their structures and functions, cellulases are classified into several categories (endoglucanases, exoglucanases or cellobiohydrolases, β-glycosidases or cellobiases). However, these different categories work in synergy to ensure the hydrolysis of celluylosic material in plant walls (Galiotou-Panayotou 2007; Obeng et al. 2007).

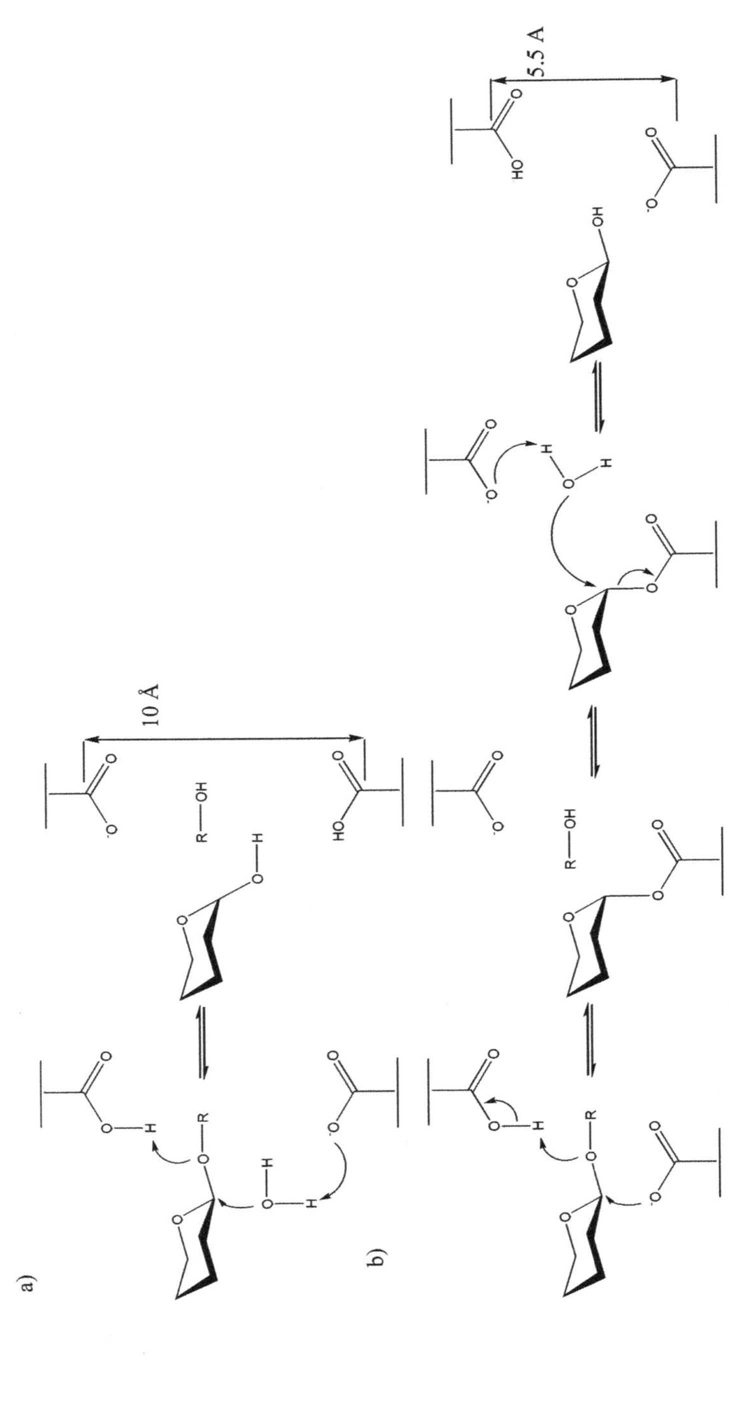

Figure 3.1. *Catalytic mechanisms of glycoside hydrolases: (a) inversion; (b) retention (according to Obeng et al. (2017))*

These enzymes catalyze the cleavage of the glycosidic bond, by "retention" or "inversion" mechanisms (Obeng et al. 2017). These two catalytic mechanisms are shown in Figure 3.1 (according to Obeng et al. 2017).

Hemicellulose is broken down by hemicellulases. These enzymes catalyze the hydrolysis of glycosidic bonds (α-glucuronidases, endo-1,4-β-xylanases, α-arabinofuranosidases, α-galactosidases and β-xylosidases).

Other enzymes may also be involved (acetyl or feruloyl-esterases), leading to the hydrolysis of the ester bonds linking acetic and ferulic acids to the side groups of the plant wall structure (Obeng et al. 2017; Sharma and Sharma 2017).

Hemicellulases exhibit similar activities to cellulases due to the 1,4 glycosidic bond that exists in most molecules containing hemicelluloses. Their modes of action are shown as a diagram in Figure 3.2 (according to Sharma and Sharma 2017).

In most cases, hemicellulases are used in combination with cellulases in lignocellulosic biomass depolymerization processes so as to produce fermentable sugar solutions (C5 and C6 sugars). The glucose formed is catabolized through the EMP pathway, while xylose is converted to xylulose 5-phosphate, either through the action of xylose isomerase, or under the effect of xylose reductase (XR) and xylitol dehydrogenase (XDH) (Narisetty et al. 2022b). The xylulose 5-phosphate thus generated is either metabolized through the pentose phosphate pathway (Valdés et al. 2020; Narisetty et al. 2022a) or by the action of phosphoketolases, which convert it directly into glyceraldehyde-3-phosphate and acetyl-phosphate (Sonderegger et al. 2004; Sarris and Papanikolaou 2016). The reaction catalyzed by phosphoketolase is a more efficient reaction for the microbial cell than the pentose phosphate pathway (Papanikolaou and Aggelis 2011).

Laccases (benzenediol: oxygen oxidoreductases; EC 1.10.3.2) are enzymes that are capable of catalyzing the oxidation of various substrates, such as the phenolic and non-phenolic sub-units of lignin (Obeng et al. 2017).

Laccases contain four copper atoms in their active site. Depending on the distribution of these copper atoms, a distinction is made between type-1, type-2 and type-3 laccases (Dwivedi et al. 2011).

These copper atoms act as catalytic metals and reducing agents for the oxidation of various carbons in the polymeric structure of lignin (C-1, C-4 and C-6) (Dwivedi et al. 2011; Obeng et al. 2017). Ascorbate oxidase, ferroxidases, and ceruloplasmin and bilirubin oxidases are also members of the multi-copper protein family (Obeng et al. 2017).

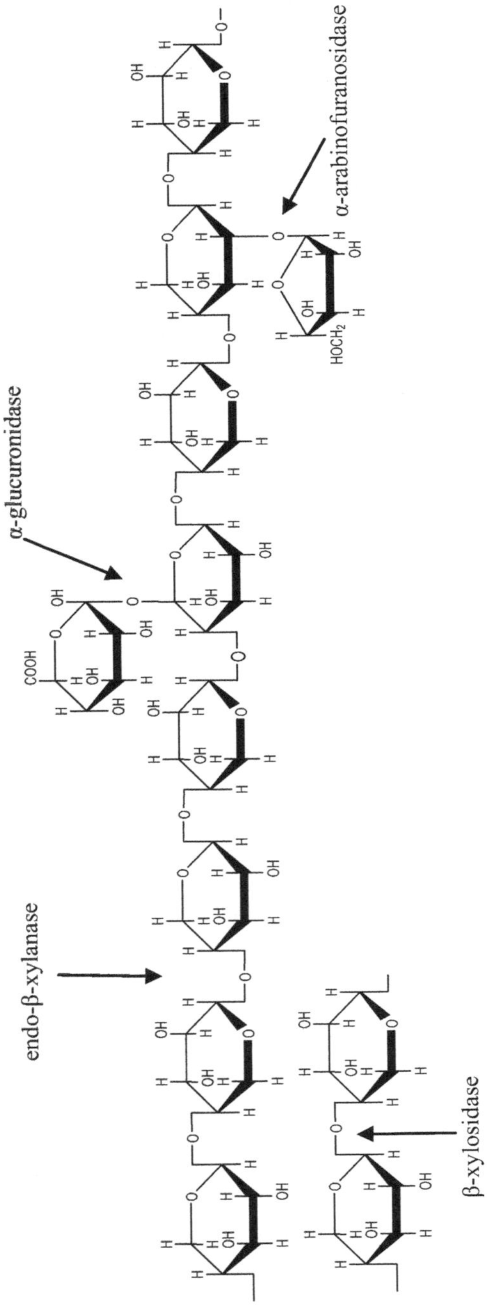

Figure 3.2. *Mechanisms of action of xylanases (according to Sharma and Sharma 2017)*

The degradation of lignocellulosic material involves three stages: delignification by laccases to liberate cellulose and hemicellulose from lignin, depolymerization to produce free sugars (C5 and C6) and, lastly, fermentation of the hexoses and pentoses that are liberated, to obtain a range of value-added compounds (Koutinas et al. 2014; Narisetty et al. 2022a).

3.3. Enzymes involved in the degradation of pectin-type substrates

Pectins are polysaccharides that are exclusively of plant origin, found in the primary walls of dicotyledons, particularly in the plant walls of many fruits and vegetables (Galiotou-Panayotou 2007). They are heteropolysaccharides characterized by a more or less branched framework, mainly composed of α-D-galacturonic acid and, to a lesser extent, of α-L-rhamnose.

Other sugars can also be associated with pectic structures, such as xylose, galactose and arabinose. The carboxyl group of galacturonic acid can be esterified with methanol (Galiotou-Panayotou 2007).

Pectinases refer to all types of so-called pectolytic enzymes, as they are capable of catalyzing the breakdown of pectin present in plant cell walls. Under the effect of these enzymes, the plant structure loses some of its rigidity, as observed during fruit ripening (Balatsouras 2006). These enzymes also ensure the decomposition of various types of plant residues in nature and contribute to the organic improvement of the soil (Jayani et al. 2005; Balatsouras 2006).

According to the classification of the International Union of Biochemistry, pectolytic enzymes either belong to the hydrolase or lyase group (Fogarty and Kelly 1983; Galiotou-Panayotou 2007). The predominant classification of pectinases is based on the type of catalytic reaction performed, as well as on the structure and type of substrate (pectin, pectic acid, etc.); a distinction is therefore made between pectinesterases, protopectinases/polygalacturonases and lyases. A classification can also be established according to the position in relation to which these enzymes are active (intra- or extra-molecular) (Table 3.1).

Enzyme	EC number	Systematic name	Mechanism e of action	Type of action	Primary substrate	Product of reaction
Esterase						
Pectinomethyl esterases	3.1.1.11	–	Hydrolysis	Random	Pectin	Pectic acid + methanol
Hydrolases (selection)						
Protopectinases	–	–	Hydrolysis	Random	Protopectin	Pectin
Endo-polygalactorunase	3.2.1.15	Poly-(1-4)-α-D-galactosiduronate glycanohydrolase	Hydrolysis	Random	Pectic acid	Oligogalacturonates
Exo-polygalactorunase	3.2.1.67	Poly-(1-4)-α-D-galactosiduronate glycanohydrolase	Hydrolysis	Terminal bond	Pectic acid	Monogalacturonates
Exo-polygalacturonan-digalacturonohydrolase	3.2.1.82	Poly-(1-4)-α-D-galactosiduronate digalacturonohydrolase	Hydrolysis	Penultimate bonds	Pectic acid	Digalacturonates
Oligogalacturonate hydrolase	–	–	Hydrolysis	Terminal bond	Trigalacturonate	Monogalacturonates
Lyases (selection)						
Endo-polygalacturonase lyase	4.2.2.2	Poly-(1-4)-α-D-galactosiduronate lyase	Trans-elimination	Random	Pectic acid	Unsaturated oligogalacturonates
Exo-polygalacturonase lyase	4.2.2.9	Poly-(1-4)-α-D-galactosiduronate exolyase	Trans-elimination	Penultimate bonds	Pectic acid	Unsaturated digalacturonates
Endo-polymethyl-D-galactosiduronate lyase	4.2.2.10	Poly-(methyl galactosiduronate) lyase	Trans-elimination	Random	Unsaturated poly-methyl-D-digalacturonates	Unsaturated methyl-oligogalacturonates
Exo-polymethyl-D-galactosiduronate lyase	–	–	Trans-elimination	Terminal bond	Unsaturated poly-methyl-D-digalacturonates	Unsaturated methyl-oligogalacturonates

Table 3.1. *Classification of pectolytic enzymes (according to Fogarty and Kelly (1983) and Jayani et al. (2005))*

Pectinesterases or pectin methylesterases catalyze the hydrolysis of ester bonds established between the carboxyl group of galacturonic acid and methanol, forming pectins with a low degree of methylation (Figure 3.3) (Jayani et al. 2005; Aggelis 2007; Galiotou-Panayotou 2007). Pectin methylesterases play a key role in plant cell growth mechanisms (Jayani et al. 2005).

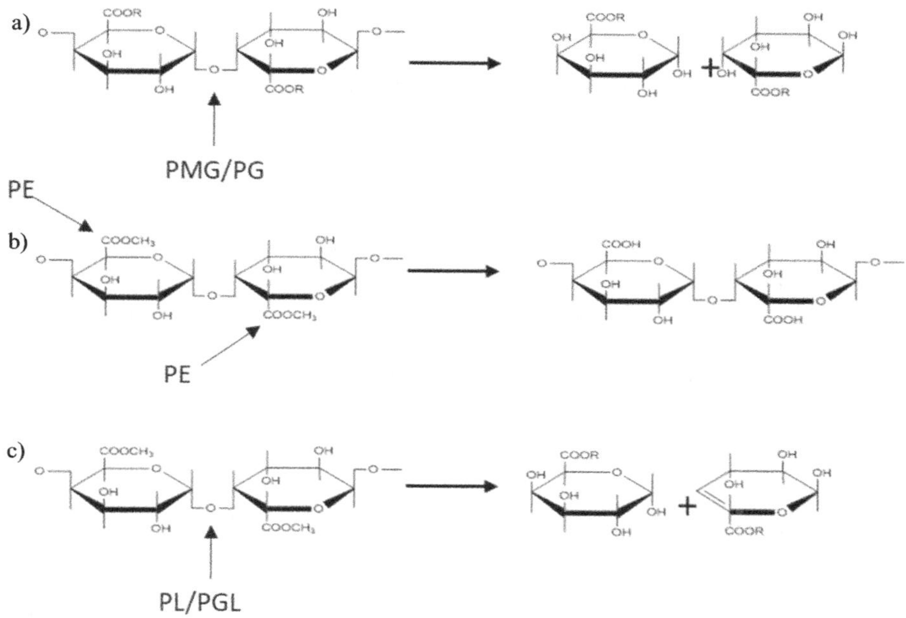

Figure 3.3. *Modes of action of pectinases: (a) R = H for PG and R = CH_3 for PMG; (b) PE; and (c) R = H for PGL and R = CH_3 for PL. PMG: polymethylgalacturonases; PG: polygalacturonases; PE: pectinesterases; PL: pectin lyases (according to Jayani et al. 2005)*

Protopectinases catalyze the conversion of insoluble protopectin into water-soluble pectin. Two types of protopectinase have been identified. The first acts on the internal bonds, while the second acts on the external bonds of the pectic structure (Fogarty and Kelly 1983).

Polygalacturonase enzymes act in a similar way to protopectinases. Depending on the site of action on the polygalacturonic acid chain, a distinction is made (Figure 3.1) between endo-polygalacturonase (endo-PG, EC 3.2.1.15) and exo-polygalacturonase (exo-PG, EC 3.2.1.67) (Jayani et al. 2005; Amin et al. 2019). Polygalacturonases cause the disorganization, decomposition and ultimate destruction of the compact structure of plant tissues; this contributes, for example, to the ripening of fruits, vegetables and other associated products (Amin et al. 2019).

Other enzymes that act on the same bonds as polygalacturonases, but using different substrates, are described in the literature. These include oligogalacturonases, endo- and exo-polymethylgalacturonases (Fogarty and Kelly 1983; Jayani et al. 2005).

Within the category of pectin-degrading enzymes, there are also lyase-type enzymes. These enzymes cause the cleavage of glycosidic bonds, without the addition of water molecules (Jayani et al. 2005). More specifically, these enzymes break the bonds between the C4 and C5 carbons of the sugar, removing a hydrogen molecule and leading to the formation of a double bond at the nonreducing end of the products. Depending on the substrate and the site of action, a distinction is made between endo- and exo-polygalacturonase lyases (endo-PGL, EC 4.2.2.2; exo-PGL, EC 4.2.2.9), as well as endo- and exo-polymethyl-D-galactosiduronate lyases (endo-PMGL, EC 4.2.2.10; exo-PMGL) (Figure 3.3). The mode of action of the various enzymes involved in pectin degradation is shown in Figure 3.3.

3.4. Enzymes involved in the degradation of lipid substrates

Lipids are mainly composed of triacylglycerols (TAGs). The enzymes involved in the degradation of these substrates are lipases (glycerol ester hydrolases; EC 3.1.1.3). These water-soluble enzymes catalyze the hydrolysis of ester functions, thus transforming glycerides (TAGs, diglycerides (DAGs) and monoglycerides (MAGs)) into glycerol and free fatty acids (Macrae 1983). They represent a sub-class of esterases.

TAG hydrolysis is a selective process. Indeed, most microbial lipases exhibit regional specificity toward sn-1,3 positions. Consequently, fatty acids (FAs) esterified at the external positions of TAGs are first released, while the remaining β-monoacylglycerols are then hydrolyzed after their isomerization into α-isomers (Fickers et al. 2005; Metzger and Bornscheuer 2006; Tzirita et al. 2018; Vasiliadou et al. 2018). In addition to this regional specificity, microbial lipases can also exhibit specificity toward released FAs. These enzymes can recognize specific aliphatic chains based on their number of carbon atoms and degree of unsaturation (Fickers et al. 2005; Metzger and Bornscheuer 2006).

The reaction diagram for the hydrolysis of a fat composed of TAG is shown in Figure 3.4.

a) TAG → DAG + AG → MAG + AG → Glycerol + AG

b)

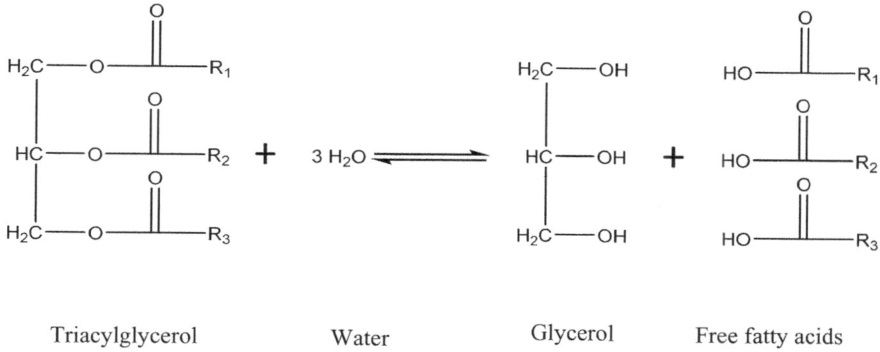

Figure 3.4. *Schematic representation of triacylglycerol (TAG) hydrolysis, catalyzed by extracellular lipases. (a) Sequential reaction; (b) total reaction (according to Macrae 1983)*

3.5. Enzymes involved in protein degradation

Peptidases, proteases or proteolytic enzymes catalyze the cleavage of peptide bonds in proteins. This is known as proteolysis. The latter involves the intervention of a water molecule for each broken peptide bond, which classifies these enzymes as hydrolases (Balatsouras 2006).

Proteases have a wide range of biological functions. They can be involved in protein maturation, food digestion, blood coagulation, tissue remodeling during organism development and wound healing (Balatsouras 2006; Aggelis 2007; Galiotou-Panayotou 2007). Some proteases are toxins, such as the botulinum toxin (Balatsouras 2006). Given their multiple biological functions, proteases are used in a wide range of food applications (meat, charcuterie, cheese, brewing, etc.) and nonfood applications (detergents, biomedical technology, nonconventional catalysis, agro-industrial waste) (Balatsouras 2006; Galiotou-Panayotou 2007; Aggelis 2017). In addition, some proteases are produced as inactive precursors, known as zymogens (proenzymes). Zymogens are usually activated by a proteolytic cleavage, which releases the functional enzyme and triggers proteolytic activity.

Zymogens are therefore protein precursors which, after appropriate activation, can produce an active enzyme (Ward 1983; Balatsouras 2006).

Depending on their site of action, proteases are divided into two groups: intra-cellular and extra-cellular proteases (Fogarty and Kelly 1990; Galiotou-Panayotou 2007; Chander and Puri 2019). Proteases can also be differentiated according to their mode of action; we can thus distinguish between endopeptidases and exopeptidases.

Exopeptidases are proteases that catalyze the disruption of peptide bonds located near the N- or C-terminus of proteins (Galiotou-Panayotou 2007). There are various sub-classes of exopeptidases (Ward 1983; Fogarty and Kelly 1990; Galiotou-Panayotou 2007): aminopeptidases (3.4.11), dipeptidases (3.4.13), peptidyl dipeptide hydrolases (3.4.15) and carboxypeptidases (3.4.16 and 3.4.17).

Aminopeptidases are the most important exopeptidase enzymes. They act close to the N-terminus of proteins. They release a single amino acid, a dipeptide or a tripeptide. Some aminopeptidases are specific to proteins with an N-terminus methionine residue (Rao et al. 1998). On the other hand, carboxypeptidases are exopeptidases that act at the C-terminus of proteins. They release a single amino acid or dipeptide (Rao et al. 1998).

Most enzymes used in the industry are endopeptidases. They are characterized by their hydrolytic action on a specific site of the polypeptide chain far from the N-terminus and C-terminus. The presence of a free amine or carboxyl group can have an inhibitory effect on the activity of these proteases (Fogarty and Kelly 1990; Rao et al. 1998). Endopeptidases are divided into four classes according to their catalytic mechanism: serine proteases (3.4.21), cysteine proteases (or thiol proteases) (3.4.22), aspartic proteases (or acid proteases) (3.4.23) and metalloproteases (3.4.24) (Ward 1983; Fogarty and Kelly 1990; Chander and Puri 2019).

Serine proteases are an industrially important sub-class of endopeptidases. These proteases are characterized by the presence of a serine residue at their active site, and by aspartate and histidine residues forming the catalytic machinery (Fogarty and Kelly 1990). They are found in viruses, bacteria and eukaryotes. These proteases generally operate at an optimum pH of 10, but this can vary between 7 and 11 (Rao et al. 1998; Chander and Puri 2019). Their activity is inhibited by the presence of diisopropyl fluorophosphate (DFP) and phenylmethylsulphonyl fluoride (PMSF). Similarly, some proteins in this sub-class are inhibited by the presence of thiol agents and p-chloromercuribenzoate (PCMB) (Fogarty and Kelly 1990). It has also been reported that some serine proteases require the presence of Ca^{2+} ions for their stability (Rao et al. 1998). Unlike serine proteases, cysteine proteases are rarely used on an industrial scale. These proteases are found in both prokaryotes and eukaryotes. Most proteases in this class are only activated in the presence of reducing agents such as cysteine or hydrocyanic acid (Chander and Puri 2019). Their activity is not affected by the presence of DFP and chelating agents, but is inhibited by the presence of sulfhydryl agents such as PCMB (Rao et al. 1998; Chander and Puri 2019).

Papain, a plant-based protein, is the only protease in this family to be significantly used on an industrial scale (Balatsouras 2006).

4

Enzyme Production for Food Applications

Mohamed Ghoul[1] and Seraphim Papanikolaou[2]

[1] *LRGP, CNRS, Université de Lorraine, Nancy, France*

[2] *LFMB, Agricultural University of Athens, Greece*

Most industrial enzymes are produced by wild or genetically modified microbial strains. The most widely used are hydrolytic enzymes, which mainly include amylases, pectinases and cellulases/hemicelluloses.

Other plant and animal-based sources of enzymes are also depicted. Examples include papain, rennet and brewing enzymes. Papain is a cysteine protease (thiol protease) that catalyzes the cleavage of peptide bonds. This enzyme is found in the latex of papaya plants (*Carica papaya* and *Vasconcellea cundinamarcensis*) and is used in the food, cosmetics and detergents industries. An example of animal-based enzymes is rennet, a mixture of proteolytic enzymes (chymosin and pepsin). This enzymatic cocktail is extracted from the abomasum of young ruminants and is widely used in the milk coagulation process (Balatsouras 2006). In terms of reaction, rennet acts on kappa casein at the Phe^{105}-Me^{106} position, triggering a specific cleavage that destabilizes the entire casein system and, consequently, the entire milk structure, inducing coagulation (Balatsouras 2006; Galiotou-Panayotou 2007).

Despite the effectiveness of plant- and animal-based enzymes, the majority of industrial processes use enzymes of microbial origin.

This chapter therefore presents developments in the field of microbial enzyme production. The strains used for the production of each enzyme category and the associated optimum operating conditions will be presented and discussed.

Implementation of Enzymatic Processes and Lactic Bacteria in the Food Industries, coordinated by Mohamed Ghoul.

4.1. Amylases

Amylases are used in many processes to transform starch and other polysaccharides. There are three types of amylase, which are as follows: α-amylase, β-amylase and γ-amylase. They are further divided into three categories: endoamylases, exoamylases and debranching enzymes. The production of α-amylase is obtained from mold, yeast and bacterial strains (*Candida tsukubaensis*, *Filobasidium capsuligenum*, *Lipomyces* sp., *Saccharomycopsis capsularis*, *Saccharomycopsis fibuligera*, *Schwanniomyces alluvius*, *Schwanniomvces casteilli*, *Trichosporon pullulans* and *Candida isikubaensis*). The optimum pH and temperature for α-amylase production from yeast are, respectively, in the range of 4.5–5.5 and 50–70°C. In contrast, production of these enzymes by bacteria (*Bacillus amyloliquefaciens*, *Bacillus acidocaldarius*, *Bacillus polymyxa*, *Bacillus subtilis*, *Bacillus stearothermophilus*, *Bacillus licheniformis*, *Clostridium thermosulfuricum*, etc.) takes place at a variable pH between 4.0 and 6.5, and at a temperature between 60 and 70°C (Balatsouras 2006; Galiotou-Panayotou 2007; Djekrif et al. 2016; Aggelis 2017). It should be noted that *Bacillus licheniformis*, *Clostridium thermosulfuricum* and, above all, *Bacillus stearothermophilus*, involved in the production of α-amylases, are thermoresistant strains with an optimum temperature of 90–95°C. The fungi most commonly used for α-amylase production are strains of *Aspergillus niger*, *Aspergillus oryzae*, *Calvatia gigantea*, *Aureobasidium pullulans*, *Mucor pusillus*, *Penicillium expansum*, *Penicillium chrysogenum* and *Penicillium roqueforti*. The optimum pH and temperature are 4.0–6.0 and 40–60°C, respectively (Galiotou-Panayotou 2007; Djekrif et al. 2016; Gopinath et al. 2017). In addition to optimal temperature and pH conditions, the production of the α-amylase enzyme is promoted by the presence of Ca^{2+} ions (Balatsouras 2006; Gopinath et al. 2017).

β-Amylase enzymes are mainly extracted from plants and animals. However, they can also be produced by microorganisms, mainly prokaryotes, and especially strains belonging to the *Bacillus polymyxa*, *Bacillus megaterium* and *Pseudomonas* sp. species/genera (Balatsouras 2006; Aggelis 2017). The strains involved in the production of this enzyme require a pH in the range of 6.0–6.5 and a temperature of 50 ± 5°C (Galiotou-Panayotou 2007). The glucoamylase enzyme is synthesized by fungi (*Aspergillus niger*, *Aspergillus oryzae*, *Rhizopus* sp. and *Mucor* sp.), yeast strains (*Saccharomycopsis capsularis*, *Trichosporon pullulans*, *Schwanniomyces castelli*, *Lipomyces kononenkoae* and *Candida guilliermondii*) and bacteria (*Enterobacter aerogenes* and *Streptococcus* sp.) (Balatsouras 2006; Aggelis 2007; Djekrif et al. 2016).

Table 4.1 summarizes the main microbial sources of starch decompositioin, as well as the type of enzymes produced by these microorganisms.

Microbial source	Type of amylase	Cleaved bond
	a) Bacteria	
Bacillus polymyxa	β-Amylase, α-amylase	α-1,4
Pseudomonas sp.	β-Amylase	α-1,4
Bacillus acidocaldarius	α-Amylase	α-1,4
Bacillus amyloliquefaciens	α-Amylase	α-1,4
Bacillus megaterium	β-Amylase	α-1,4
Bacillus subtilis	α-Amylase	α-1,4
Bacillus stearothermophilus	α-Amylase	α-1,4
Bacillus licheniformis	α-Amylase	α-1,4
Clostridium thermosulfurogenes	α-Amylase	α-1,4
Clostridium thermosulfuricum	α-Amylase	α-1,4
Enterobacter aerogenes	Glucoamylase	α-1,4; α-1,6; α-1,3
Streptococcus sp.	Glucoamylase	α-1,4; α-1,6; α-1,3
	b) Yeast	
Saccharomycopsis capsularis	α-Amylase, glucoamylase	α-1,4; α-1,6; α-1,3
Saccharomycopsis fibuligera	α-Amylase	α-1,4
Lipomyces kononenkoae	Glucoamylase	α-1,4; α-1,6; α-1,3
Candida tsukubaensis	α-Amylase	α-1,4
Candida guilliermondii	Glucoamylase	α-1,4; α-1,6; α-1,3
Trichosporon pullulans	α-Amylase, glucoamylase	α-1,4; α-1,6; α-1,3
Microbial source	**Type of amylase**	**Cleaved bond**
	c) Mushrooms	
Aspergillus niger	α-Amylase, glucoamylase	α-1,4; α-1,6; α-1,3
Aspergillus oryzae	α-Amylase, glucoamylase	α-1,4; α-1,6; α-1,3
Calvatia gigantea	α-Amylase, glucoamylase	α-1,4; α-1,6; α-1,3
Mucor sp.	α-Amylase, glucoamylase	α-1,4; α-1,6; α-1,3
Rhizopus sp.	Glucoamylase	α-1,4; α-1,6; α-1,3
Aureobasidium pullulans	α-Amylase, pullulanase	α-1,4; α-1,6

Table 4.1. *Production of amylases by microorganisms (according to Balatsouras (2006), Aggelis (2007), Galiotou-Panayotou (2007), Djekrif et al. (2016), Gopinath et al. (2017) and Saini et al. (2017))*

The biosynthesis of extracellular amylases is affected by several factors, the most significant of which is the nature of the carbon source used as substrate. Indeed, some work has indicated that microbial amylase production is induced by the presence of starch in the fermentation medium (Kumari et al. 2019). On the other hand, other work has reported that amylase synthesis is not subjected to catabolic repression by monosaccharides present in the fermentation medium, and that the latter can instead constitute a substrate that is favorable to the production of this enzyme (Gopinath et al. 2017; Kumari et al. 2019). Other factors, such as the age and initial amount of inoculum, the source and initial amount of nitrogen, and the incubation temperature, also appear to play an important role in the process of amylase biosynthesis (Gopinath et al. 2017; Kumari et al. 2019). Amylase production by various fungal strains (belonging to the *Calvatia gigantea* and *Aspergillus oryzae* species, etc.) is in the range of 20–40 U/mL (1 U = the amount of enzyme releasing 1 μmol of maltose per minute). This is considered a satisfactory production (Gopinath et al. 2017). In the case of using starch as the sole carbon source by wild strains such as *Cunninghamella echinulata* and *Mortierella* sp., the amount of α-amylases synthesized is significantly lower (≈0.5–2.5 U/mL) (Dyal and Narine 2005; Papanikolaou et al. 2007a).

Research over the last 20 years has shown that the main strains used for amylase production are genetically modified strains of the yeast *Yarrowia lipolytica*. These strains have undergone genetic transformation in order to construct mutant phenotypes capable of constitutively secreting amylases, while simultaneously producing very large quantities of dry yeast biomass (>60 g/L) and extracellular α-amylase (>25 U/mL). These strains are grown on inexpensive carbon sources (i.e. glycerol, that is the principal by-product of biodiesel production) in perfectly stirred bioreactors (Kim et al. 2000; Celińska et al. 2017).

4.2. Pullulases

The pullulanase enzyme is a debranching enzyme that hydrolyzes the α-1,6 bonds of amylopectin, releasing dextrins that only contain α-1,6 bonds (Saini et al. 2017). This enzyme is produced by various bacteria and archaea.

4.3. Lignocellulases

Cellulases are classified into three major categories (endoglucanases, exoglucanases, or cellobiohydrolases, and β-glycosidases or cellobiases)

(Galiotou-Panayotou 2007; Obeng et al. 2007). The vast majority of cellulase preparations are produced by *Trichoderma reesei*-type strains. These strains have been the subject of a great deal of research concerning both the mechanisms of cellulase production and the degradation processes of lignocellulosic biomass carried out by these very enzymes. The knowledge acquired through the various studies carried out on *Trichoderma reesei* strains has been exploited, so as to develop new cellulase-producing strains using various engineering methods. This has been particularly applied in strains of the non-conventional yeasts *Yarrowia lipolytica* and *Pichia pastoris* (Celińska et al. 2021).

Although a large proportion of the cellulases used for lignocellulosic biomass hydrolysis are derived from fungi (*Trichoderma reesei*), new strains of prokaryotes with cellulase production potential have been characterized in recent years and are now widely exploited (Obeng et al. 2017). In several bacteria, cellulases are mainly present in extracellular form in aggregated structures attached to cells in extracellular complexes. These include cellulosomes produced by bacteria, belonging to the *Clostridium* and *Thermotoga* genera. These bacteria are highly stable against various extreme values of temperature and pH (Obeng et al. 2017).

Cellulolytic enzymes are enzymes whose biosynthesis can either be induced or suppressed by the presence of various components in the fermentation medium. Indeed, in the case of fungus-like microorganisms, studies have shown that the presence of cellulose, cellobiose, sophorose or lactose in the culture medium induces cellulase synthesis, whereas the presence of glucose completely suppresses this very process (Enari 1983). Given that cellulose is not very soluble in the medium, the most influential inducers are those resulting from the hydrolysis of this biopolymer. Thus, cellobiose seems to be the most significant inducer for the synthesis of these enzymes. Table 4.2 summarizes the induction and/or repression of cellulase synthesis (according to Enari (1983)).

Enzyme	Inducer	Catabolic repressor	Inhibitor
Cellobiohydrolase	Cellulose, cellobiose	Glucose, galactose, and others	Cellobiose
Endoglucanase	Sophorose, lactose	Easily assimilated carbon sources	Glucose

Table 4.2. *Cellulase biosynthesis control in Trichoderma reesei and Phanerochaete chrysosporium (according to Enari (1983))*

Hemicellulases consist of a group of enzymes that are capable of hydrolyzing glycosidic bonds (α-glucuronidases, endo-1,4-β-xylanases, α-arabinofuranosidases, α-galactosidases and β-xylosidases) and enzymes catalyzing the hydrolysis of ester bonds involving acetic and ferulic acids (acetyl or, alternatively, feruloyl esterases) (Obeng et al. 2017; Sharma and Sharma 2017). Xylanases can be produced by bacteria and actinomycetes (*Bacillus* sp., *Pseudomonas* sp., *Streptomyces* sp.). They are effective over a wide range of pH levels (from 5.0 to 9.0) and temperature (35°C and 60°C) (Sharma and Sharma 2017). Nevertheless, the most important sources of xylanases are fungi (strains of *Fusarium* sp., *Aspergillus* sp., *Pleurotus* sp., etc.). These xylanases are more active than those produced by bacteria. However, their range of action regarding pH or temperature is limited (Sharma and Sharma 2017).

Cellulases are used in the degradation process of various substrates, either as extracts or as whole cells (Diamantopoulou and Philippoussis 2001; Philippoussis et al. 2001, 2011; Philippoussis 2009; Knob et al. 2014; Economou et al. 2017, 2020; Sharma and Sharma 2017; Mejias et al. 2018).

The production of cellulases essentially involves solid-state fermentation processes. Under these conditions, production ranges between 0.5 and 400 U/g dry matter for cellulolytic enzymes, and 30 and 40,000 U/g dry matter for xylanases (Philippoussis 2009; Philippoussis et al. 2011; Knob et al. 2014; Mejias et al. 2018). These very significant differences can be attributed to the types of strains used (Philippoussis et al. 2011; Mejias et al. 2018), the presence of nitrogen and the initial C/N ratio of the culture medium (Enari 1983; Philippoussis et al. 2001, 2011), as well as the types of carbon substrates used (Philippoussis 2009; Philippoussis et al. 2011; Knob et al. 2014). They may also be explained by the fact that for xylanases, there is no standard substrate for measuring the reducing sugars released, and that for pure xylan, other substrates have been used (Knob et al. 2014).

In general, fungal sources are the most important producers of cellulase/hemicellulase enzymes. Table 4.3 shows the fungal sources of cellulase/hemicellulase production, the type of substrate used, and the enzyme activity obtained (adapted from Knob et al. (2014) and Mejias et al. (2018)).

Cellulases can also be produced by bacteria. Table 4.4 summarizes the microbial sources of cellulase/hemicellulase production, as well as several characteristics of the substrate used to produce these enzymes.

Fungal source	Type of enzyme	Production (U/g)	Substrate
Aspergillus niger	Xylanase	5,662	Apple pomace
Aspergillus niger	Xylanase	42	Barley fiber
Aspergillus niger	Xylanase	106.42	Citrus peel
Aspergillus foetidus	Xylanase	3,065	Corn cob
Trichoderma reesei	Cellulase	10.78	Soy hulls and wheat fiber
Aspergillus oryzae	Endoglucanase	100.67	Soy hulls and wheat fiber
Trichoderma asperellum	β-Glycosidase	384.4	Agave atrovirens fiber
Trichoderma reesei	Cellulase	5.74	Soy hulls
Trichoderma reesei	Endoglucanase	58.6	Soy hulls
Aspergillus foetidus	Xylanase	100–40,000	Agri-food waste
Aspergillus oryzae	Endoglucanase	132.3	Rice straw and wheat fiber
Aspergillus oryzae	Cellulase	35.8	Rice straw and wheat fiber
Trichoderma reesei	Cellulase	22.8	Wheat fiber
Fungal source	**Type of enzyme**	**Production (U/g)**	**Substrate**
Aspergillus fumigatus	Cellulase	98.1	Corn semolina
Aspergillus fumigatus	Endoglucanase	341.3	Corn semolina
Fusarium oxysporum	Xylanase	1,840	Corn stalk
Humicola lanuginosa	Xylanase	610	Canola flour

Table 4.3. *Comparative results of cellulase/hemicellulase production by fungal species (adapted from Knob et al. (2014) and Mejias et al. (2018))*

Laccases are enzymes that are widely distributed in plants, fungi and bacteria. They exhibit different functions and stabilities, depending on the source from which they are derived (Galiotou-Panayotou 2007). Studies carried out on the molecular structure of these enzymes and the assessment of their electrochemical potential have made it possible to classify laccases according to their high, medium or low redox potential (Obeng et al. 2017). Laccases of plant and/or bacterial origin are considered low-redox potential enzymes, while laccases of fungal origin belong to the high or medium-redox potential enzymes (Obeng et al. 2017). Consequently, laccases of fungal origin exhibit high activity towards wood depolymerization, and are widely distributed among ascomycetes, deuteromycetes and basidiomycetes. The species known to be most efficient at secreting and producing laccases are the "white-rot" fungi (*Pleurotus* sp., *Lentinula edodes*, *Ganoderma* sp., etc.) (Diamantopoulou and Philippoussis 2001; Philippoussis et al. 2001; Aggelis 2007, 2017; Philippoussis 2009; Dwivedi et al. 2011). There are also bacterial strains capable of secreting laccases (*Bacillus subtilis*, *Anabaena azollae*, *Streptomyces cyaneus* and *Azospirillum lipoferum*) (Kefalogianni and Aggelis 2006; Tsagou and Aggelis 2006; Dwivedi et al. 2011; Obeng et al. 2017).

Microbial source	Type of enzyme	Origin
	Bacteria	
Paenibacillus polymyxa	Cellulase	Compost
Bacillus cereus, B. subtilis	Cellulase, β-glycosidase	Soil
Pseudomonas fluorescens	Cellulase, β-glycosidase, β-xylosidase	Soil, activated sludge
Pseudomonas putida	Cellulase	Soil, activated sludge
Oenococcus oeni	β-Glycosidase	Grape must
Alicyclobacillus acidocaldarius	β-Glycosidase	Compost
Pyrococcus furiosus	β-Glycosidase	Compost
Clostridium phytofermentans	Cellulase	Soil
Clostridium acetobutylicum	Endo-1,4-β-xylanases	Soil
	Fungi	
Aspergillus niger, Trichderma resei	Cellulase, β-glycosidase	Wheat fiber
Aspergillus terreus	Cellulase, β-glycosidase	Wheat straw
Aspergillus oryzae, Trichderma resei	Cellulase, β-glycosidase, xylanases	Soy hulls
Aspergillus foetidus	Xylanase	Agricultural residues
Trichoderma asperellum	Cellobiohydrolases, xylanases	Agave fiber
Pleurotus ostreatus	Cellulase, xylanases	Corn silage digestate

Table 4.4. *Production of cellulases/hemicellulases by microorganisms (Obeng et al. 2017; Sharma and Sharma 2017; Mejias et al. 2018)*

Fungi have the ability to degrade cellulose, hemicellulose and lignin, either simultaneously or sequentially (Elisashvili et al. 2009; Philippoussis 2009; Philippoussis and Diamantopoulou 2012). They can be used directly as catalysts to produce molecules of interest (Diamantopoulou and Philippoussis 2001; Elisashvili et al. 2009; Philippoussis 2009; Philippoussis and Diamantopoulou 2012; Diamantopoulou et al. 2020).

Figure 4.1 illustrates the ability of these fungi to degrade the phenolic compounds in pomace derived from olive oil extraction.

Figure 4.1. *Removal of phenolic compounds during fermentation of olive mill wastewater by Ganoderma resinaceum strains. Kinetic results obtained according to Diamantopoulou et al. (2021). For a color version of this figure, see www.iste.co.uk/ghoul/implementation.zip*

In order to improve the laccase production process, several genes involved in the biosynthesis of these enzymes were cloned from different microorganisms, and were heterologously expressed in yeasts (*Saccharomyces cerevisiae, Pichia pastoris, Pichia methalonica, Yarrowia lipolytica* and *Kluyveromyces lactis*), filamentous fungi (*Aspergillus niger, Aspergillus oryzae* and *Trichoderma reesei*) and bacteria (*Escherichia coli* and *Streptomyces lividans*) (Dwivedi et al. 2011).

4.4. Pectinases

Pectolytic enzymes are present in plant cells and are naturally endogenous enzymes. Pectinases can also be produced by numerous phytopathogenic microorganisms (Aggelis 2007, 2017). The most suitable microorganisms for pectolytic enzyme synthesis are in order of priority: fungi, bacteria and yeasts

(Lowe 2001; Galiotou-Panayotou 2007). On an industrial scale, these enzymes are mainly produced by culturing various types of fungi. In most cases, these enzymes are produced extracellularly, mostly in submerged liquid cultures. The composition of the culture medium is the most significant factor affecting the production of these enzymes. Indeed, most pectinases are produced "inducibly"; in other words, in the presence of a carbon source (mainly pectins) (Fogarty and Kelly 1983; Lowe 2001; Galiotou-Panayotou 2007). The best-known suppliers of pectolytic enzymes are as follows: Novozymes (Denmark); Novartis (Switzerland); Roche (Germany); Biocon (India) (Gummadi and Panda 2003). According to the classification of the International Union of Biochemistry, pectolytic enzymes are either hydrolases or lyases (Fogarty and Kelly 1983; Galiotou-Panayotou 2007). These include protopectinases, polygalacturonases, lyases and pectinesterases.

Pectinesterases or pectinomethylesterases hydrolyze the ester bonds formed between the carboxyl of galacturonic acid and methanol. These enzymes exist in the plant tissues of several plants (*Carica papaya*, *Lycopersicum esculentum*, *Prunus malus*, *Vitis vinifera*, *Citrus* sp., etc.) (Jayani et al. 2005). They can also be synthesized by fungal, yeast and bacterial strains (*Phytophthora infenstans*, *Saccharomyces cerevisiae*, *Aspergillus niger*, *Aspergillus japonicus*, *Aspergillus sojae*, *Aspergillus saito*, *Xanthomonas* sp., *Penicillium digitatum*, *Penicillium frequetans* and *Penicillium chrisogenum*) (Fogarty and Kelly 1983; Jayani et al. 2005; Aggelis 2007, 2017).

Protopectinases catalyze the conversion of insoluble protopectin to water-soluble pectin. Protopectinase synthesis has been identified in some yeasts (*Kluyveromyces fragilis* and *Galactomyces reesei*) and bacteria (*Bacillus* sp.) (Jayani et al. 2005).

Polygalacturonases hydrolyze the α-1,4 glycosidic bonds between biopolymer building blocks. They are divided into endo-polygalacturonases (endo-PG, EC 3.2.1.15) and exo-polygalacturonases (exo-PG, EC 3.2.1.67) (Jayani et al. 2005; Amin et al. 2019). The production of endo-polygalacturonases has been achieved by several microorganisms, including filamentous fungi and bacteria (*Rhizoctonia solani*, *Thermomyces lanuginosus*, *Aspergillus* sp., *Rhizopus arrhizus*, *Erwinia carotorova*, *Penicillium expansum*, *Penicillium chrysogenum*, etc.) (Fogarty and Kelly 1983; Jayani et al. 2005). In contrast, the production of exo-polygalacturonases is less widespread among microorganisms and is mainly carried out by bacteria (*Bacillus* sp., *Erwinia aroideae*, *Pseudomonas* sp.) (Fogarty and Kelly 1983).

Polygalacturonases are produced by biotechnological methods, involving a large number of microorganisms. The main enzyme-producing systems are filamentous fungi belonging to the *Aspegillus* sp. genus (Galiotou-Panayotou et al. 1993; Nighojkar et al. 2006; Amin et al. 2019). In most cases, liquid-state fermentation has been used for the production of these enzymes (Amin et al. 2019). Nevertheless, in recent years, numerous publications have highlighted the performance of solid-state fermentation for obtaining these enzymes (Silva et al. 2005; Wong et al. 2017; Amin et al. 2019).

Lyases catalyze the cleavage of glycosidic bonds, without the addition of water molecules. This is a β-type detachment.

Table 4.5 shows the microbial sources of pectolytic enzyme production, with some characteristics of the fermentations carried out.

Microorganism	Activity	Substrate	Type of culture	Reference
Aureobasidium pullulans LV10	22 U/mL	Apple pectin	SmF (*)	(Manachini et al. 1988)
Aspergillus niger VTT-D-86267	180 U/mL	Sugar beet pectin	SmF/bioreactor	(Bailey and Pessa 1990)
Aspergillus niger NNRL-364	14.5 U/mL	Citrus pectin	SmF (*)	(Galiotou-Panayotou et al. 1993)
Aspergillus niger	4 U/mL	Sucrose	SmF (*)/ bioreactor	(Friederich et al. 1994)
Aspergillus sp. ATHUM-3482	4.3 U/mL	Citrus pectin	SmF (*)	(Galiotou-Panayotou et al. 1997)
Rhizopus stolonifer	4 U/mL	Citrus pectin	SmF (*)	(Naidu and Panda 1998)
Aspergillus kawachii IFO 4308	12 U/mL	Soy flour	SmF (*)/ bioreactor	(Kojima et al. 1999)
Aspergillus niger M21	13.6 U/mL	Wheat–soy fiber	SSF (**)	(Castilho et al. 2000a)
Aspergillus niger	19 U/mL	Wheat–soy fiber	SSF (**)	(Castilho et al. 2000b)
Aspergillus niger 3T5B8	30.75 U/mL 4.20 U/mL	Wheat fiber Mango peel	SSF (**)	(Couri et al. 2000)
Penicillium occitanis CT1	221 U/mL exo-PG 86 U/mL endo-PG	Apple pectin	SmF (*)	(Hadj-Taieb et al. 2002)
Aspergillus niger V. Tieghem	11 U/mL	Apple pectin	SmF (*)	(Fawole and Odunfa 2003)
Trichoderma reesei Rut C-30	1.1 U mL	Cellulose–cane sugar	SmF (*)/ bioreactor	(Olsson et al. 2003)
Penicillium viridicatum RFC3	8.9 U/mL	Citrus bagasse–wheat fiber	SSF (**)	(Silva et al. 2005)

Microorganism	Activity	Substrate	Type of culture	Reference
Aspergillus niger	0.938 U/mL 0.963 U/mL	Citrus peels Citrus pectin	Immobilization of filaments/ SmF	(Nighojkar et al. 2006)
Mortierella isabellina	1.20 U/mL	Apple pectin	SmF (*)	(Papanikolaou et al. 2007a)
Cunninghamella echinulata	1.62 U/mL	Apple pectin	SmF (*)	(Papanikolaou et al. 2007a)
Aspergillus terreus NCFT4269.10	~6,500 U/g	Banana peel	SSF (**)	(Sethi et al. 2016)
Coriolus versicolor	~90 U/g	Wheat fiber	SmF (*)	(Amin et al. 2017a, 2017b)
Penicillium notatum	~110 U/g	Wheat fiber	SmF (*)	(Amin et al. 2017a, 2017b)
Aspergillus fumigatus R6	565 U/g	Wheat fiber	SSF **	(Wong et al. 2017)

SmF (*): submerged fermentation; SSF(**): solid-state fermentation.

Table 4.5. *Comparative results for the production of pectolytic enzymes by microbial species*

4.5. Lipases

Lipases (glycerol ester hydrolases; EC 3.1.1.3) are water-soluble enzymes that are capable of hydrolyzing ester functions and are specialized in the transformation of glycerides (triglycerides (TAGs), diglycerides (DAGs) and monoglycerides (MAGs)) into glycerol and free fatty acids (FFAs) (Macrae 1983). The study of fatty acid transfer and, in general, the degradation of fatty (hydrophobic) materials, mostly by the unconventional polymorphic yeast *Yarrowia lipolytica* has been the subject of several articles (Aggelis et al. 1997; Papanikolaou et al. 2001, 2003; Papanikolaou and Aggelis 2003, 2010; Beopoulos et al. 2009; Tzirita et al. 2018). In the case of the growth of microorganisms on substrates composed of TAGs, FFAs liberated after lipase-catalyzed hydrolysis of the mentioned TAGs, are incorporated inside the cells and a portion of them is stored, again, as TAGs, and to a lesser extent, steryl-esters. This is the so-called ex novo process of intracellular lipid accumulation. Other FFAs, liberated as mentioned above through medium TAGs hydrolysis, can be metabolized through β-oxidation, providing the intracellular carbon source for the biosynthesis of cellular constituents (Beopoulos et al. 2009; Papanikolaou and Aggelis 2010, 2019; Athenaki et al. 2018).

Microorganisms can only grow on media where the carbon source is TAG-type, if these very strains used are capable of producing lipases (Macrae 1983; Ratledge 1994; Beopoulos et al. 2009; Papanikolaou and Aggelis 2010). Indeed, in the case of the non-conventional polymorphic yeast *Yarrowia lipolytica*, its culture on TAG-type substrates is accompanied by the secretion of an extracellular lipase called Lip2p, encoded by the LIP2 gene (Pignède et al. 2000; Beopoulos et al. 2009). The secreted lipase has been reported to be a 301 amino acid glycosylated polypeptide, belonging to the TAG hydrolase family (EC 3.1.1.3) (Pignède et al. 2000; Beopoulos et al. 2009; Papanikolaou and Aggelis 2019). Other intracellular lipases (Lip7p, Lip8p) can also be produced and secreted into the culture medium. These enzymes show different specificities toward extracellular fatty acids, with maximum activity toward Δ9C18:1 (oleic), C6:0 (caproic) and C10:0 (capric) acids (Fickers et al. 2005; Beopoulos et al. 2009; Papanikolaou and Aggelis 2010).

Lipase secretion into the culture medium can be considered as a physiological response of *Yarrowia lipolytica*, as well as other yeast and fungal strains, to the presence of long-chain FFAs or TAGs in the medium (Hadeball 1991; Sharma et al. 2001; Fickers et al. 2005). Microbial lipase production has been improved by the addition of organic nitrogen to the fermentation medium (peptone, urea, etc.) (Pereira-Meirelles et al. 1997; Dominguez et al. 2003; Montesinos et al. 2003; Papanikolaou et al. 2007b; Papanikolaou and Aggelis 2010). Similarly, lipase production is strongly influenced by the addition of various nutrients and surfactants to the medium, by the agitation and aeration rates imposed, and by the use of perfluorocarbons as oxygen carriers (Dominguez et al. 2003; Alonso et al. 2005; Amaral et al. 2006). Several wild strains of microorganisms (*Yarrowia lipolytica*, *Candida rugosa*, *Aspergillus niger*, *Aspergillus* sp., *Penicillium* sp., *Mucor* sp., etc.) are also capable of producing lipases on hydrophobic media at a rate between 1.0 and 60 U/mL (Pereira-Meirelles et al. 1997; Dominguez et al. 2003; Montesinos et al. 2003; Alonso et al. 2005; Kamzolova et al. 2005; Szczęsna-Antczak et al. 2006; Papanikolaou et al. 2007b, 2011).

Improvements in lipase production have been achieved through the use of genetic engineering tools and random mutagenesis, notably in the *Yarrowia lipolytica* yeast (Pignède et al. 2000; Nicaud et al. 2002; Fickers et al. 2004, 2005; Madzak 2021). An expression system was therefore developed, thanks to which strains containing multicopy integration of expression cassettes carrying the LIP2 gene, expressed under the control of the oleic acid-inducible POX2 promoter, were obtained (Pignède et al. 2000; Nicaud et al. 2002). Through the optimization of culture conditions, lipase concentrations of around 2,000 U/mL were achieved in shake flasks, and up to 11,500 U/mL in batch bioreactors. These trials were carried out with glucose as the sole carbon source. Bioreactor cultures of genetically modified *Yarrowia lipolytica* strains in fed-batch mode produced very high yields (90,500 U/mL). These fermentations were carried out with glucose as the sole carbon source. During these fermentations, biomass reached concentrations approaching 80 g/L (Nicaud et al. 2002).

4.6. Proteases

Peptidases (or proteases, or so-called "proteolytic" enzymes) are enzymes that catalyze the cleavage of peptide bonds in proteins. On an industrial scale, proteases are among the most widely used hydrolases. Proteases are divided into two groups according to their mode of action: intracellular and extracellular proteases (Fogarty and Kelly 1990; Galiotou-Panayotou 2007; Chander and Puri 2019).

Intracellular proteases are involved in several cellular processes, such as sporulation, differentiation, hormone maturation and maintenance of the cellular protein pool (Aggelis 2017). However, this type of protease is of moderate interest, as their use requires a cell lysis stage to ensure their extraction (Galiotou-Panayotou 2007).

Extracellular proteases enable the hydrolysis of proteins in the cell's external environment, helping the cell to absorb and utilize the products of this hydrolysis, such as amino acids (Ward 1983; Aggelis 2007, 2017; Galiotou-Panayotou 2007). These enzymes are therefore more interesting and easier to use on an industrial scale than intracellular proteases (Balatsouras 2006; Galiotou-Panayotou 2007). Recovery of these enzymes involves a centrifugation stage in the case of bacteria and yeasts, or a filtration stage in the case of filamentous fungal mycelia (Galiotou-Panayotou 2007). Within this group of enzymes, a distinction is made between endopeptidases and exopeptidases. Aminopeptidases and carboxypeptidases are the most important exopeptidase enzymes. This group acts close to the N-terminus of proteins. There are over 150 different species of microorganisms (bacteria, yeasts and molds) capable of producing exopeptidases (Ward 1983). On an industrial scale, endopeptidases are the most widely used. They are characterized by their hydrolytic action at a specific site on the polypeptide chain, far from the N-terminus and C-terminus (Fogarty and Kelly 1990; Rao et al. 1998).

Table 4.6 shows the most important protease classes and sub-classes, as well as the microbial sources involved in the production of these enzymes (according to Ward (1983), Fogarty and Kelly (1990) and Rao et al. (1998)).

Type of protease	Class and sub-class	Microbial source
Exopeptidases (3.4.11-17)	α-Aminopeptide hydrolases (3.4.11)	–
	Aminopeptidase (cytosolic) (3.4.11.1)	Yeasts, bacteria
	Proline iminopeptidase (3.4.11.5)	*Escherichia coli*
	Aminopeptidase P (3.4.11.9)	*Escherichia coli*
	Thermophilic aminopeptidase (3.4.11.12)	Bacteria
	Clostridium histolyticum aminopeptidase	*Clostridium histolyticum*
	Dipeptide hydrolases (3.4.13)	–
	α-Glutamyl-glutamate dipeptidase (3.4.13.7)	Bacteria
	β-Aspartyldipeptidase (3.4.13.10)	*Escherichia coli*
	Dipeptidase (3.4.13.11)	*Mycobacterium phlei*
	Peptidyl dipeptide hydrolases (3.4.15)	–
	Dipeptidyl carboxypeptidase (3.4.15.1)	*Escherichia coli*
	Serine carboxypeptidases (3.4.16)	–
	Carboxypeptidase (3.4.16.1)	Yeasts, *Penicillium* sp.
	Metallo-carboxypeptidases (3.4.17)	–
	Glycine carboxypeptidase (3.4.17.4)	Yeasts
	Aspartate carboxypeptidase (3.4.17.5)	*Pseudomonas* sp.
	Alanine carboxypeptidase (3.4.17.6)	Bacteria
	Acylmuramoil-alanine carboxypeptidase (3.4.17.7)	*Bacillus subtilis*
Endopeptidases (3.4.21-24)	Serine proteases (3.4.21)	–
	Elastase (3.4.21.11)	*Pseudomonas* sp.
	Myxobacter α-lytic protease (3.4.21.12)	*Mycobacterium sorangium*
	Microbial serine proteases (3.4.21.14)	*Bacillus subtilis*, *Tritirachium album*, *Pseudomonas aeruginosa*, *Staphylococcus aureus*, *Yarrowia lipolytica*, etc.

Type of protease	Class and sub-class	Microbial source
Endopeptidases (3.4.21-24)	Thiol proteases (3.4.22)	–
	Clostripain (3.4.22.8)	*Clostridium histolyticum*
	Yeast protease B (3.4.22.9)	Yeasts
	Acid proteases (3.4.23)	–
	Microbial carboxyl proteases (3.4.23.6)	*Aspergillus niger*, *Aspergillus oryzae*, *Rhizopus* sp., *Endothia parasitica*, *Mucor pusillus*, *Mucor miehei*, *Sacchromyces cerevisiae*, *Rhodotorula glutinis*, *Tetrahymena pyriformis*, etc.
	Metallo-proteases (3.4.24)	–
	Clostridium histolyticum colagenase (3.4.24.3)	Similar enzyme in many bacteria
	Microbial metallo-proteases (3.4.24.4)	*Streptomyces thermophilus*, *Streptomyces diacetilactis*, *Sarcina* sp., *Bacillus subtilis*, *Pseudomonas aeruginosa*, *Citrobacter freundii*, *Bacillus thermoproteolyticus*, *Aspergillus oryzae*, *Aspergillus niger*, *Penicillium roqueforti*, *Serratia marescens*, etc.
	Achromobacter iophagus colagenase (3.4.24.8)	*Achromobacter iophagus*
	Trichophyton mentagrophytes keratinase (3.4.24.9)	*Trichophyton mentagrophytes*

Table 4.6. *Classification and protease-producing microbial sources (according to Ward (1983), Fogarty and Kelly (1990) and Rao et al. (1998))*

Strains of the *Bacillus* sp. genus are the most significant source of proteases. The proteases produced by these strains require relatively neutral pH values (5.0–8.0) and have limited stability at high temperatures (Chander and Puri 2019). Their low thermal tolerance is advantageous for controlling their activity when manufacturing food hydrolysates with a low degree of hydrolysis. Some of the neutral proteases belong to the metallo-proteases and require divalent metal ions for their activity, whilst others are serine proteases, whose activity is unaffected by chelating agents (Chander and Puri 2019). Unlike the proteases synthesized by the *Bacillus* sp. genus, proteases produced by strains of *Pseudomonas* sp. show significant activity, even at high pH values (9.0–10.0) and temperatures (50–55°C) (Chander and Puri 2019).

Protease production by molds has been observed in several species (*Aspergillus* sp., *Trichoderma* sp., *Rhizopus* sp., etc.). Production operating conditions (pH, temperature) vary according to the species/strain used (Chander and Puri 2019).

In addition to pH and temperature, protease production by microbial strains is affected by a number of factors, such as carbon source, nitrogen source, agitation/aeration, etc. The use of glucose has been extensively studied. It has been established that glucose inhibits protease production through a catabolic repression mechanism in *Bacillus* sp. bacteria (Ferrero et al. 1996; Gupta et al. 2002). Indeed, during fermentation using glucose as a carbon source, protease production only begins after glucose depletion. In contrast, other carbon sources (lactose, fructose, etc.) can stimulate protease production in *Bacillus licheniformis* when used in relatively low concentrations (2–4% w/v) (Chander and Puri 2019). Similarly, low protease productions are obtained using inorganic nitrogen as the sole nitrogen source. The use of corn maceration liquor, soy flour and casein hydrolysates as a nitrogen source promotes protease synthesis in bacteria and molds (Chander and Puri 2019). On the other hand, the synthesis of these enzymes is inhibited by the presence of inorganic nitrogen sources that can be rapidly metabolized, such as amino acids and ammonium ions (Kumar et al. 1999). During the fermentation stage, oxygen transfer to the culture medium is an important factor (Calik et al. 1998, 2000). Calik et al. (2000) have demonstrated that control of oxygen transfer through the development of appropriate agitation and aeration processes significantly increases the activity of proteases produced by *Bacillus licheniformis* strains in synthetic media. Protease production has also been achieved on various residues employed as microbial substrates, including, but not limited to, residual corn liquors (Chander and Puri 2019), fish residue hydrolysates (Ellouz et al. 2001, 2003), whey (De Souza et al. 2001), sunflower meal (Kachrimanidou et al. 2013), rapeseed meal (Chatzifragkou et al. 2014), etc.

4.7. Conclusion

In this chapter, we have reviewed state-of-the-art sources and substrates used for enzyme production. For each enzyme category, the optimum conditions in terms of temperature, pH, culture medium composition, agitation, oxygen transfer and bioreactor operation have been reported. For some enzymes, genetic engineering tools are used to make production compatible with industrial requirements in terms of cost, stability, specificity and productivity. These improvements are described below.

Industrial-scale production is carried out using different bioreactor operating modes (batch, fed-batch, continuous) in liquid and solid media. The performance of these processes varies according to the induction or repression mechanisms that are specific to each enzyme family. It is therefore important to consider these aspects when developing an industrial enzyme production strategy.

Today, purified enzymes and enzyme cocktails can be found on the market that have been prepared specifically for the substrates to be hydrolyzed or functionalized. Thanks to this progress, the use of enzymes in the transformation of raw materials has become a standard operation and has been applied to many industrial fields, both food and non-food.

5

Ways to Improve Enzymatic Processes

Isabelle CHEVALOT and Catherine HUMEAU
LRGP, CNRS, Université de Lorraine, Nancy, France

5.1. Enzyme engineering

Enzymes are the catalysts of choice because of their specificity, chemo-, regio- and enantio-selectivity. However, these characteristics can also be a hindrance to their industrial use, including to specific non-natural substrates or reactions. Furthermore, like all proteins, enzymes are sensitive to environmental factors. pH, temperature, ionic concentration and the presence of organic solvents are all factors that can affect enzyme activity during a process. Lastly, some enzymes of interest display catalytic activity suited to the targeted reaction, but this is insufficient activity for their use on a large scale.

This requires enzymes to be adapted to production conditions, which are often far removed from natural ones, as well as to unnatural substrates or reactions (Figure 5.1). Furthermore, with a view to industrial efficiency, the catalytic potential of these enzymes needs to be optimized (Zhang et al. 2019; Bilal and Iqbal 2020; Contesini et al. 2020). These phases of adaptation and/or optimization of enzyme activity are inspired by the widely described concept that the structure of an enzyme determines its function(s). Thus, the three-dimensional structure of an enzyme includes at least one region in which recognition of the substrate(s) takes place, as well as a set of residues ensuring catalytic functions. In addition, specific structural

For a color version of all the figures in this chapter, see www.iste.co.uk/ghoul/implementation.zip.

Implementation of Enzymatic Processes and Lactic Bacteria in the Food Industries,
coordinated by Mohamed GHOUL.

elements guarantee the enzyme's stability in its environment and facilitate its access to the substrate. Any modification – even a minor one – of an enzyme's three-dimensional structure can alter its activity. This can be harmful, leading to a reduction or loss of activity; however, it can also be extremely beneficial, enabling the enzyme to bypass the limitations described above.

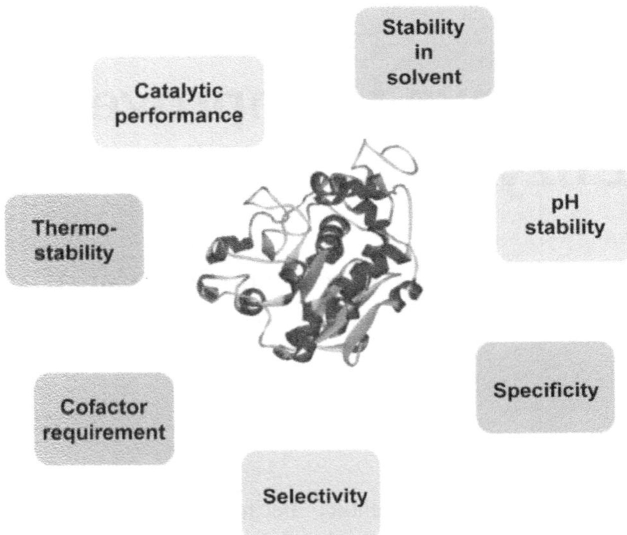

Figure 5.1. *Main features of enzymatic catalysts that need to be improved for industrial-scale use*

5.1.1. *Enzyme engineering methods*

Various approaches to enzyme modification have been developed. Some consider amino acids with functional groups to be potential reaction sites on which structural modifications can be made using classical chemistry. Others, which will be discussed in this chapter, propose to modify the set of amino acids involved in enzyme structure and function using protein engineering tools. Various engineering methods have been developed. Some are based on a random search for mutants that perform well in specific reactions, substrates and reaction conditions; others are more rational, using tangible elements to guide mutations. The choice of engineering method depends on several factors, such as the amount and type of information available concerning the enzyme's structure and catalytic mechanism, but also the ability to generate a large number of mutants and test their catalytic activity under given conditions. A lot of the time, a combination of different methods makes it possible to benefit from their advantages while avoiding their disadvantages, limiting the risk of failure in the search for mutants of interest. In addition, the work cited below as an example will show that bioinformatics tools have high potential for guiding engineering strategy (Sharma et al. 2021; Zhang et al. 2019; Bilal and Iqbal 2020).

5.1.1.1. *Directed evolution*

The natural diversity of biological systems is generated by a natural evolutionary process, leading to the production of enzymes that contribute to the survival of microorganisms in their natural environment. In the course of evolution, living organisms have had to adapt to various natural environments, such as salt-saturated waters, hot springs in volcanic regions, the cold waters of Antarctica and the high-pressure, low-temperature environments of the seabed. This slow evolution has been accompanied by the production of enzymes that are adapted to a wide variety of biological molecule transformations, according to reaction patterns that are often highly complex, under specific and sometimes extreme conditions. Directed evolution uses the principles and mechanisms of natural evolution in an accelerated way in a laboratory (Arnold 2018; Chowdhury and Maranas 2020). This method is used to study fundamental questions of enzyme evolution, with the aim of optimizing the genes of a living organism and, consequently, the enzymes it produces. In this case, directed evolution consists of modifying the genetic sequence encoding the enzyme to be improved, with the aim of creating banks of randomly generated mutants that target a specific characteristic.

A major challenge of this approach is the ability to assess the desired activity of these mutants, so as to identify sequences of interest. This approach therefore combines a process for generating genetic diversity with an efficient mutant selection process. The population of mutants generated is led to evolve by selecting mutants that have integrated the trait of interest (notion of selection pressure). The corresponding genetic information is then isolated and amplified. The mutation/selection cycle can be repeated several times in succession, so as to accumulate favorable mutations and gradually orientate the enzyme's spectrum of activity (Figure 5.2). It is also possible to combine genes that have been improved in previous evolutionary cycles. This approach has led to the development of DNA shuffling methods, which involve fragmenting first-generation selective genes, isolating mutation-carrying fragments and assembling them to reconstruct whole genes. Many researchers report that the resulting combinations of mutations allow the beneficial effects of independent mutations to be cumulated. These methods form the basis for the development of other methodological approaches, aimed at the creation of genetic diversity and the screening of activities.

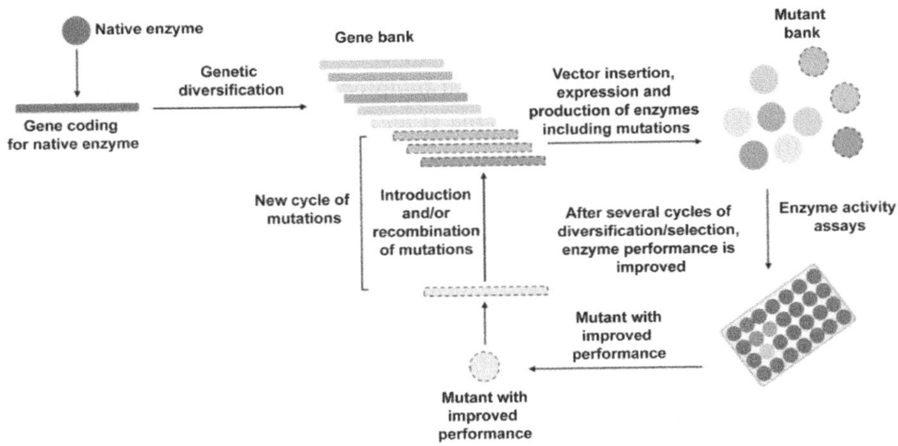

Figure 5.2. *Schematic representation of the main stages in directed evolution. An iterative process combining genetic diversification and the selection of mutants of interest, with respect to the desired enzymatic characteristics, is used to improve enzyme performance*

New or optimized enzymes, obtained by directed evolution, contain mutations distributed throughout structure, and not just strategically localized in regions for the catalytic activity or stability of the protein; the effect of these mutations on enzyme performance does not have a rational explanation, even in hindsight. This shows the potential of such an approach, compared with enzyme engineering methods that are based on a rational choice of modifications to be made. Indeed, if the starting gene is correctly selected and the activities of a large number of mutants can be screened, directed evolution can widen the field of possibilities by exploring mutation sites throughout the protein, without any preconceived notion of their potential involvement in the enzyme's properties.

Another advantage of directed evolution is that the structure or catalytic mechanism of the enzyme of interest does not need to be known. Indeed, directed evolution assumes that a randomly introduced mutation can have a beneficial effect that is difficult to predict. Furthermore, directed evolution provides access to a great deal of genetic diversity in a short space of time, compared with natural evolution. Nevertheless, the potential of this approach may prove useless in the absence of a means to screen the mutants generated, which would make it possible to identify sequences of interest with regards to specific implementation conditions or a given property. High-throughput analysis methods are therefore required for measuring the effects of a large number of random mutations. The development of such methods is often highly specific to the preservation of a particular activity, and therefore difficult to transfer to the study of other reaction systems. However, this stage is indispensable in the implementation of an enzyme improvement strategy through directed evolution.

5.1.1.2. *Site-directed mutagenesis*

In contrast to the random approach described above, a rational approach can be considered by directed mutagenesis. The latter aims to induce one or more mutations, accurately and deliberately (Kapoor et al. 2017; Korendovych 2018). It therefore requires in-depth knowledge of the structure of the enzyme of interest, its active site, the catalytic mechanism and the role of the different residues in its functions, stability and substrate recognition. Specific sequence changes are made, targeting particular regions of the protein involved in the property to be improved. There are three kinds of sequence change: substitution, involving the replacement of one or more residues by an identical number of residues; insertion, resulting in the addition of one or more residues to the sequence; or deletion, leading to the deletion of one or more residues from the sequence. A library containing a limited number of mutants, whose catalytic performance is to be determined, is thus generated (Figure 5.3). Unlike directed evolution, site-directed mutagenesis does not necessarily imply the availability of high-throughput tools for assessing enzymatic activity.

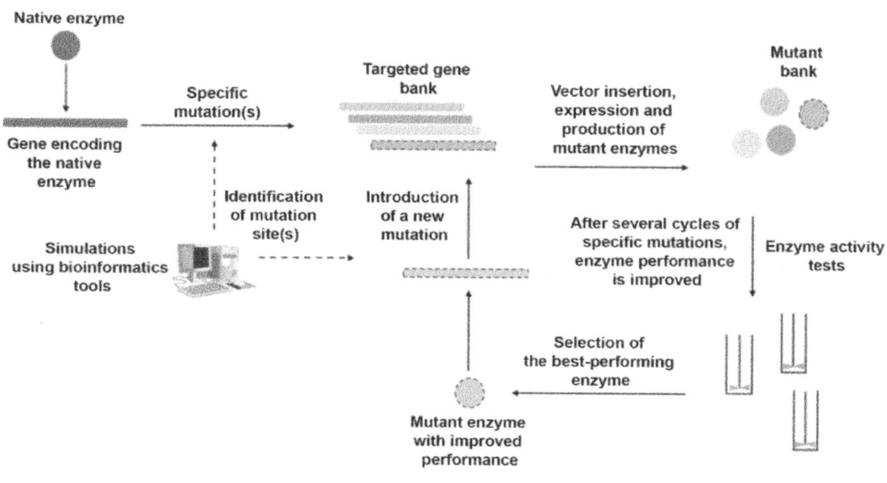

Figure 5.3. *Schematic representation of the main stages in enzyme improvement by site-directed mutagenesis. Digital simulations guide the choice of mutation sites of interest and the nature of the mutations*

Widely used to test hypotheses concerning the role of specific residues in enzyme function and, ultimately, to establish structure/function relationships, site-directed mutagenesis is also a powerful engineering method for improving biocatalysts. This approach is accompanied by simulations using bioinformatics tools to analyze the structure of the enzyme of interest to explain experimental biocatalysis results, such as selectivity or substrate specificity, and to understand and/or predict the effect of mutations on biocatalyst characteristics. When the structure of the enzyme has not been clarified, models based on the structure of homologous enzymes can be constructed (homology modeling).

Numerous studies using site-directed mutagenesis tools have shown that the mutation of a single residue can significantly alter the properties of an enzyme, either positively or negatively. Examples of frequently encountered mutations are the replacement of one or more hydrophobic residues by polar residues (or vice versa) to modify interactions with the substrate, or the replacement of one or more bulky residues by small ones (or vice versa) to modulate the volume of the catalytic site, with the introduction of residues favoring the establishment of disulfide bridges; the addition of polar – or, on the contrary, apolar – residues reinforce the network of electrostatic or hydrophobic interactions on the protein surface or in the environment surrounding the active site, etc. Once the mutation has been made, the new enzyme is produced and its properties are assessed. If these are not satisfactory, a new mutation cycle can be applied.

Despite its effectiveness in improving biocatalysts and generating knowledge about the structure/function relationships of enzymes, site-directed mutagenesis presents a number of limitations. A precisely and deliberately introduced mutation may be accompanied by unpredictable and undesirable effects, such as destabilization of the protein and subsequent loss of activity. Furthermore, a mutation may not have the desired effect, with insufficient or no gain in activity.

Lastly, the beneficial effect of certain mutations is sometimes difficult to predict, even when the structure and mechanism of the enzyme are well known; the rational nature of the approach therefore limits the field of investigation, at the risk of overlooking structural elements that are potentially involved in the desired characteristic.

5.1.1.3. *Combinatorial approaches*

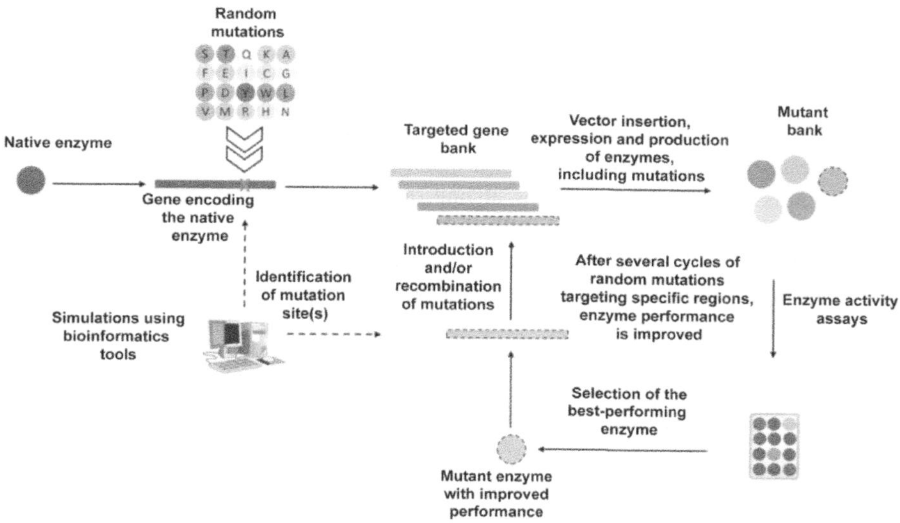

Figure 5.4. *Schematic representation of the main stages in site saturation mutagenesis. Random mutations are induced at one or more specific sites on the protein. Digital simulations guide the choice of mutation sites*

Directed evolution and site-directed mutagenesis involve a set of protein engineering methods that can lead to the improvement of enzymatic catalysts. As mentioned above, both approaches have their advantages and limitations. The choice of one or the other depends on a number of factors. The first is the availability of the three-dimensional structure of the enzyme of interest, or, failing that, of the structure of an enzyme that presents a high degree of sequence homology with the enzyme of interest. While this factor is essential for a rational strategy of site-directed mutagenesis, it is not necessary for random mutations. The second factor is the level of knowledge of the catalytic mechanism and, more generally, of the structure/function relationships characterizing the enzyme of interest. Any element relating to these different aspects will be extremely useful in guiding the choice of mutation sites and the type of mutation within the framework of a site-directed mutagenesis strategy. The third factor is the availability of high-throughput analysis tools and methods for assessing the enzymatic activity of a large mutant library. If this is the case, an exclusively random approach can be considered, without necessarily satisfying the first two factors. The random and rational approaches described above can be implemented in combination, exploiting their respective advantages while circumventing their main limitations (Chica et al. 2005; Kapoor et al. 2017;

Bornscheuer and Höhne 2018; Korendovych 2018; Zhang et al. 2019; Dadwal et al. 2020; Soni 2021). Described as semi-rational, the methods derived from this strategy involve identifying regions of interest for mutations based on enzyme sequence, structure and/or three-dimensional models, then inducing these mutations by generating a limited mutant library (Figure 5.4). Lastly, the catalytic activity of these mutants is studied. An example of this type of approach is site saturation mutagenesis. This is a random mutagenesis technique used to create a mutant library that contains all possible mutations at one or more targeted positions in the gene encoding the enzyme. This approach makes it possible to determine the effect of substituting one amino acid for all possible amino acids at a specific position. When this strategy is extended to several codons, the number of possible combinations of mutations leads to a considerably enlarged bank of mutants that need to be tested. This difficulty can be circumvented by limiting the number of synonymous codons or by iteratively and systematically exploring the different mutation sites, so as to identify possible synergistic effects (Galanie et al. 2020).

In conclusion, the choice of an engineering strategy must take into account the information available involving the enzyme of interest, as well as the ability to assess the property to be improved (Chica et al. 2005). There are several possible scenarios. In the absence of structural information and when high-throughput analysis tools are available, the random approach is preferred. This leads to the construction of a large bank of mutants, whose activity is tested. Mutations are carried out without any preconceived ideas regarding their location or possible effects. The risk of not finding any mutants with the desired characteristics exists; this risk can be limited by enlarging the mutant bank through one or more iterative genetic diversification campaigns. When the enzyme is structurally and catalytically well known, and the ability to test mutants for activity is limited, a rational approach combined with digital simulations is an appropriate choice. A targeted library of mutants is generated; mutations concern specific regions of the protein. There is a real risk of missing mutation sites of interest, as only mutations with an a priori effect are considered. However, numerous studies have shown that the effect of mutations on enzyme properties is not always predictable. If structural data on the enzyme of interest and high-throughput activity screening tools are unavailable, a preliminary stage with the objective of acquiring knowledge about the catalyst is required. This may include one or more random point mutation campaigns leading to a limited number of mutants, whose catalytic properties are analyzed. In such a context, the construction of homology models providing even partial information on the structure of the protein, from its sequencing, may prove useful. Lastly, a wealth of information on the enzyme and tools for high-throughput activity screening are available. In this ideal case, mutations targeting particular regions of the protein are often effective in improving the enzyme's properties, while multiple random mutations limit the risk of overlooking regions of interest for the desired property. It should also be noted that the choice of an appropriate engineering strategy also depends on the nature of the improvements to be made. If the aim is to modify the interactions between the enzyme and the substrate, to adapt the catalytic site to a non-natural ligand, or to modulate the enzyme's selectivity, the engineering work will primarily focus on the active site and its immediate environment. If the aim is to increase the enzyme's thermal stability, the regions of interest for mutations are more difficult to predict and potentially concern the entire protein structure.

5.1.2. *Improving enzyme stability*

Enzymes are, by nature, long chains of amino acids that form three-dimensional arrangements, whose maintenance depends on numerous electrostatic, hydrophobic, and hydrogen-bonding interactions. Under the effect of factors such as temperature, pH, ionic concentration and the nature of the surrounding environment, this set of complex interactions can be compromised, leading to enzyme denaturation and a loss of activity (Kazlauskas 2018).

5.1.2.1. *Improving thermostability*

The production of thermostable enzymes is one of the biggest challenges on an industrial scale. Several categories of microorganisms can be defined, depending on the temperature range that promotes their growth. There are hyperthermophilic organisms (above 80°C), thermophilic organisms (45–80°C), mesophilic organisms (20–45°C) and psychrophilic organisms (below 20°C). By extension, the enzymes produced by these microorganisms can be similarly classified according to the temperature range in which they maintain their activity. For example, thermophilic enzymes are highly active and stable at high temperatures. These attract considerable interest due to their high application potential in sectors such as detergent production, starch processing and the production of high-fructose corn syrup. However, natural diversity does not always provide the thermophilic enzymes needed to catalyze industrial transformations. For this reason, strategies have been developed to improve the thermostability of enzymes with the desired spectrum of activity.

Numerous studies have been carried out to identify the factors contributing to protein stability under extreme temperature conditions. Structural studies on thermophilic enzymes have helped define the molecular basis of thermostability.

Thus, it has been determined that the intensity of interactions, as well as their localization within the protein structure, play an essential role in its stability (Miotto et al. 2019). More specifically, the presence of a large number of disulfide bridges and a dense network of hydrogen bonds involving the side chains of protein residues are two features common to most thermophilic enzymes (Kumar et al. 2000).

In addition, the number of thermolabile cysteine and serine residues is lower in the sequence of these enzymes than in mesophilic enzymes, while the number of arginine and tyrosine residues is higher. High hydrophobicity, a compact structure with little flexibility, the absence of loops and the presence of short loops, as well as small and few cavities, are other structural elements that seem to promote thermostability (Mahmoudi et al. 2016; Rigoldi et al. 2018; Han et al. 2019; Miotto et al. 2019; Sang et al. 2020).

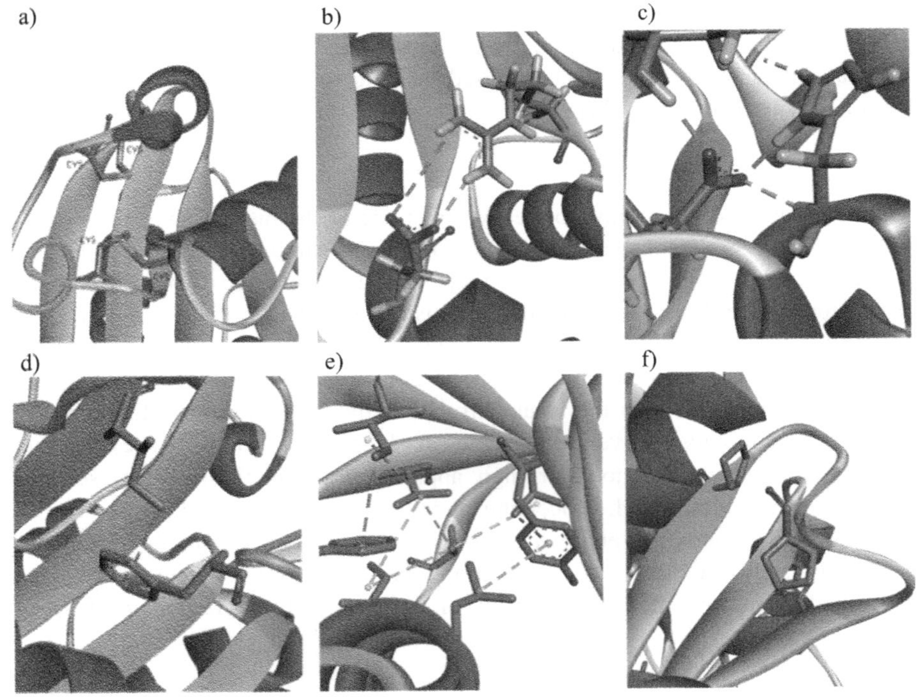

Figure 5.5. *Strategies for improving the thermostability of enzymes: (a) addition of disulfide bridges; (b) addition of ionic bridges; (c) reinforcement of the network of hydrogen bond interactions; (d, e) reinforcement of the network of hydrophobic interactions; (f) enrichment of the sequence with proline residues*

Although protein thermostability is a difficult characteristic to predict, the information reported in the literature greatly contributes to a better understanding of the factors influencing this property and forms the basis of strategies implemented to improve enzymes (Kazlauskas 2018; Rigoldi et al. 2018).

Therefore, one of the most widely used approaches to increasing the thermostability of an enzyme is the introduction of disulfide bridges, accompanied by stabilization of protein chain folding. Other strategies are reported in the literature, such as reinforcing the network of hydrogen bonds and salt bridges on the protein surface, which reduces the free energy of the system. It has also been shown that adding proline residues to the mobile loops of proteins stabilizes their structure by reducing the relative entropy in the denatured state. Lastly, we can mention the importance of hydrophobic interactions in maintaining the structure (Figure 5.5). These factors provide ways for the rational improvement of enzymes.

5.1.2.1.1. Improving the thermal stability of α-amylases

In the food industry, the main applications for α-amylases are in starch processing and the manufacture of bakery products. The various stages leading to starch hydrolysis, also known as liquefaction, and to the production of fructose and glucose syrup are carried out at temperatures between 60°C and 105°C. Furthermore, in industrial baking, α-amylases are used at temperatures between 30°C and 50°C (Mehta and Satyanarayana 2016). The thermal stability of these enzymes is a decisive factor for their use in these industrial sectors.

The structure of α-amylases has three distinct conserved domains (Machius et al. 2005). Figure 5.6 shows the structure of *Bacillus licheniformis* α-amylase as an example. The central domain A is the largest and contains the active site. It consists of eight parallel β-sheets and eight alternating α-helices (Figure 5.6). It takes on the appearance of a (β/α) barrel. The sheets and helices are connected by irregular, more or less complex loops numbered from 1 to 8. The catalytic and calcium ion-binding sites are located on the C-terminal side of the domain. The catalytic triad, consisting of two aspartate residues and a glutamate residue, as well as the amino acids involved in substrate binding, are located at the loops linking the β-sheets to adjacent α-helices. One of these loops, located between the third sheet and third helix, forms a protuberance considered as a separate domain (domain B). This contains amino acids that function as substrate-binding sites. It is also involved in the binding of calcium Ca^{2+} ions, which play a role in maintaining the protein's correct conformation. Domain C is located after the (β/α)8 barrel in the C-terminal part of the sequence. It is organized in a Greek key pattern. Its β-sheet sandwich structure also contributes to the correct folding of the polypeptide chain.

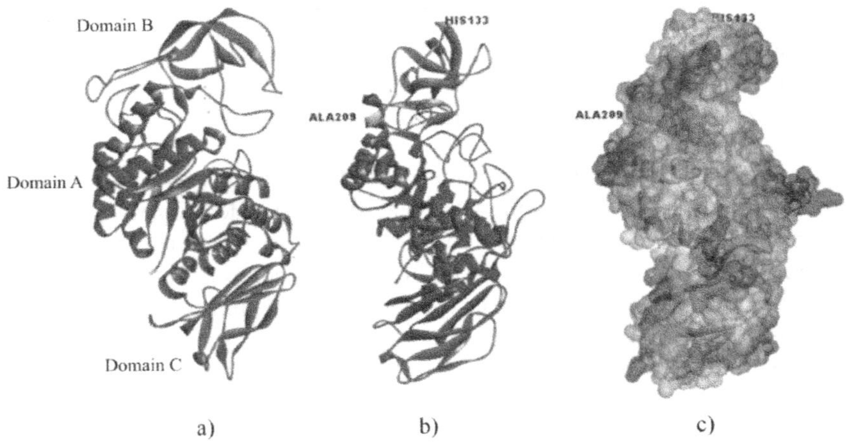

Figure 5.6. *Three-dimensional structure of α-amylase from* Bacillus licheniformis – *Protein Data Bank reference 1BLI*

COMMENT ON FIGURE 5.6.– *(a) Domains A, B and C are colored, respectively, in red, green and blue. Calcium ions appear as small green spheres; sodium ions as a small blue sphere; (b) mutation sites of interest are colored yellow; (c) the Van der Waals surface of the protein is colored according to the hydrophobicity of the residues; blue, red and white designate polar residues, hydrophobic residues and residues of intermediate polarity, respectively.*

A great deal of work has been carried out with the aim of obtaining α-amylases with improved enzymatic efficiency that are compatible with industrial conditions of use (Declerck et al. 2002; Mehta and Satyanarayana 2016). In order to achieve this goal, three main challenges need to be addressed. The first is to increase high-temperature stability. The second is less dependence on calcium. The last challenge is to maintain high catalytic activity in acidic environments. Random mutagenesis studies carried out on α-amylase from *Bacillus licheniformis* have demonstrated the key role of two positions (133 and 209), where mutation of a residue strongly decreases or increases the enzyme's thermal stability (Declerck et al. 2002). Site saturation mutagenesis experiments identified several mutants, whose inactivation rate at 80°C is three times lower than that of the native enzyme. Once the crystallographic structure of the protein had been clarified, three-dimensional models were built to understand the effect of these mutations on enzyme stability. Position 209 is located at the N-terminal of helix 3 of the domain A at the bottom of a small cavity. Replacing alanine 209 with a bulkier hydrophobic residue, such as a valine, led to a threefold increase in the enzyme half-life compared to the native enzyme at 90°C, without affecting amylolytic activity. Three-dimensional models have led to the hypothesis that there is greater structural compaction in the A209V mutant, leading to greater structural stability than in the native enzyme. Residue 133 is located in domain B, on the surface of the protein, close to the N-terminal of a β-sheet. The replacement of histidine by hydrophobic residues mostly leads to an increase in the enzyme stability, which may be correlated to the ability of the mutated residues to form a sheet, rather than to their hydrophobic nature. The role of these two residues, shown in a large-scale screening of mutants, could hardly have been predicted, demonstrating the value of the randomized approach.

Knowledge of the enzyme structure and mode of operation also makes it possible to modify its structure in a targeted way, which improves its properties. Thus, based on the observation that the region forming the interface between domain A and domain B of the enzyme is rich in charged residues, forming salt bridges, or that it is involved in the binding of calcium and sodium ions, a hypothesis was made that it plays a predominant role in protein stability. A site-directed mutagenesis approach, targeting this part of the enzyme, has shown that any mutation affecting this complex network of interactions strongly compromises the protein thermostability. On the contrary, the introduction of aromatic residues enriching the network of hydrophobic interactions leads to increased stability. Lastly, multiple mutations combining the beneficial effects of random mutations in the protein and targeted mutations were introduced. The mutant enzyme retains 50% of its activity at 106°C, compared with 83°C for the native enzyme, demonstrating the benefits of combining random and site-directed mutagenesis approaches.

As previously mentioned, improving the thermostability of α-amylases is a major challenge for the starch products industry. Among the enzymes of particular interest on an industrial scale, the α-amylases produced by psychrophilic microorganisms exhibit optimal activity at low temperatures, which is a clear advantage for the development of energy-efficient processes. However, the main drawback of these enzymes is their elevated structural flexibility, which makes them sensitive to temperature rises and limits their large-scale use (Yang et al. 2017). Structural modifications through site-directed mutagenesis can be used to improve the thermostability of these enzymes. Possible strategies include introducing hydrophobic residues into the core of the enzyme to promote protein compaction, enriching the structure with localized proline residues on the surface to stiffen the structure, increasing the number of disulfide and salt bridges to strengthen the electrostatic interaction network, or reducing the length of loops to limit the associated entropy (Isaksen et al. 2016). These different options for improvement can be used alone or in combination, leading to mutations in different regions of the protein. For example, the half-life of *Bacillus licheniformis* α-amylase at 95°C was increased ninefold by mutating three residues to stabilize a long labile loop belonging to the enzyme B-domain (Li et al. 2017). Another approach involving the introduction of disulfide bridges near the catalytic site of α-amylase has been described (Liu et al. 2014). Based on a comparison of the sequences of a mesophilic α-amylase and a psychrophilic α-amylase, a low-temperature-active enzyme resistant to temperature rises was developed using a site-directed mutagenesis approach targeting, on the one hand, the region surrounding the active site and, on the other hand, regions on the protein surface. The developed strategy combines enrichment of the domain A and domain B loops of the psychrophilic enzyme with proline residues, and the replacement of valine residues involved in non-polar interactions with catalytic residues with polar residues, enabling the establishment of hydrogen bonds. Activity tests revealed that the mutants produced had a half-life twice as long as the native enzyme at 50°C, and a kcat/Km value that was 56% higher at 35°C. These results can be explained by a stiffening of the protein labile regions and a reinforcement of the interaction network in the vicinity of the catalytic site.

5.1.2.1.2. Improving the thermal stability of lipases

Over 75% of the enzymes used on an industrial scale are hydrolases, including lipases and phospholipases, which play a major role in fat modification and flavor production (Prakash et al. 2013; Rao and Sobha 2020).

As previously stated, the disulfide bridges present within protein structures greatly contribute to their maintenance over a wide temperature range. This can be exploited to improve the thermostability of lipases by increasing the number of disulfide bridges. There are open-access applications for predicting the possibility of increasing the number of disulfide bridges within a given structure, after replacement of one or more residues by a cysteine (DbD2, SSBOND, MODIP, BridgeD, etc.) (Sowdhamini et al. 1989; Dombkowski 2003; Pellequer and Chen 2006; Gao et al. 2020). This strategy has inspired numerous works aimed at improving the thermostability of lipases. *Rhizomucor miehei* lipase naturally carries three disulfide bridges. The addition of a disulfide bridge to stabilize a loop covering the enzyme active site has led to a significant reduction in the protein free energy, thereby enhancing its conformational stability (Han et al. 2009) (Figure 5.7). No loss of catalytic activity was observed, despite the increased rigidity of the protein structure.

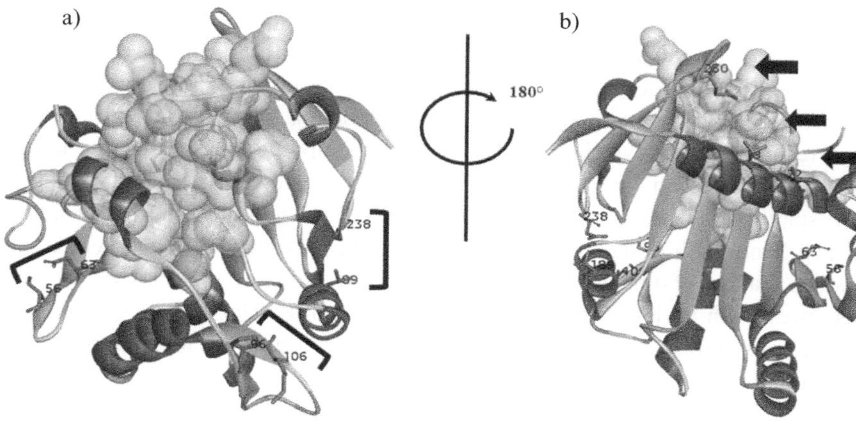

Figure 5.7. *Strategies for improving the thermostability of* Rhizomucor miehei *lipase (based on the work of Han et al. (2019) and Li et al. (2018)). The surface of residues within 4 Å of the catalytic triad is shown in blue*

COMMENT ON FIGURE 5.7.– *(a) Mutation sites for disulfide bridges are indicated by black brackets (S56/N63; V189/D238; P96/K106); (b) mutation sites for increased lipase thermostability (T18, T22, E230) are indicated by black arrows.*

More recently, new mutations made on the beta-sheets and between an alpha-helix and a loop of the enzyme – leading to the formation of new disulfide bridges – have made it possible to increase the protein melting temperature from 1 to 4°C (Li et al. 2018). These mutations are localized in flexible regions on the protein surface, thereby maximizing entropic effects, leading to structure stabilization without affecting the organization of the active site. This rational approach of increasing the protein rigidity at flexible surface-localized regions can be combined with a random search for additional mutations that also lead to improved stability. Three mutation sites were thus identified (T18, T22, E230), each leading to an improvement in the enzyme thermostability. The combination of these beneficial and synergistic effects resulted in a 15°C increase in the protein melting temperature and a 40% increase in catalytic efficiency at 70°C.

As each enzyme has its own structural peculiarities and specific characteristics, transposing an improvement strategy from one protein to another may not be relevant. While the above-mentioned studies underline the benefit of cumulating several structural modifications, other studies have highlighted the significant effect of a single mutation on protein stability. For example, the thermostability of the *Staphylococcus epidermidis* lipase was increased by 25°C thanks to a single mutation (Veno et al. 2019). In this case, a totally random directed evolution approach, followed by activity assays, identified a mutant with increased stability, compared to the native enzyme. Comparison of the structural characteristics of the native enzyme and the variant by molecular modeling showed that replacing a glycine residue with a cysteine residue in the amphiphilic flap of the lipase has several side-effects: an increase in the hydrophobic nature of the flap, giving it greater stability, a strengthening of interactions involving neighboring residues (hydrogen bonds, salt bridges) and a reduction in the flexibility of the structure. An interesting feature of this work is the increased stability of the enzyme at high temperatures, with no loss of activity at low temperatures. The authors explain this result as a compromise between increasing the structure rigidity at high temperatures and maintaining its plasticity at low temperatures.

In conclusion, our work shows that protein stability depends on many factors, such as the nature of the amino acids making up the protein chain and their proportions, chain folding and, in particular, the organization within the protein and on its surface, the ability of the chain to form regions organized as helices or sheets, the network of intramolecular interactions, etc. Although the structure/properties relationships of enzymes are becoming increasingly well known, improving the stability of these protein constructs remains a complex problem that requires a strategy combining rational and random approaches. Bioinformatics tools are ideal devices in this approach, enabling us to predict the effect of specific mutations, identify mutation sites of potential interest for enhancing protein stability, reduce the size of mutant libraries to be tested, and accelerate the discovery of enzymes with high industrial potential.

5.1.2.2. *Improving pH stability*

Various factors can lead to partial or total loss of enzyme activity. Among these, pH is a limiting factor for the use of biocatalysts on an industrial scale. Indeed, industrial processes often require pH conditions far from the optimum pH of the enzyme that is catalyzing the reaction of interest. Under these conditions, there is a significant risk of protein denaturation due to structural changes (Iyer and Ananthanarayan 2008). α-Amylases are particularly concerned by this problem. The pH of native starch milk takes on values between 3 and 4.5, while the optimal pH value of most α-amylases is close to 6. This leads to pH adjustment stages during starch liquefaction processes. These stages could be avoided by using enzymes with enhanced stability over a wider range of pH values. Scientific studies show that structural modifications to improve the thermal stability of these enzymes are very often accompanied by an increase in their resistance to acidic pH. However, specific strategies have been proposed to address this issue, which is of vital importance on an industrial scale. Some enzymes produced by acidophilic microorganisms are naturally resistant to acidic pH; their structures have been analyzed to identify the determining factors of this stability and implement them in enzymes in need of improvement (Parashar and Satyanarayana 2018). A high density of negative charges on the surface of acidophilic proteins, contributed by glutamate and aspartate residues, has been identified as a key determinant of their stability. This phenomenon can be explained by an increase in the protonation state of these residues at acidic pH levels. This leads to a reduction in the repulsion forces between negative charges on the protein surface and an increase in its stability. Another theory has been put forward, explaining the particular stability of acidophilic α-amylases. This is based on the pKa values of the residues involved in the catalytic mechanism. It has been shown that small variations in these pKa values can strongly modify the activity profile of the enzymes, as a function of pH. The residues of the active site must be in a specific protonation state in order to play their role effectively. Thus, the proton donor catalytic glutamate must be in a protonated state, while the nucleophilic aspartate must be deprotonated (Figure 5.8). In an effort to explain why certain enzymes remain active over a wide range of pH values, some authors have hypothesized that the pKa values of these key residues change with pH, induced by a change in the electrostatic field surrounding them.

Based on this information, site-directed mutagenesis work has been undertaken, targeting residues in the catalytic domain with the aim of lowering their pKa value (Yang 2017; Wang 2018a). Various strategies have been developed, such as replacing histidine residues with aspartate residues in the vicinity of the active site, so as to reinforce the network of hydrogen bonds and salt bridges. An increase in the stability of the mutants at acid pH was observed, without any loss of thermostability. These results can be explained by a decrease in the pKa of the catalytic acid glutamate and nucleophilic aspartate, and hence a greater ability to yield protons in acidic conditions.

This rational approach to protein modification can be based on results from models with known enzyme structures (Wang et al. 2018a). Simulations can be used to predict a number of system properties, such as the pKa of the proteic residues. A study carried out on α-amylase from *Bacillus subtilisis* identified mutants of interest using models and calculations of pKa values, limiting the number of unsuccessful experiments. The most promising mutants were produced and their catalytic activity was characterized. This approach, which combines modeling tools and experimental enzyme engineering techniques, produced a double mutant with a pH optimum two units lower than that of the native enzyme, with four times the catalytic activity at acidic pH.

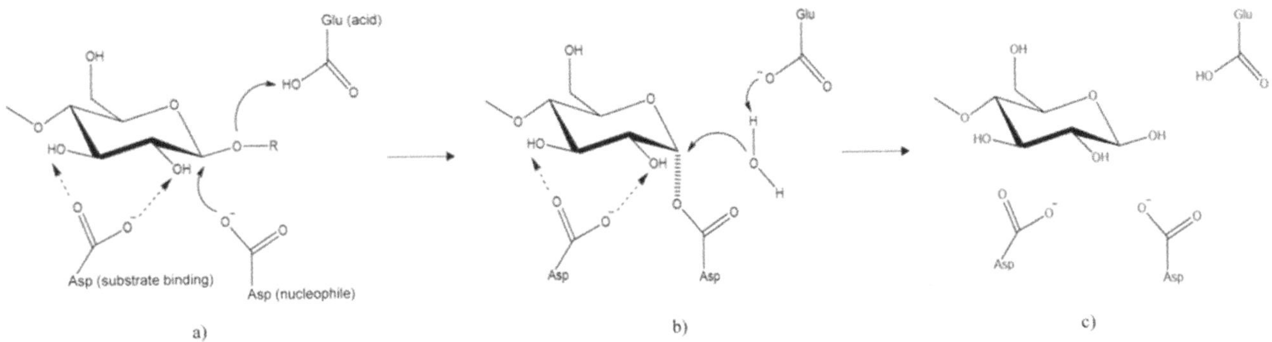

Figure 5.8. *Catalytic mechanism of glycosyl hydrolases. (a) Proton transfer from glutamic acid to glycosidic oxygen and attack of the C1 position of glucose by nucleophilic aspartate; stabilization of the substrate by hydrogen bonding with an aspartate residue; (b) activation of a water molecule and breaking of the C1-Asp covalent bond; (c) regeneration of the initial protonation state of the enzyme*

5.1.2.3. *Improving stability in solvent*

Water is the natural medium for enzyme-catalyzed transformations. However, many biocatalyzed industrial processes involve other, more favorable media for the targeted reaction, whether for reasons of substrate solubilization, product recovery, or shifting the reaction equilibrium when a synthesis reaction must be favored over a hydrolysis reaction. Under these conditions, the enzyme/medium combination needs to be optimized in order to ensure smooth catalyst operation and high yields. To achieve this, three main strategies can be implemented, either alone or in combination. The first involves adapting the medium to limit its denaturing effect on the enzyme; the second exploits natural diversity to identify an enzyme that catalyzes the reaction of interest and is sufficiently robust in order to retain its activity in the chosen medium; the third involves modifying the enzyme catalyzing the reaction of interest to increase its stability in the chosen medium (Stepankova et al. 2013). The latter strategy is based on two major approaches: immobilizing the enzyme on (or in) a medium and modifying the enzyme structure to make it more resistant to its environment.

Modification of the enzyme structure is preferably achieved by genetic and enzymatic engineering, and relies on classic stability-enhancing strategies such as the addition of disulfide bridges, reinforcement of the protein hydrophobic core, modulation of charge distribution on the protein surface, reduction of flexible regions, etc. The mechanisms by which enzymes are destabilized in the presence of solvents are complex, not always well understood and little studied compared with the phenomenon of thermal denaturation. Molecular dynamics simulations have enabled us to highlight two important effects of organic solvents such as 1,4-dioxane, dimethyl sulfoxide and 2,2,2-trifluoroethanol on *Bacillus subtilis* lipase A activity: the trapping of water molecules on the protein surface and the entry of solvent molecules into the substrate binding site, leading to conformational changes and enzyme inhibition (Cui et al. 2020). In addition, activity assays were carried out on the 3,439 variants, resulting from saturation mutation of the 181 residues of the protein chain. This work enabled us to define two strategies for improving lipase stability in organic media. The first consists of increasing the number of positive charges on the protein surface to promote the formation of a hydration layer and limit contact with solvent molecules. The second involves introducing non-polar residues at the substrate binding site, so as to limit the entry of solvent molecules.

The value of combining digital simulations with genetic engineering methods to improve enzymes has also been demonstrated, in a study aimed at increasing the solvent resistance of a promising metalloprotease for the synthesis of peptides of interest (Zhu et al. 2020). Directed evolution has identified two mutants that are stable in the presence of acetonitrile and acetone, with a greater capacity to catalyze casein hydrolysis and dipeptide synthesis than the native enzyme. Molecular modeling simulations explain the results by a strengthening of the interaction network within the protein and an increase in its compactness, when only two polar residues are replaced by two aromatic hydrophobic residues. A similar approach was used to improve the stability of the *Geobacillus stearothermophilus* lipase used for biodiesel synthesis from poultry fat (Gihaz et al. 2018). A rational site-directed mutagenesis approach increased the enzyme stability in methanol through the introduction of aromatic hydrophobic residues. New π-π and CH_3-π interactions, strengthening the hydrophobic core of the protein, were revealed by solving the crystallographic structures of the mutants.

5.1.3. *Improving enzyme specificity and selectivity*

Specificity and selectivity are the main advantages enzymes have over most chemical catalysts in facilitating the transformation of sometimes complex substrates, according to well-defined reaction schemes.

Within a given family, enzymes catalyze the transformation of a broader or narrower spectrum of substrates. Highly specific enzymes catalyze the conversion of a single substrate or a small number of substrates, while others are able to accommodate substrates of various structures in their active site.

Furthermore, for a given substrate, different enzymes catalyzing the same transformation can lead to different products, depending on their selectivity.

Developing an enzymatic process on an industrial scale often requires an improvement of native enzymes, with a view to adapting their specificity and selectivity properties to the product of interest.

5.1.3.1. *Improving lipase specificity*

Lipases form a large family of enzymes that naturally catalyze the hydrolysis of triglycerides into glycerol and fatty acids.

Some are also capable of hydrolyzing phospholipids, cholesterol esters and synthetic esters.

In low-water environments, these enzymes are capable of catalyzing the reverse reaction of hydrolysis, leading to the establishment of ester bonds.

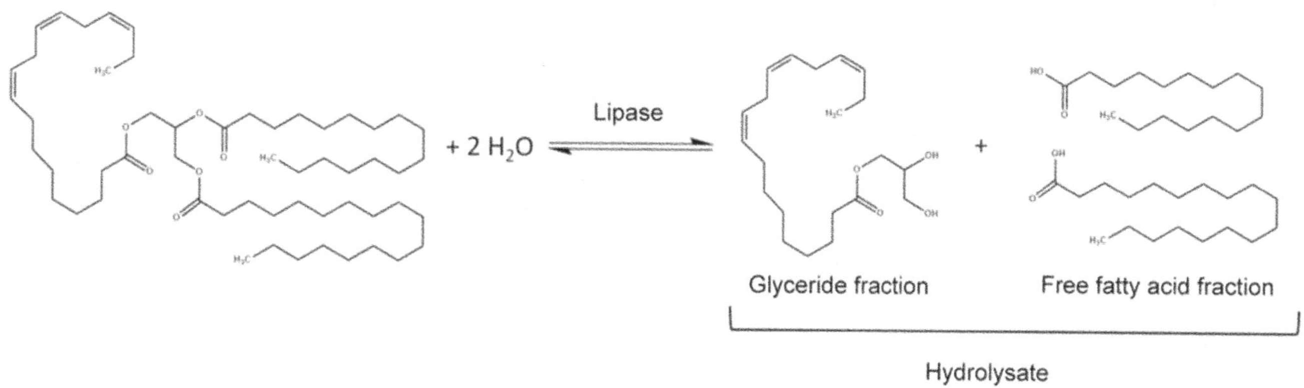

Figure 5.9. *Principle of oil enrichment in unsaturated fatty acids by lipase-catalyzed triglyceride hydrolysis. A specific lipase promotes the hydrolysis of saturated fatty acids, thereby enriching the glyceride fraction with unsaturated fatty acids*

Lastly, a number of studies have reported the ability of lipases to catalyze reactions involving nucleophiles other than hydroxyl groups, such as amine or thiol functions. These characteristics make these enzymes highly sought-after catalysts for the production of fatty acids, esters of interest, structured lipids and polymers.

Among the fats sought after by the food industry, unsaturated lipids from the omega-3 series are being extensively researched for their nutritional value.

The growing demand for industrial products enriched with these lipids justifies the search for specific lipases in order to produce concentrated extracts that can be incorporated into food matrices.

To achieve this objective, an oil hydrolysis strategy can be implemented, leading to the release of fatty acids other than those required (Figure 5.9).

This approach requires the use of suitable catalysts whose specific properties can be optimized by enzyme engineering, based on the hypothesis that structural complementarity between the ligand, the active site and interactions within enzyme/substrate complexes are the main determinants of specificity (Albayati et al. 2020). For example, lipase from *Geobacillus thermoleovorans* has been mutated to increase its specificity in favor of saturated fatty acid release, leading to enrichment of the hydrolysate glyceride fraction with unsaturated fatty acids (Moharana and Rao 2020). A rational approach was developed based on the hypothesis that a reduction in the size of the catalytic cavity is likely to discourage access to sterically unsaturated acids, which are bulkier than their saturated counterparts (Figure 5.10). Two residues in the cavity have been replaced by bulkier ones, resulting in a narrower access tunnel for substrates to the active site. Hydrolytic activity measurements on anchovy oil revealed an enrichment of the glyceride fraction in unsaturated fatty acids, notably eicosapentaenoic acid (EPA) and docosahexaenoic acid (DHA), while affinity calculations applied to models showed the double mutant affinity for unsaturated fatty acids to be lower than that of the native enzyme.

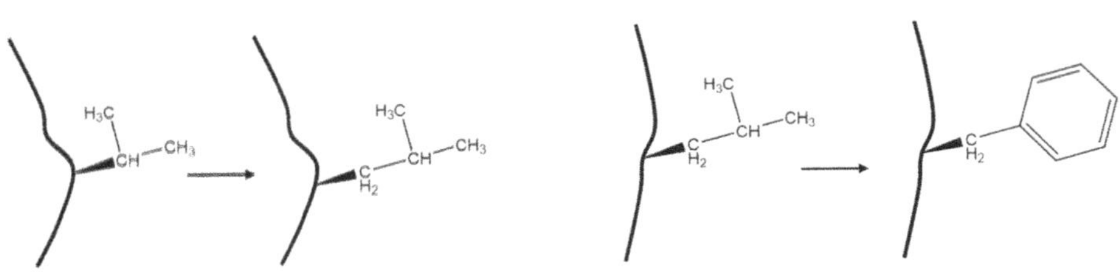

Figure 5.10. *Schematic representation of mutations that restrict access of unsaturated fatty acids to the active site of Geobacillus thermoleovorans lipase*

5.1.3.2. *Improving the enantioselectivity of lipases*

Structural complementarity between the enzyme active site and its substrates is also a determining factor in catalyst selectivity. Lipases are widely used to resolve racemic mixtures. The ability of a lipase to select between two enantiomers is assessed by the enantiomeric excess obtained after a hydrolysis or synthesis reaction applied to a racemic mixture. This enantioselective property is of great interest to the industry for the production of compounds with controlled stereochemistry. Much work has been carried out on lipases to improve or modulate their ability to catalyze the transformation of chiral compounds, inspired by the work of Kazlauskas et al. (1991). According to this author, the enantioselectivity of lipases is explained by the preferential orientation of a chiral substrate, guided by steric constraints at the active site. For example, residue Trp104 of *Candida antarctica* lipase B is thought to be responsible for the enzyme selectivity toward secondary alcohols by creating a steric hindrance (Figure 5.11) (Melais et al. 2016).

It is therefore possible to modulate the enantioselective properties of enzymes by modifying the shape of their active site. Structurally, lipases share a number of common elements: a catalytic triad consisting of a serine residue, an aspartate or glutamate residue and a histidine residue; an oxyanion hole formed by residues that stabilize the intermediate formed during catalysis by establishing hydrogen bonds; a non-polar pocket interacting preferentially with acyl-donor substrates (fatty chains of triglycerides in the case of hydrolysis, or carboxylic acid in the case of esterification); a hydrophilic cavity establishing interactions with nucleophilic, acyl-acceptor substrates (water molecules in the case of hydrolysis, or alcohols in the case of esterification); and a hydrophobic cavity accommodating the substituents of secondary alcohols. The geometrical peculiarities of these different substrate binding sites lead to preferential orientations of the substrates in the catalytic cavity and, consequently, to selectivity. Although it is difficult to generalize, it would appear that the replacement of bulky residues by smaller ones in the region hosting the acyl-donor substrate and, to a lesser extent, in the oxyanion hole or hydrophobic pocket, increases the selectivity of lipases toward carboxylic acids with an R-configuration (Maldonado et al. 2021). With regards to the resolution of racemic alcohols, mutations located in the hydrophilic and hydrophobic cavities of lipases are of particular interest for increasing, or even reversing, enantioselectivity. This work shows that the nature of the chiral substrate must be considered in the development of a strategy to improve selectivity. The nature of the reaction is another important consideration. In an ester hydrolysis reaction, the substrate interacts simultaneously with the hydrophobic cavity and the acyl chain binding site, making these two regions promising mutation sites, regardless of the location of the chiral center (Figure 5.12). On the other hand, during an esterification reaction, the acyl-donor substrate is the first to interact with its binding site. Mutations in this region therefore seem particularly relevant for improving the enzyme ability to determine chiral carboxylic acids. All this work, which represents a major advance in understanding the structure/function relationships of lipases, led authors to come up with the term KRIE, for *key regions influencing enantioselectivity*, specifically designating mutation sites of interest for improving enantioselectivity.

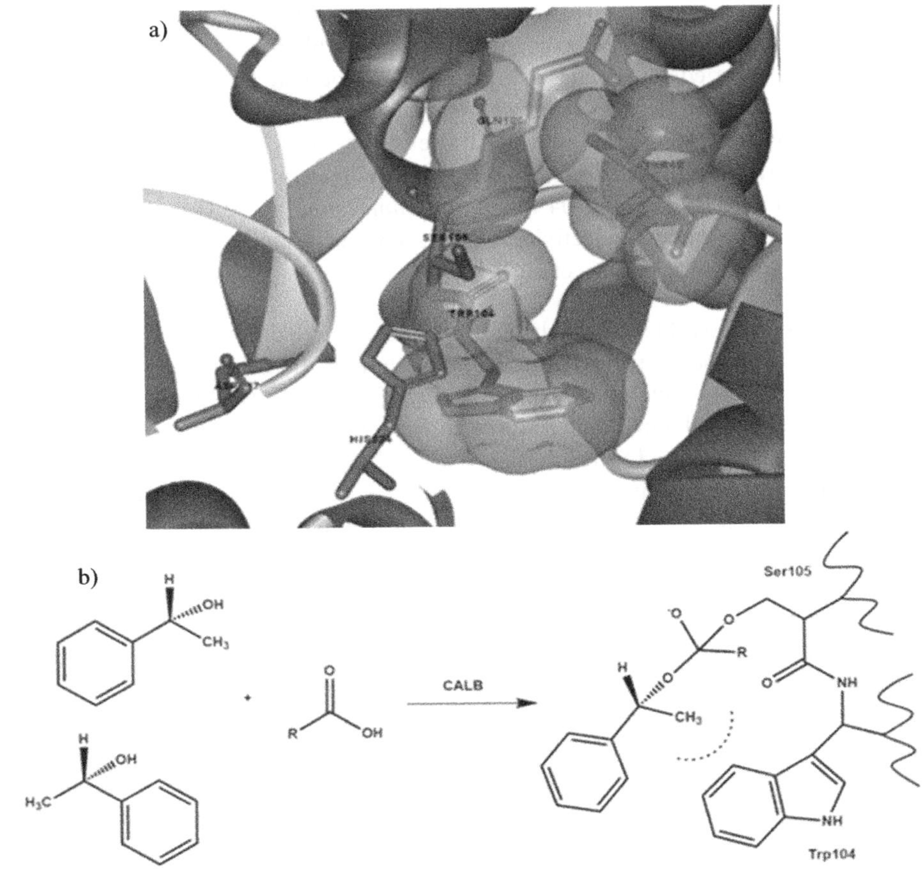

Figure 5.11. *Illustration of Kazlauskas' work*

COMMENT ON FIGURE 5.11.– *(a) Three-dimensional structure of the catalytic site of* Candida antarctica *lipase B. Catalytic residues (Ser105, His224, Asp187) and the Trp104 residue are shown as sticks; the surface indicating the obstruction of the Trp104 residue is shown in purple. Residues constituting the oxyanion hole are shown as sticks (Thr40, Glu10); their surface is colored in orange; (b) acylation of 1-phenylethanol; schematic representation of the steric hindrance caused by Trp104 residue, leading to lipase enantioselectivity favoring acylation of the R-configuration stereoisomer.*

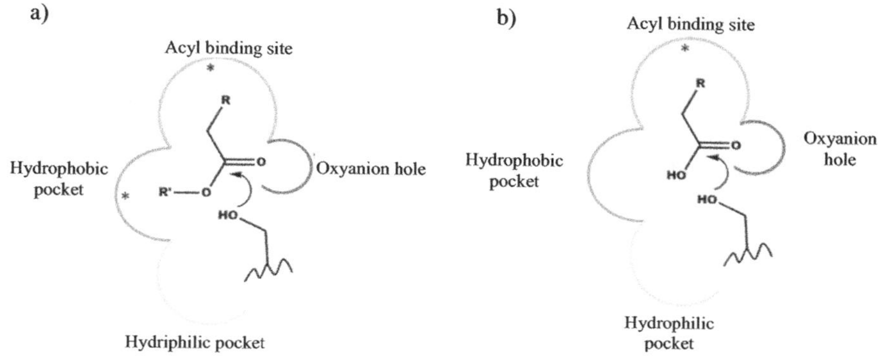

Figure 5.12. *Schematic representation of the different lipase substrate binding sites*

COMMENT ON FIGURE 5.12.– *The acyl binding site is colored green. The hydrophobic pocket interacting with secondary alcohol substituents is colored blue. The hydrophilic pocket hosting the nucleophilic group of the acyl-acceptor substrate is colored yellow. The oxyanion hole stabilizing the reaction intermediate is colored red. (a) Case of a chiral ester hydrolysis reaction; (b) case of a chiral carboxylic acid esterification reaction. The symbol * shows the regions of interest for mutations to improve enantioselectivity.*

5.1.4. *Improving catalytic performance*

Some of the enzymes used to catalyze transformations of interest to the food industry have the disadvantage of a high cost, which limits their use. This high cost is sometimes compounded by low activity on recalcitrant substrates, requiring the use of large quantities of catalyst or catalyst mixtures. Additionally, some enzymes are capable of catalyzing several reactions, one of which may be predominant under specific conditions. This lack of specific action leads to product mixtures and a lack of efficiency for a given reaction.

In this context, researchers and industrialists have developed various strategies for improving enzymes in order to make them more efficient and more selective in catalyzing the desired transformation.

5.1.4.1. *Improving the catalytic efficiency of cellulases*

Cellulases are a class of enzymes that are widely used on an industrial scale in various sectors, including the paper, detergent, animal feed and fruit juice industries. A great deal of work has been carried out to improve the properties of these enzymes, with the aim of increasing their tolerance to pH changes, their thermostability and their catalytic efficiency (Dadwal et al. 2020). Another difficulty inherent in the industrial use of these enzymes is their inhibition by glucose, the product of catalysis, which limits the cellulose degradation process and leads to the accumulation of cellobiose and polysaccharides.

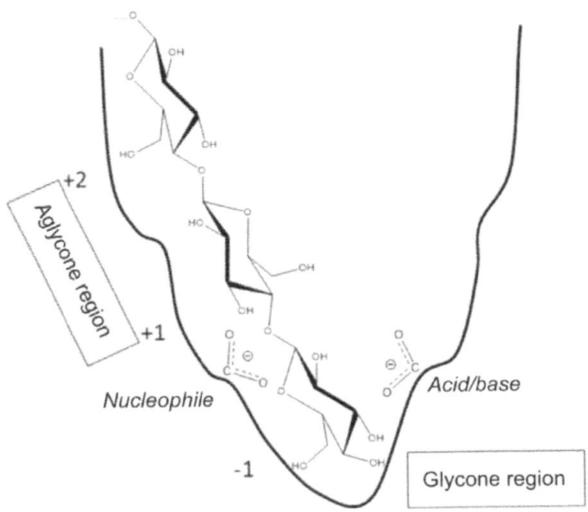

Figure 5.13. *Schematic representation of the beta-glucosidase catalytic cavity*

COMMENT ON FIGURE 5.13.– *The different areas of interaction between the oligosaccharide substrate and the cavity walls are designated by –1, +1 and +2. The –1 sub-site hosts the glucose entity beyond which hydrolysis takes place, with the +1 and +2 sub-sites interacting with the rest of the molecule. Catalytic residues acting as acid and nucleophile are shown.*

This phenomenon of inhibition by glucose varies in significance, depending on the beta-glucosidase. These different levels of tolerance could be explained by elements of the protein structure, such as the hydrophobic nature of the active site or the crowding of the tunnel leading to the active site (Giuseppe et al. 2014; Goswami et al. 2017). It has been shown that substrate hydrolysis catalyzed by beta-glucosidases takes place between two regions of the protein, one referred to as a glycone and the other as an aglycone. The glycone region, also known as sub-site −1, hosts the glucose entity, after which hydrolysis takes place. The aglycone region, on the other hand, interacts with the leaving group initially bound to the glucose entity of the substrate; this region, which is rich in hydrophobic residues and sometimes called the antechamber, is composed of several substrate-binding sub-sites known as +1, +2 (Figure 5.13).

Structural and modeling studies have revealed that glucose can act as a competitive inhibitor, limiting substrate access to the active site. In some cases, however, glucose can stimulate enzymatic activity by participating in transglycosylation reactions. Depending on the nature of the residues making up the catalytic cavity of enzymes, the preferred site of interaction of glucose varies, explaining the different effects observed (Yang et al. 2015). Various interactions (hydrogen bonds with hydroxyl groups, hydrophobic interactions with aliphatic hydrogen atoms) typically stabilize glucose at the entrance to the catalytic cavity, in the medial region leading to the active site, or within the active site itself (Figure 5.14). These differentiated zones of interaction, known as sub-sites, explain the diverse effects of glucose on catalytic activity.

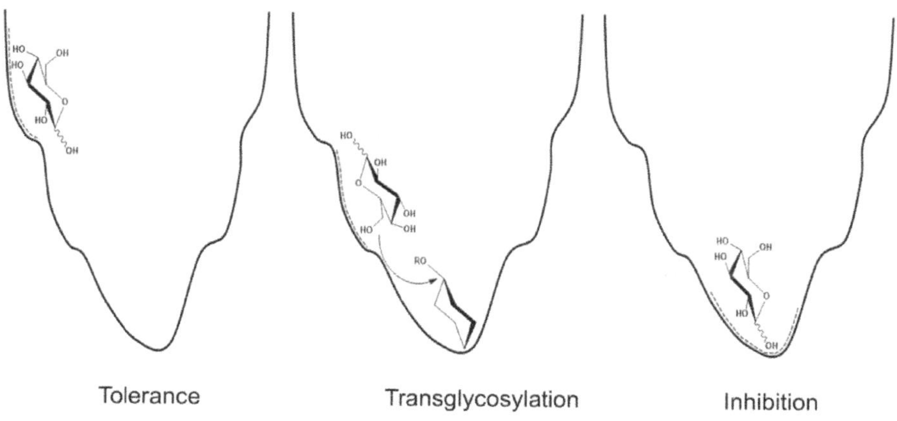

Figure 5.14. *Schematic representation of the different glucose interaction sites in the catalytic cavity of a beta-glucosidase*

These discoveries, which are very important for understanding the structure/function relationships of these enzymes, have served as the basis for numerous enzyme engineering projects aimed at developing new biocatalysts with improved properties.

A first strategy consisting of modifying the ability of different regions of the catalytic cavity to interact with polysaccharide substrates or glucose has been followed by many researchers. For example, replacing two residues (L167W and P172L) located at the entrance to the catalytic cavity of *Trichoderma reesei* β-glucosidase increased the tolerance of the enzyme by promoting hydrophobic interactions between glucose and the +2 sub-site of the enzyme (Guo et al. 2016). Other work has shown the potential of directed evolution to generate a series of *Humicola insolens* beta-glucosidase mutants with improved catalytic properties (Meleiro et al. 2017). Structural analyses of these mutants led to the identification of determining factors of catalytic activity. The hydrolytic activity of the enzyme is predominant in the absence of a monosaccharide. In the presence of a monosaccharide, competition between hydrolysis and transglycosylation activity can be observed. The predominance of one or the other of these transformation pathways can be modulated by tweaking the affinity of monosaccharides and the substrate for the glycone (sub-site −1) and aglycone (sub-sites +1 and +2) regions. In particular, mutations in residues 235 and 237 near sub-sites +1 and +2 can modulate the balance between the hydrolytic and transglycosylation activities (Figure 5.15).

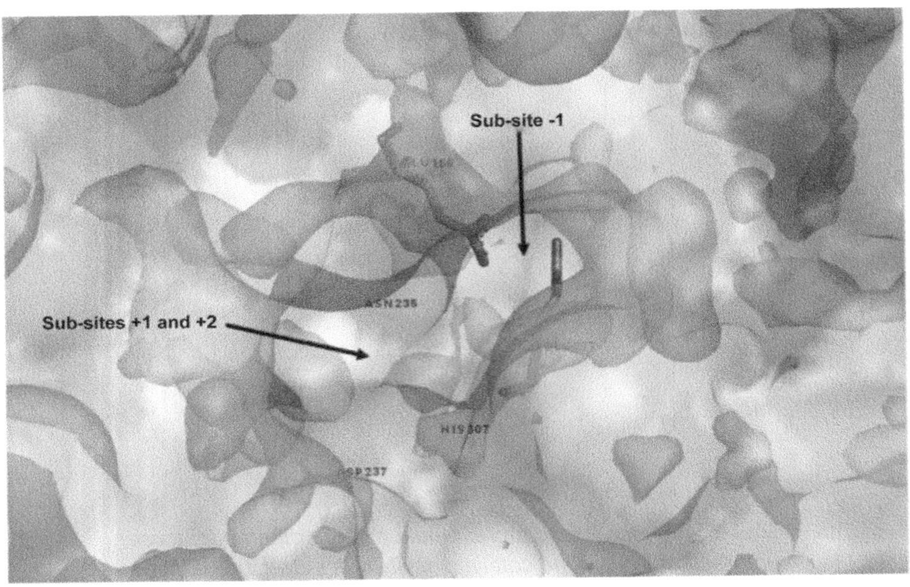

Figure 5.15. *Structure of the catalytic cavity of Humicola insolens beta-glucosidase. Catalytic residues Glu166 and Glu377 are shown as sticks. The surface of mutated residues is shown in magenta*

This understanding of the mechanisms underlying enzyme specificity can be used to improve the efficiency of enzymes catalyzing biopolymer degradation, as well as to develop methods for the production of disaccharides of interest from glucose, by optimizing the transglycosidase activity. Based on the hypothesis that a decrease in the hydropathy index of the active site of *Trichoderma reesei* beta-glucosidase should increase its ability to catalyze the transglycosylation reaction, authors have replaced three hydrophobic residues of the active site with three hydrophilic residues (Niu et al. 2020). Compared with the native enzyme, the mutant increases the concentration of total disaccharides by a factor of three and a half. Conversely, the replacement of an asparagine residue by a leucine residue leads to a loss of the enzyme ability to catalyze the transglycosylation reaction.

A second strategic approach is to modulate the shape of the catalytic cavity and, in particular, accessibility to the active site. A highly glucose-tolerant beta-glucosidase has been developed using a site-directed mutagenesis approach to shrink the catalytic cavity by 2 Å at sub-site +1 (Santos et al. 2019) (Figure 5.16). Activity studies revealed significantly greater glucose release during lysis of lignocellulosic compounds catalyzed by the mutant, compared with the native enzyme, which demonstrates significantly improved glucose tolerance.

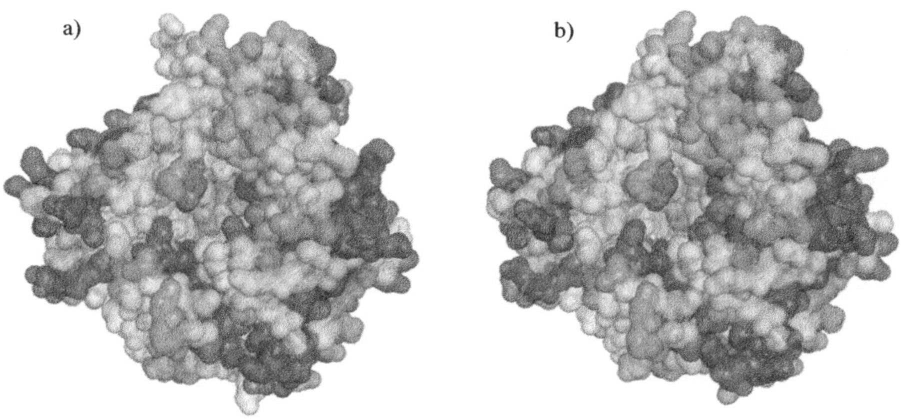

Figure 5.16. *Three-dimensional structures of beta-glucosidase from Trichoderma harzianum (PDB5BWF entry) and its double mutant L167W/P172L (PDB6EFU entry). The surface of residues 338, 167 and 172 contributing to the topology of the catalytic cavity at sub-site +1 is shown in magenta*

5.1.4.2. *Modulation of lipase catalytic activity*

The ability of lipases to catalyze hydrolysis or ester synthesis makes them powerful tools for the industry. The synthesis activity of lipases can be enhanced by carrying out reactions in media with low water content or low water activity. However, these conditions are difficult and costly to implement on an industrial scale, which led researchers to develop alternative enzyme engineering solutions. The strategy is to minimize the involvement of water as a nucleophilic substrate in favor of the acyl-acceptor substrate, so as to promote ester synthesis. Numerous specific tunnels allowing substrate access to the active site have been identified in enzymatic catalysts. The hypothesized existence of a tunnel specifically used by water molecules in the structure of *Candida antarctica* lipase B formed the basis of an engineering strategy, aimed at modulating the hydrolytic/synthetic activity balance (Larsen et al. 2010). A library of mutants resulting from random mutations at positions 42, 47 and 106 was generated (Figure 5.17). Enzyme activity tests in the presence of water and butanol, as nucleophilic agents, showed that the S47L mutant, in which the size of the water access tunnel is considerably reduced compared with the native enzyme, is characterized by a synthesis activity 14 times greater than the hydrolysis activity. In contrast, the Q46A mutant, characterized by a larger water access tunnel than the native enzyme, exhibits increased hydrolytic activity and decreased synthetic activity.

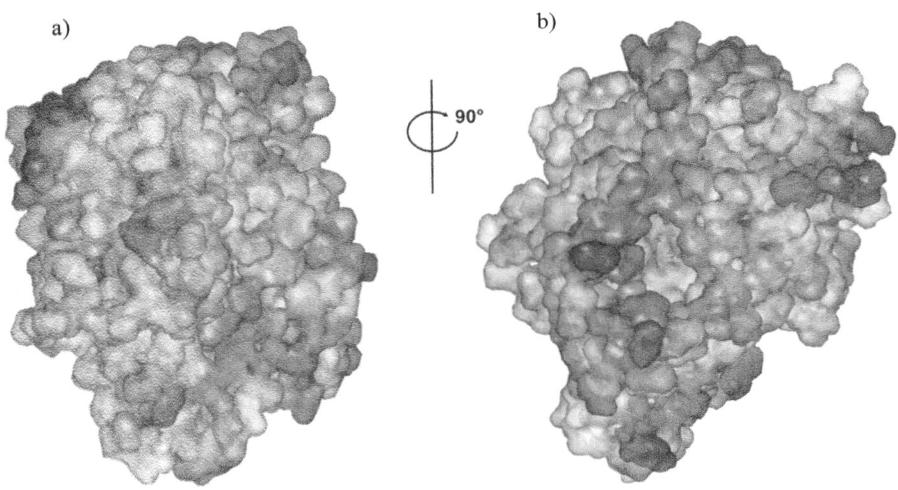

Figure 5.17. *Three-dimensional structure of Candida antarctica lipase B*

COMMENT ON FIGURE 5.17.– *The protein surface is colored according to the hydrophobicity of the residues (red: hydrophobic residues, blue: hydrophilic residues); the surface of mutated residues Thr42, Ser47, Gln106 is colored in yellow. (a) Side view showing the entrance to the water access tunnel; (b) front view showing the entrance to the catalytic cavity.*

5.2. Process engineering

5.2.1. *Introduction*

The relatively high cost of enzymes and their sometimes-insufficient stability have prompted many researchers to search for new technologies, in order to improve their efficiency, stability and enable reuse, or the recycling of these catalysts (Bashari et al. 2016). Several avenues have been proposed, including the immobilization of these enzymes, and the use of technologies, such as microwaves and ultrasound. More recently, intensification methodologies have been considered to increase reaction yields through the development of microstructured, microfluidic or controlled-flow reactors.

5.2.2. *Enzyme immobilization technology*

Enzymes are widely used in a variety of food transformation processes – for example, in beer, wine and cheese production – but their industrial application is often limited by their low stability and the difficulty of recovering them for reuse. These drawbacks can be mitigated by using a variety of methods to immobilize enzymes. Immobilized enzymes have been shown to be far more resistant than free enzymes. Numerous reviews thus report the immobilization of enzymes by various methods, depending on industrial needs (Basso and Serban 2019; Bilal and Iqbal 2020; Taheri-Kafrani et al. 2020).

5.2.2.1. *Impact of immobilization*

Enzyme immobilization is generally defined as the physical confinement of biocatalysts to a certain region of space, stabilizing their catalytic activity for a longer period than without confinement. Enzyme immobilization is generally considered a necessary compromise in industrial biocatalysis processes. The high production, purification and development costs of enzymatic processes are the main factors that sometimes make it necessary to reuse biocatalysts. However, the operating costs of immobilized enzyme production can be substantial. Additionally, conventional enzyme immobilization techniques can reduce catalytic activity, as shown by catalytic parameters such as Vm and/or Km. The development of new media and techniques at the molecular level has improved enzyme stability.

5.2.2.2. *Choice of media*

The choice of media is crucial, conditioning the type of interaction with the enzyme and conferring specific mechanical, chemical, biochemical and kinetic properties to the immobilized enzyme (Sheldon 2007). The properties of a medium depend on its hydrophilicity, inertia toward the enzyme, biocompatibility, resistance to microbial attack, resistance to compression, availability and cost.

Although there is no universal media suitable for all enzymes and applications, certain characteristics of the medium material need to be taken into account, such as high protein affinity, availability of reactive functional groups, mechanical stability, feasibility of regeneration, non-toxicity and biodegradability (Khan and Alzohairy 2010).

Some materials for immobilizing enzymes, notably those based on silica, acrylic resins, synthetic polymers, active membranes and ion exchange resins, have drawbacks, such as the cost of the materials, as well as the functionalization methods required for proper enzyme bonding.

Various immobilization media have been developed, offering a wide variety of physical and chemical properties (variable pore size, hydrophilic/hydrophobic balance and modified surface chemistry) favoring enzyme bonding. Media can be classified into three groups: organic, hybrid or inorganic, depending on their chemical composition, and can be further subdivided into natural and synthetic polymers (Figure 5.18).

The most commonly cited media are carboxymethylcellulose, starch, collagen, modified sepharose, ion-exchange resins, activated carbon, silica (Wu et al. 2012), clay, aluminum oxide, titanium, diatomaceous earth, hydroxyapatite, ceramic, celite (Khan et al. 2006), agarose (Brena and Batista-Viera 2006) or treated porous glass, as well as certain polymers and biopolymers (Gorecka and Jastrzebska 2011). One of the main desirable characteristics of the substrate is its mesoporosity, enabling it to offer large surface areas. A large number of pores leads to higher enzyme loading per unit mass, and the immobilized enzyme is also better protected from the environment. Pore characteristics, such as particle size, determine the total surface area and thus strongly influence the interaction capacity between the enzyme and the medium.

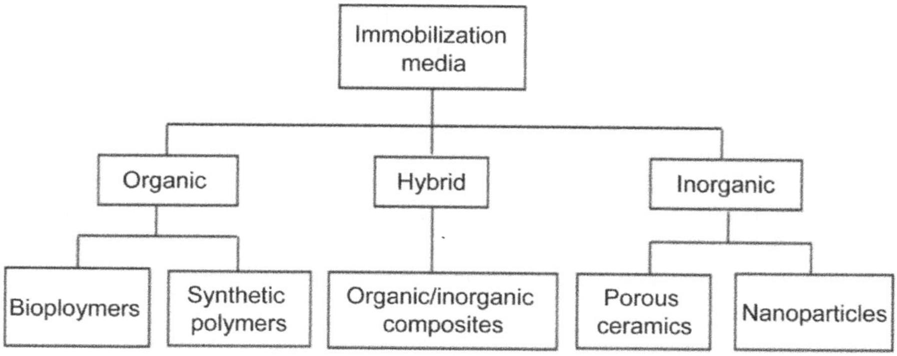

Figure 5.18. *Classification of materials used as media for enzyme immobilization*

Materials of organic and inorganic origin can also be used together to produce a wide range of hybrid media for enzyme immobilization. The most frequently used inorganic precursors for hybrid production are inorganic oxides of silica, titanium and zinc, as well as magnetic nanoparticles, carbonaceous materials and minerals.

These can be combined with synthetic polymers, such as polyethyleneimine, polyacrylonitrile or polyvinyl alcohol, as well as with biopolymers, such as chitosan, alginate and lignin (Sha et al. 2013). These hybrid materials have been used for physical adsorption, covalent bonding, as well as immobilization by encapsulation of different types of enzymes. In these hybrid materials, the inorganic precursor provides high stability, inertness and mechanical strength, while the organic part provides high affinity for the enzyme due to the existence of several functional groups that can easily interact with the enzyme (Zhang et al. 2015). Consequently, the major advantage of these hybrid materials lies in their functionality and stability, which make them suitable for a variety of applications. For example, several studies have described the use of silica/chitosan-based media as suitable media for the covalent immobilization of lipases and carboxypeptidases (Cazaban et al. 2018; Manzo et al. 2018).

More recently, nanostructured systems, such as nanoparticles, nanofibers, nanotubes and nanocomposites, have been described as media for enzyme immobilization and stabilization (Ahmad and Sardar 2015). Although these media turn out to be quite costly, they are robust and are excellent materials for enzyme immobilization, as they have a large surface area and good mechanical properties that allow a large amount of enzyme immobilization with minimal diffusion limitation. The properties of carbon nanotubes also enable easy separation and reuse of the biocatalyst by simple filtration or with the help of a magnetic field (Peijun et al. 2010; Suganthi et al. 2019). Compared with conventional immobilization techniques, nanoparticle-based immobilization offers three major advantages: (i) easy synthesis of high-solids nano-enzyme particles, without the use of surfactants and toxic reagents; (ii) possible adaptation of particle size; (iii) obtaining a homogeneous, well-defined level of nanoparticles with a thick enzyme layer (Ansari and Husain 2012). In addition, with the growing interest in enzymatic cascade reactions, the co-immobilization of multienzyme systems on these nanoparticles can be achieved (Suganthi et al. 2019).

5.2.2.3. *The different methods of immobilization*

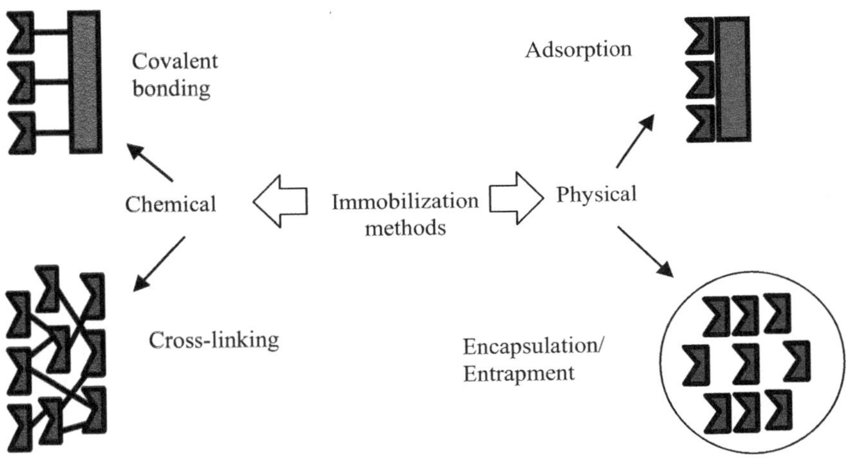

Figure 5.19. *The different methods of immobilization*

Enzyme immobilization can be achieved using a variety of techniques: generally speaking, they are classified into two methods: chemical and physical. In physical methods, various weak interactions are established between enzymes and media, while in chemical methods, enzymes and media are linked by covalent bonds. Loss of enzyme activity can be minimized by using an appropriate immobilization technique, which can avoid changes in the chemical structure of the enzyme and the reactive groups present at the enzyme's active site. There are several common methods of enzyme immobilization, such as adsorption, encapsulation, covalent coupling, cross-linking and entrapment (Cantone et al. 2013) (Figure 5.19). These immobilization methods and the nature of the enzyme media are very important, as the stability and long-term use of an enzyme depends on the immobilization method chosen (Asgher et al. 2014).

Enzyme co-immobilization can also be considered, enabling co-factor recycling and the combination of several biocatalysts for the synthesis of molecules of interest. Selective compartmentalization of enzymes during immobilization can offer a number of advantages. For example, starch-converting enzymes, α-amylase and glucoamylase, have been immobilized on surface-modified media, and it has been shown that these enzymes, immobilized on hydrophilic silica gel and DEAE-cellulose entrapped in alginate beads, exhibited 92.3% and 88.9% residual activity, even after 10 cycles of reuse (Park et al. 2005).

5.2.2.3.1. Immobilization by adsorption

Adsorption is the simplest immobilization method. It involves a variety of non-covalent interactions, including Van der Waals forces, hydrophobic interactions and hydrogen bonds. The procedure involves contacting the media, which can be immersed in an enzyme solution for physical adsorption based on the principles of the enzyme-ligand interaction. This method is inexpensive, requires no activating reagent and also protects enzymes from proteolysis and aggregation. The main advantages are easy contact between the media and the enzyme, enabling immobilization in a matter of minutes, and therefore minimal changes in enzyme conformation, which generally leads to the maintenance of high enzyme activity.

Disadvantages of adsorption include desorption of the enzyme from the media due to weak bonds between the media and the enzyme, limiting the enzyme's ability to recycle. In some cases, physical adsorption can also lead to denaturation of the enzyme depending on the surface chemistry of the media material.

5.2.2.3.2. Covalent immobilization

The formation of covalent bonds is responsible for stable interactions between the enzyme and the media, which are formed during the reaction with functional groups (e.g. imidazole, indolyl, hydroxyl, amino and thiol) that are present on the enzyme surface. Enzymes can be covalently immobilized by two methods involving (i) direct bonding of the enzyme to the surface of the media material using the functional group or (ii) the addition of a spacer arm for bonding of the enzyme to the media. In the latter case, the functional groups present on the material are activated with a specific reagent, then the enzymes are attached. In general, this media surface modification ensures better thermal stability of the covalently bonded enzyme (Hartmann and Kostrov 2013). However, when the enzyme is covalently immobilized in porous materials, limitations on enzyme diffusion within the media may occur. In particular, at high enzyme content, diffusion limitations can occur due to the uneven distribution of the enzyme in the porous media, leading to a reduction in the pore volume available for substrate and product diffusion. Retention of activity by immobilized enzymes bound to the media is generally around 50% (Janssen et al. 2002).

5.2.2.3.3. Immobilization by entrapment

In encapsulation methods, enzymes are trapped in natural polymer beads by two methods: (i) by thermal inverse polymerization, where agar, agarose or gelatin is melted, and as it cools, the enzyme is introduced to be trapped; or (ii) by ionotropic gelation, where calcium alginate or carrageenan is extracted from a gel with ions such as Ca^{2+} (Datta et al. 2013; Kumar et al. 2013, 2014; Keerti et al. 2014). In addition, numerous synthetic polymers, such as polyvinyl alcohol hydrogel, polyacrylamide, tetramethoxysilane (TMOS) and propyl trimethoxysilane (PTMS), have been described for enzyme immobilization (Grosová et al. 2008). Mass transfer limitation is the main disadvantage of this method.

Enzyme entrapment is based on the inclusion of biocatalysts in a matrix. Two main strategies are traditionally used. The first is based on the addition of enzymes to the solution containing the matrix components, followed by polymerization, resulting in the confinement of the biocatalyst within this matrix. This strategy has been used in a variety of matrices, such as carrageenans, acrylamides, silanes and polyglycolic-lactic acid. The advantage of this system is that the enzyme is placed in a system where it is confined and protected, limiting enzyme denaturation processes.

In the second strategy, the porous media is soaked in an enzyme solution, and the biocatalyst is trapped in the matrix by a diffusion mechanism. In this system, the positioning of the enzyme depends on the molecular properties of the enzyme (e.g. molecular weight, surface charges and spatial structure), the characteristics of the matrix and the interactions between the two; thus, the enzyme can be located on the surface, inside or in both positions. However, an additional stage is often required to prevent enzyme diffusion from inside the media by applying derivatization procedures with cross-linking agents or by depositing polymer layers on the matrix surface.

5.2.2.3.4. Membrane immobilization

Different types of membrane exist for enzyme entrapment, allowing a variety of useful techniques for processes with allosteric enzymes or when the presence of co-factors is required. The main advantages of these membranes are that they offer a large surface area, enabling large quantities of enzymes to be immobilized on their surface, leading to high substrate conversion rates. However, the main disadvantage is the exposure of the enzyme to the medium, where shear stresses can limit enzyme stability and activity. A variety of hydrophobic and hydrophilic membrane materials have been developed. Hydrophobic membranes can be made of polypropylene, polytetrafluoroethylene, polyvinyl chloride, polyvinylidene difluoride or polyetherimide. Hydrophilic membranes are made from cellulose, acrylic, polyamide or polyacrylonitrile (Gupta et al. 2013). For example, lipase from *Mucor miehei* was immobilized by adsorption on microporous, asymmetric hollow-fiber membranes and implemented to hydrolyze two different oils, namely palm oil and olive oil, with superior adsorption of the enzyme on hydrophobic membranes (Shamel et al. 2007).

5.2.2.3.5. Immobilization by enzyme cross-linking

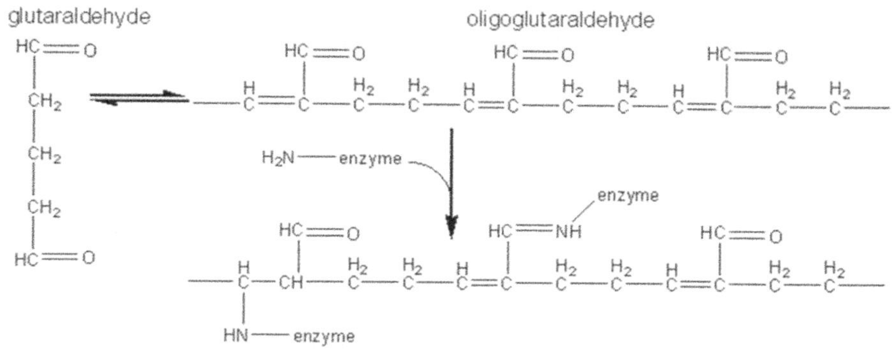

Figure 5.20. *Enzyme cross-linking with glutaraldehyde*

This enzyme immobilization technique is also known as copolymerization and is based on a cross-linking of enzymes, which are attached to each other at several points without the use of media (Datta et al. 2013). These types of enzyme aggregates are known as cross-linked enzymes (CLE). They range in size from 1 to 100 µm and have high reactivity due to the high concentration of enzymes. The enzyme solution is, however, rather viscous; although this method involves covalent bonds, enzymes immobilized using this method often undergo conformational changes; consequently, it is not uncommon for there to be a loss of enzyme activity. Glutaraldehyde is commonly used as a cross-linking agent because of its bifunctional groups (Figure 5.20).

In CLEs, the main advantage is the absence of media, which implies the absence of interactions with other molecules that could potentially have a negative effect on enzymatic activity. The procedure involves cross-linking the biocatalyst with a bifunctional reagent (such as glutaraldehyde). Numerous variants of this technique have been developed to achieve high catalytic activity. The system is advantageous because even labile and/or allosteric enzymes can display high productivity and long-term use.

However, there are disadvantages to this approach, including problems of substrate and product diffusion, as well as steric hindrance and protein flexibility. Since the development of this technique by Professor Roger A. Sheldon, many different procedures have been developed, some of which have been shown to achieve excellent enzyme stabilization. The most popular techniques are cross-linked enzyme crystals (CLEC) and cross-linked enzyme aggregates (CLEA).

The vast potential of CLEA at the industrial level means that this technology can be explored for numerous applications in the food industry, such as winemaking and citrus juice clarification.

5.2.2.4. *Types of reactors for immobilized enzymes*

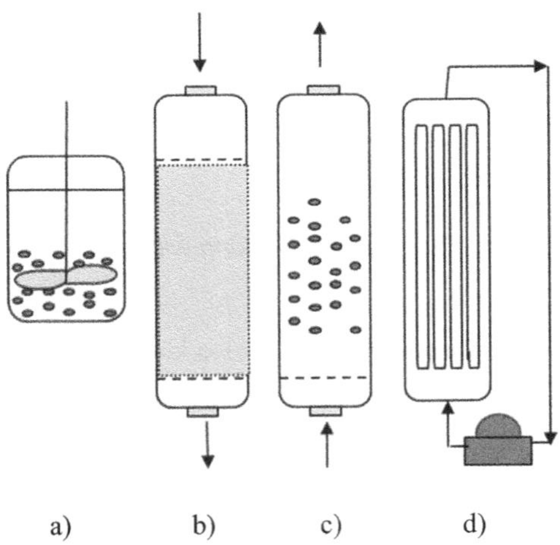

Figure 5.21. *Different reactor configurations for immobilized enzymes*

Various bioreactor configurations have been developed for processes with immobilized enzymes. Four main configurations have been proposed: the mechanically stirred reactor, the fixed-bed reactor, the fluidized-bed reactor and the membrane reactor (Figure 5.21).

5.2.2.4.1. Mechanically stirred reactor

The most currently commonly used enzyme reactor is the mechanically stirred reactor. The operating principle of stirred bioreactors is relatively simple: substrates and enzyme are introduced into the reactor, which can be fitted with fixed baffles to improve stirring efficiency and mass transfer. Temperature and pH control are always necessary in this type of reactor, to ensure optimum conditions for enzyme activity. The stirred reactor is ideal for industrial applications – particularly for processing high-viscosity or low-solubility substrates – and can offer both low capital and operating costs. However, this type of reactor has certain disadvantages, such as low reaction efficiency, high energy input required for agitation and the need to remove enzymes from the products, once the desired reaction has taken place (Williams 2002).

5.2.2.4.2. Fixed-bed reactor

The fixed-bed reactor is cylindrical in shape and filled with immobilized enzymes into which the substrate enters at one end and the product flow exits at the other. The most important feature of this type of reactor is that the reaction solution flows through the reactor in the form of a piston-type flow; the reactor is therefore also referred to as a piston reactor. The absence of a stirring device limits fluid mixing in the reactor; flow properties will vary in both longitudinal and radial directions, but the variation in the radial direction is always much smaller than that in the longitudinal direction. In order to produce an ideal piston flow in the bed-fixed reactor, a turbulent flow regime is preferred to a laminar flow, enabling improved mixing and heat transfer. A significant advantage of the fixed-bed reactor is that high concentrations of immobilized enzymes can be used. With the enzyme immobilized in the reactor bed, the substrate solution to be converted is transformed into a product that can be continuously collected at the bioreactor outlet. This type of reactor is also well-suited to reactions involving product inhibition. The disadvantages of the fixed bed reactor are the absence of possible pH control by the addition of acids or bases, and difficult temperature control in the event of excessive heat release – a problem which can be particularly prevalent in large-scale reactors (Martin and Christopher 1990). Moreover, immobilized enzymes are easily clogged by the formation of precipitates, which presents cleaning problems and may require the replacement of immobilized enzymes.

5.2.2.4.3. Fluidized-bed reactor

The fluidized-bed reactor consists of a bed of immobilized enzyme, fluidized by the rapid upward flow of substrate alone. The substrate passes through the bed of immobilized enzyme at a speed high enough to lift the particles. The enzyme is immobilized on small particles that move with the fluid. There is a minimum fluidization speed required to achieve bed expansion, which depends on the size, shape, porosity and density of the particles, as well as the density and viscosity of the solution. The particle size of immobilized enzymes is an important factor in the formation of a homogeneous fluidized bed. Immobilized enzyme particles should be small enough to achieve a high biocatalytic surface area. These particles must be sufficiently dense, relative to the substrate stream, so as to prevent them from being carried out of the reactor. The advantages of the fluidized bed reactor are its ease of pH and temperature control, and its excellent heat and mass transfer characteristics, which make it ideal for highly exothermic reactions. However, the fluidized-bed reactor has a number of disadvantages, such as the need for high input power to fluidize the bed, the difficulty of scaling up, the potential loss of biocatalysts associated with very high flow rates, and the susceptibility to destruction of immobilized enzymes due to particle fluidization.

5.2.2.4.4. Membrane reactor

The usual choice of membrane reactor for enzymatic conversion is the hollow-fiber reactor. This reactor consists of a module that contains hundreds of thin tubular fibers, each with an internal diameter of 200–300 μm and an external diameter of around 300–900 μm. Substrate diffusion through the tubular wall enables it to come into contact with the enzyme and be converted into product. Subsequent diffusion of the product facilitates its separation for recovery. Under the influence of differential pressure along the tube wall, the product flows inside the tubes. Advantages of the membrane reactor are the ease with which these systems can be set up; they are often used for small-scale production (grams to kilograms), particularly when a multienzyme system or coenzyme regeneration is required; they allow easy enzyme replacement in processes involving particularly labile enzymes; and they can also be used for biphasic reactions. The main disadvantages of this type of reactor are the cost of the membranes, their clogging, and the need to change them at regular intervals.

5.2.2.5. *Enzyme kinetics of immobilized enzymes*

The rate laws are of the same form as those established for free enzymes, with a coupling between the biochemical reaction and the liquid–solid matter transfer. When the enzyme is bonded to the surface of solid particles, the substrate must be transferred from the liquid phase to the liquid–solid interface, where it is transformed by the reaction. A quasi-stationary state is reached, corresponding to a transfer rate equal to the enzymatic reaction rate. Considering a reaction according to the Michaelis–Menten law, the rate equality leads to the following relationship:

$$k_s(S - S_S) = \frac{V_{max} \cdot S_S}{K_m + S_S} \qquad [5.1]$$

with:

k_s: mass transfer coefficient $m^3. kg^{-1} s^{-1}$;

V_{max}: maximum specific reaction rate per unit mass of media ($mol \cdot s^{-1} \cdot kg^{-1}$);

S: substrate concentration in the liquid (mol/m^3);

S_s: substrate concentration at the liquid–solid interface (mol/m^3);

K_m: Michaelis and Menten constant (mol/m^3).

This relationship can be written in the following dimensionless form:

$$\left(\frac{K_m}{(S)} + \frac{(S_S)}{(S)}\right)\left(1 - \frac{(S_S)}{(S)}\right) = Da\frac{(S_S)}{(S)} \qquad [5.2]$$

$$Da = \frac{V_{max}}{k_S.S} \qquad [5.3]$$

Da is the Damköhler number: it represents the ratio between the maximum rate of reaction and the maximum rate of matter transfer. A Damköhler number that is much greater than 1 means that the rate of reaction is greater than the rate of transfer, and corresponds to a so-called diffusion regime, since it is the transfer of matter that is the limiting step. Conversely, when the Damköhler number is less than 1, the transfer rate is greater than the reaction rate, which is the limiting step, and corresponds to a so-called kinetic regime.

5.2.2.6. *Examples of the use of immobilized enzymes in the food industry*

Immobilized enzymes have many applications in the food industry, such as the production of flavors, syrups, confectionery, dairy products, alcohols and fruit drinks, yeasts for bakery products and whey lactose hydrolysates (Nisha et al. 2012). The use of immobilized enzymes in the dairy industry is very important: for example, immobilized lactase, which catalyzes the hydrolysis of lactose, enables the production of lactose-free milk for people with lactose intolerance. Immobilized enzymes are also used for clarifying fruit juices and removing bitterness from citrus juices (Adhikari 2019). Biosensors have also been developed to analyze food composition and monitor product quality (Sharma and Leblanc 2017). Immobilized enzymes also have applications in active food packaging to extend food shelf life and improve quality (Nestorson et al. 2008). In the food industry, immobilized enzymes cannot only be used in batch processes, but also in continuous processing, which is one of the advantages of immobilized enzymes over soluble enzymes. Some examples of immobilized enzymes for different food applications are shown in Table 5.1.

Enzymes	Media	Immobilization technique	Applications	References
Candida rugosa lipase	Membrane PS	Cross-linking	Olive oil hydrolysis	(Wang et al. 2008)
β-Galactosidase (lactase)	Chitosan	Cross-linking	Production of lactose-free products	(Zawawi et al. 2020)
α-Amylase	TiO_2 nanoparticles	Adsorption	Starch hydrolysis	(Ahmad et al. 2014)
Thermolysin	Organic media	Adsorption and cross-linking	Aspartame production	(Nagayasu et al. 1994)
Lipase	Celite	Adsorption	Production of substitutes, butter, cocoa	(Mojovic et al. 1995)
Urease	Calcium alginate	Entrapment	Milk quality control (test)	(Das et al. 1998)

Table 5.1. *Example of the use of immobilized enzymes in the food industry*

5.2.3. *Microwave and ultrasound-assisted enzyme processes*

5.2.3.1. *Microwave-assisted processes*

Recent advances in the implementation of emerging food technologies have paved the way for increased use of microwave energy. Generally speaking, the microwave applications described for food processing include drying, tempering and sterilization. This technology offers advantages such as energy savings through volumetric heating, reduced processing time and operational cost (Chandrasekaran et al. 2013).

Microwaves are electromagnetic waves with frequencies ranging from 300 MHz to 300 GHz. In general, to balance efficiency and cost, the domestic microwave frequency is 2.45 GHz, while the industrial microwave frequency is between 915 MHz and 2.45 GHz.

The first studies reported in the literature focused on the impact of microwave irradiation heating on various enzymes used in the food processing industry, such as oxidases, oxygenases and phenol oxidases. This work has shown that microwaves deactivate these enzymes, which are responsible for oxidation, browning and flavor degradation (Matsui et al. 2007; Keying et al. 2009).

Microwaves make two contributions:

– a purely thermal contribution, resulting from molecular agitation caused by the inversion of dipoles with alternating electric fields;

– a specific contribution of electrostatic origin, resulting from dipole–dipole interactions between the polar molecules and the electric field, which can induce the appearance of charges in the transition state with, for example, the formation of a tetrahedral intermediate, stabilized under microwave irradiation.

Among the various applications of microwave technologies in enzymatic processes, microwaves have been reported as a contributor for enzyme immobilization: for example, Bezbradica et al. (2009) showed that the specific activity of *Candida rugosa* lipase, after microwave-assisted immobilization on the Eupergit C250L medium, was higher than that obtained after immobilization without microwaves, due to a better diffusion rate.

In general, microwave irradiation technology is used to solve problems associated with diffusion limitations. A classic example of the use of microwaves is in enzyme-assisted extraction (EAE). Indeed, EAE is an alternative to traditional extraction methods, due to its particular advantages, including reduced solvent use, increased extraction efficiency and ease of use (Puri et al. 2012; Cheng et al. 2015). EAE has already been expressed for the recovery of constituents from rapeseed, sunflower, olive and sesame, but also for fruit juice processing: for the latter, EAE made it possible to modify the integrity of the fruit cell wall, allowing the release of phenolic compounds into the juice, leading to an extended shelf life and an increase in overall product quality (Puri et al. 2012). To increase the effectiveness of EAE, one approach has been to apply microwave irradiation during EAE. The procedure, known as microwave-assisted extraction (MAE), is described as having good prospects, as the energy provided by the non-ionizing microwave radiation induces fragmentation of the cellulose matrix, thanks to the heat that is continuously supplied (Wang et al. 2007; Adetunji et al. 2017). In order to achieve good performance, various factors need to be optimized, such as the processing time and the solid–liquid ratio.

For example, MAE has been used to extract organic compounds (such as phenols, carbohydrates and lipids) from plant matrices. The use of microwaves increased the temperature inside the system and improved solvent diffusion in the solid matrix, thus enhancing the extraction of organic compounds. This combination of different processing strategies (for example, microwave, ultrasound and supercritical fluid extraction) with enzymes has proved highly effective in recovering bioactive compounds from plant matrices and their by-products (Nadar et al. 2018). This integrated strategy therefore has the potential to reduce the volume of solvents used, thus achieving greater extractability in shorter reaction times (Macedo et al. 2021).

In addition to the contribution of microwave irradiation to facilitate the enzymatic extraction of molecules from renewable matrices, there has also been reports of interest in microwaves for increasing the efficiency of enzymatic reactions. Microwave irradiation acts by delivering energy that reduces the distance between the reactive groups of the substrate molecule and the catalytic residues at the enzyme's active site. The application of microwaves thus generates an increase in enzyme reaction efficiency. For example, work by Zhang et al. (2012) has demonstrated changes in the spatial conformation of glucoamylase, following the application of microwave energy. As a result, the alpha-helix content sharply diminished and beta-sheets and kinks increased. These conformation changes may explain the increased rate of starch hydrolysis by glucoamylases, observed upon application of microwave processing.

5.2.3.2. *Ultrasound-assisted processes*

Ultrasound-assisted processes are being increasingly used in the food industry, including processing (crystallization, drying, filtration, cutting, etc.), preservation (inactivation of microorganisms and enzymes) and extraction (Misra et al. 2017). They have been detailed for the intensification of numerous physical, chemical and biological processes, including enzymatic reactions.

The principle behind ultrasound-assisted processes is to generate an acoustic wave with a frequency >20 kHz, which propagates through the medium. Much like mechanical waves, ultrasound waves can lead to multiple compression and rarefaction processes. During the rarefaction cycle caused by reduced pressure, microbubbles are formed. Acoustic cavitation is generated by the formation and transfer of these microbubbles. During cavitation, the pressure and temperature inside the bubble can reach over 1,000 atm and 5,000 °K, respectively (Wang et al. 2018b).

For many years, ultrasound has been used as an enzyme inactivation method (O'Donnell et al. 2010), but research has shown that, under mild conditions, it does not inactivate all enzymes. On the contrary, ultrasound can actually have positive effects on enzyme activity and can be used to accelerate enzymatic reactions. Ultrasonic cavitation causes changes in protein conformation, influencing enzyme activity. Generally, a short duration and low intensity ultrasound treatment is beneficial for enhancing enzyme activity, while prolonged exposure can lead to a progressive loss of enzyme stability and activity (Delgado-Povedano et al. 2015).

Ultrasound can accelerate enzymatic reactions by acting on different targets: it can modify both substrate macromolecules and enzymes, leading to an increase in enzymatic activity and product yields. First, ultrasound can induce structural changes in food materials prior to enzymatic processing. For example, they affect the surface morphology of proteins by promoting the unfolding of protein molecules and creating micropores on their surface. As a result, protein molecules tend to be more disordered and their surface areas increase, thereby improving the affinity between enzymes and substrates, and promoting an acceleration in the rate of enzymatic hydrolysis (Jin et al. 2016). Various enzymatic parameters can be improved, such as the initial reaction speed, as well as the energy efficiency (Gibbs enthalpy, entropy and free energy).

In a study of the effect of ultrasound on the aqueous enzymatic extraction of oil from perilla seeds (*Perillafrutescens L.*), it was observed that when an ultrasonic power of 250 W was applied, an increase in oil yield was observed (Li et al. 2014).

Other work has investigated the effect of ultrasound-assisted enzymatic extraction on the recovery of proteins and antioxidant compounds from sesame bran (Gorgüç et al. 2019). An increase in protein and phenol yields was obtained, as well as antioxidant capacities, compared to the standard alkaline extraction method.

Numerous works have thus demonstrated the benefits of ultrasound on the efficiency of EAE, as well as on enzymatic hydrolysis of food materials, such as corn gluten meal, rice proteins, tea residues, sunflower meal and rapeseed proteins (Jin et al. 2016; Yagoub et al. 2017; Dabbour et al. 2018).

Lastly, processes that pair microwaves and ultrasound have also been proposed; a study was carried out to assess the influence of a combined microwave and ultrasound process on dextran hydrolysis, catalyzed by dextranase (Bashari et al. 2016). This combination led to an increase of over 170% in the rate of dextran hydrolysis, compared to the traditional heating method. Furthermore, the results showed higher values of Vmax and catalytic efficiency (kcat/Km) for dextranase when both microwave and ultrasound technologies were simultaneously used, compared to when ultrasound or microwave technologies were used separately. Compared with the enzymatic reaction of conventional heating, the combined technology showed an increase in the thermal stability of dextranase, leading to accelerated hydrolysis of the dextran substrate.

5.2.4. *Process intensification*

The long-term stability of enzymes under certain temperature and pH conditions, as well as enzyme productivity, are sometimes not sufficient to meet the requirements of large-scale applications. In addition to more conventional strategies, such as enzyme immobilization and enzyme engineering techniques, process engineering and, more specifically, process intensification can improve enzyme performance and increase the cost-effectiveness of the enzyme process. The term "process intensification" comes from the world of chemistry, where it is described as "any development that leads to a smaller, cleaner and more energy-efficient technology". Although it is also often used for biological processes, the definition varies. One definition has been proposed (Vaghari et al. 2015) and describes intensification as "the development of emerging equipment and operational methods leading to smaller, cleaner plants that increase production efficiency and quality while reducing by-product formation, capital costs and energy consumption". This intensification is a key factor in increasing product yield, process productivity and/or enzyme stability. A wide range of intensification processes have been described, based on the elimination of the product formed, in order to shift the thermodynamic equilibrium toward production or avoid further processing of this product of interest, such as evaporation, extraction into another phase, membrane-based size-selective permeation or immobilization of the product by adsorption.

In terms of processes, two types of bioreactors are commonly used in industry to carry out enzymatic transformations: the stirred-tank reactor and the fixed-bed reactor or the fluidized bed reactor for immobilized enzymes. More recently, microfluidic reactors have been developed. Microfluidics is the technology that processes or manipulates small quantities (10^{-9} to 10^{-18} liters) of fluids through channels of a few dozen to a few hundred micrometers; this is used in the analysis of biochemical reactions, immunoassays, biotechnological and pharmaceutical processes. Microfluidic reactors offer a number of advantages, such as a large specific surface area and high mass transfer rates. In addition, the structure, shape and properties of the channel walls can be modulated to influence the fluid flow state, which in turn affects the mass transfer rate.

Breaking with traditional discontinuous batch processes, the concept of intensification aims to promote continuous processes with intensified operating conditions.

For enzyme immobilization, two concepts of micro-structured immobilized enzyme reactor exist. One option is to directly attach the enzyme to the surface of the microscopic channel(s). Another is to immobilize the enzyme on microparticles (usually spherical) that fill the microchannel, so as to generate a micro-bed. The pressure differential required to operate a reactor with a fixed micro-bed under continuous flow conditions could be quite high. At these microreactor scales, transfer phenomena are considered highly efficient and can satisfy certain kinetic and reaction selectivity constraints, as well as certain safety aspects (Wohlgemuth et al. 2015). However, the substitution of conventional reactors for enzymatic processes by a system based on microfluidic technology will only occur if there are clear and convincing advantages to achieving it. It is therefore necessary to identify the potential advantages of these micro-structured reactors for biocatalytic synthesis in the food sector.

Other types of reactors have more recently been described for food processes: for example, the Taylor–Couette flow reactor has been used to intensify the process of starch hydrolysis into sugar (Matsumoto et al. 2021). At the industrial level, several reactors are used in starch hydrolysis, for the various stages of gelatinization, liquefaction and saccharification. The continuous hydrolysis of starch with a single Taylor–Couette flow has enabled a satisfactory yield of starch hydrolysis into reducing sugars because of the configuration and characteristics of this reactor.

6

Main Enzymatic Processes in the Food Industry

Latifa CHEBIL and Mohamed GHOUL
LRGP, CNRS, Université de Lorraine, Nancy, France

Enzymes are increasingly being used in the food industry, playing different roles on several levels:

– improving the composition of raw materials and/or their transformation processes;

– improving the quality of existing products;

– acquiring new products.

Enzymes can correct the natural deficiencies of raw materials, enabling formulations to be standardized. They also have the capacity to act as a technological tool during a transformation process, while improving the qualities of the finished product, especially in organoleptic terms.

This chapter summarizes the main applications of enzymes in the food industry.

6.1. Baking and pastry

With an aim to replace chemical improvers, enzymes are widely used as biological improvers for bakery and pastry products.

For a color version of all the figures in this chapter, see www.iste.co.uk/ghoul/implementation.zip.

Implementation of Enzymatic Processes and Lactic Bacteria in the Food Industries, coordinated by Mohamed GHOUL.

The miller needs correction agents and the baker needs technical aids. The miller's objective is to supply the baker with flour of consistent quality and performance. The baker needs to produce a wide variety of breads and pastries efficiently on a daily basis and meet all quality requirements.

Enzyme manufacturers have developed a whole range of "tailor-made" enzyme products to meet the needs of the bakery and pastry industry. The enzymes used belong to the hydrolase and oxidoreductase families. The hydrolases used are amylases, hemicellulases, pentosanases, lipases and proteases. Oxidoreductases include lipoxygenase, glutathione dehydrogenase, ascorbic acid oxidase, catalase, glucose oxidase, sulfhydryl oxidase, phenol oxidase, laccase and peroxidase (Dahiya et al. 2020).

Table 6.1 summarizes the different enzymes used in baking and pastry-making.

Enzyme	Organoleptic/nutritional effects	Technological effects	References
Amylases (α-amylase, β-amylase)	– Modification of the bread aroma through formation of volatile compounds from secondary fermentation, as well as Maillard and caramelization reactions – Promotes crust coloring	– Promotes gas release during fermentation: bread is well-balanced and not sticky – Can reduce gas retention during baking – Anti-staling effect of bacterial amylases	(Van Dam and Hille 1993)
Hemicellulases and pentosanases (endoxylanases)	Promotes crust coloring	– Improved dough characteristics (extensibility, elasticity, stickiness) and bread characteristics (volume, appearance, crumb) – Lower energy consumption during kneading	(Rouau 1996)
Proteases	Promotes crust coloring	– Reduces elasticity and improves dough extensibility – Modulation of the Maillard reaction and therefore aroma formation – Increased water absorption rate	(Dahiya et al. 2020)

Enzyme	Organoleptic/nutritional effects	Technological effects	References
Oxidases (lipoxygenase, glutathione dehydrogenase, ascorbic acid oxidase, catalase, glucose oxidase, sulfhydryl oxidase, phenol oxidase, laccase, peroxidase)	Lipoxygenase (crumb whitening, modification of the aroma following fatty acid oxidation and hexanal production)	– Lipoxygenase (accelerates formation of the gluten network, increases bread volume, tighter crumb cell) – Glutathione dehydrogenase (glutathione oxidation, resulting in firmer dough) – Ascorbic acid oxidase and catalase (O_2 production); glucose oxidase (H_2O_2 production) – Phenol oxidase, sulfhydryl oxidase, laccase, peroxidase (enzymatic browning)	(Potus and Drapon 1997)
Phytases	Increased digestibility of calcium, magnesium and iron	–	(Kumar et al. 2010)
Lipases	– Promotes loss of essential fatty acids – Modification of the bread aroma through formation of volatile compounds resulting from oxidation of non-esterified fatty acids	– Improved rheological properties of doughs – Anti-staling effect	(Dahiya et al. 2020)

Table 6.1. *Organoleptic, nutritional and technological effects of enzymes in baking and pastry-making*

Joan and Drost-Lustenberger (2002), described the potential role of enzymes as an alternative to other "technological aids". With the addition of xylanase, the bread volume is slightly improved. When using xylanase in combination with fungal α-amylase, volume is further increased; it also improves crumb structure, so that cell walls are thinner and the crumb is silkier. When lipase is combined with the other enzymes, not only is the volume of the bread further increased, but the structure of the crumb is considerably improved too, with more uniform cells. As a result, this bread will have the softest crumb among these samples. Thus, the combination of lipase with α-amylase or xylanase can provide a fine, silky and uniform crumb structure compared to that obtained for breads made with unprocessed dough.

The main hydrolases involved in bread fermentation are summarized in Figure 6.1.

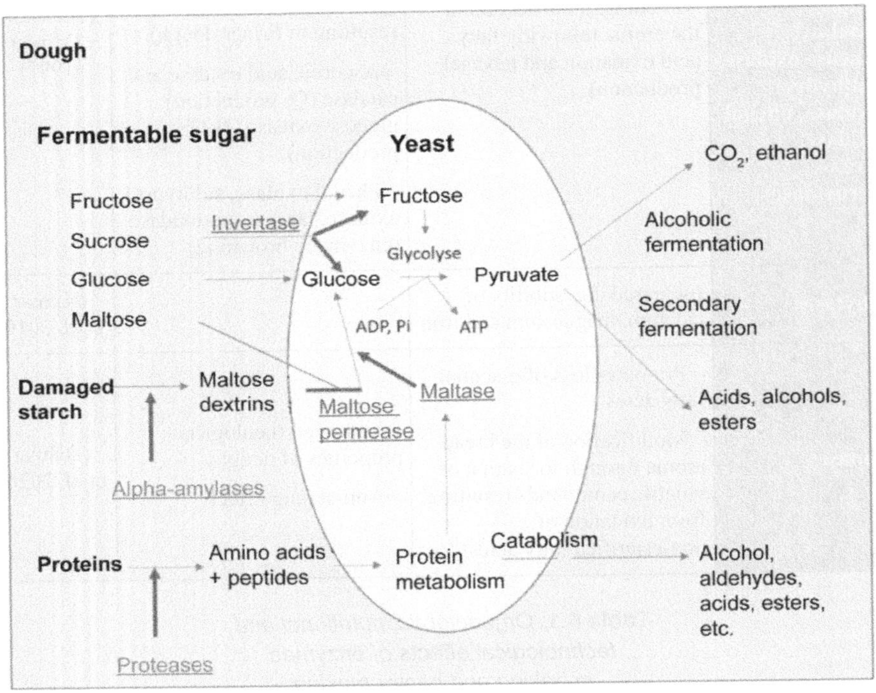

Figure 6.1. *Involvement of hydrolases in bread fermentation*

6.2. Malting and brewing

The beer-making process involves ingredients such as barley malt, hops, yeast and water, as well as technological aids such as exogenous enzymes.

A distinction is made between endogenous enzymes, supplied by the malt, and those added in response to a specific manufacturing problem, such as to supplement or replace malt enzymes, improve and standardize the beer process and quality, or produce beers with new properties (Geissinger et al. 2021).

Figure 6.2 illustrates the beer-making process and the enzymes involved in malting and brewing.

6.2.1. *Malting*

The key stage in the malting process is barley germination. During this stage, numerous endogenous enzymes are synthesized or take an active form. The enzymes involved in malting are as follows: phytases, hemicellulases, proteolytics and amylolytics.

Phytases release phosphates and metal ions required by the yeast during fermentation.

Hemicellulases play an important role in the degradation of the albumen cell wall, thanks to solubilase, which cleaves the protein-β-glucan bond and thus solubilizes β-glucans. This enzyme remains active during mixing. Subsequently, β-glucanases, synthesized during germination, degrade the β-glucans.

Proteolytic enzymes such as endopeptidases and carboxypeptidases, which are present in barley and synthesized during germination, allow amylases to access starch, but also to degrade insoluble storage proteins into soluble proteins.

Amylolytic enzymes are widely used to release the fermentable sugars needed for fermentation. These enzymes are generally α-amylases and β-amylases that are synthesized during malting and active during brewing. The α-amylases will break the α-(1-4) bonds within the amylose and amylopectin chains. β-Amylases attack the ultimate α-(1-4) bonds at the non-reducing end of starch chains. These enzymes are synthesized during germination.

In malting, exogenous enzymes such as cellulases, amylases and proteases are added during the malting process to improve the physicochemical properties of the malt. Thermostable enzymes (β-glucanases, pentosanases, etc.) are also used to remedy the problems of poor-quality wheat and barley malts.

6.2.2. *Brewing*

As described above, endogenous malt enzymes are involved in the brewing process.

These enzymes are involved in the proteolytic stage (temperature between 40°C and 50°C, and a duration between 15 and 60 minutes) and the saccharification stage (temperature between 60°C and 70°C, and a duration of 30 minutes).

In brewing, exogenous enzymes can be added to the mash tun. Either malt is added as a source of α-amylases or α-amylases from different bacterial species are added.

Exogenous α-amylases are much more stable than malt α-amylases. These enzymes involve a low water-to-grain ratio, which makes it possible to increase mash tun capacity and control temperature when mash tun corn is added to the material tun.

Filtration is a key stage in the beer-making process. The main molecules affecting this stage are high molecular weight β-glucans and hemicellulases. To control this stage and improve filtration speeds, enzymes such as α-amylases from *Bacillus subtilis* or *Aspergillus oryzae* can be added during the brewing process. β-Glucanases from *Bacillus subtilis*, either alone or in combination with other hydrolases, have been shown to be hydrolytically effective during brewing, fermentation and storage.

During beer maturation, diacetyl, which gives beer a buttery taste, can be eliminated by adding exogenous enzymes, such as α-acetolactate decarboxylase (ALDC) extracted from either *Enterobacter aerogenes* or *Bacillus subtilis*.

Another very important parameter that must be controlled during maturation is the beer's clarity and brightness. The protein-polyphenol complex is the main reaction leading to the formation of colloidal haze in beer. Papain is the most efficient proteolytic enzyme that is capable of hydrolyzing proteins into small peptides that are unable to bind to polyphenols. This enzyme is active at temperatures between 55°C and 65°C, during beer pasteurization, and lasts for two days.

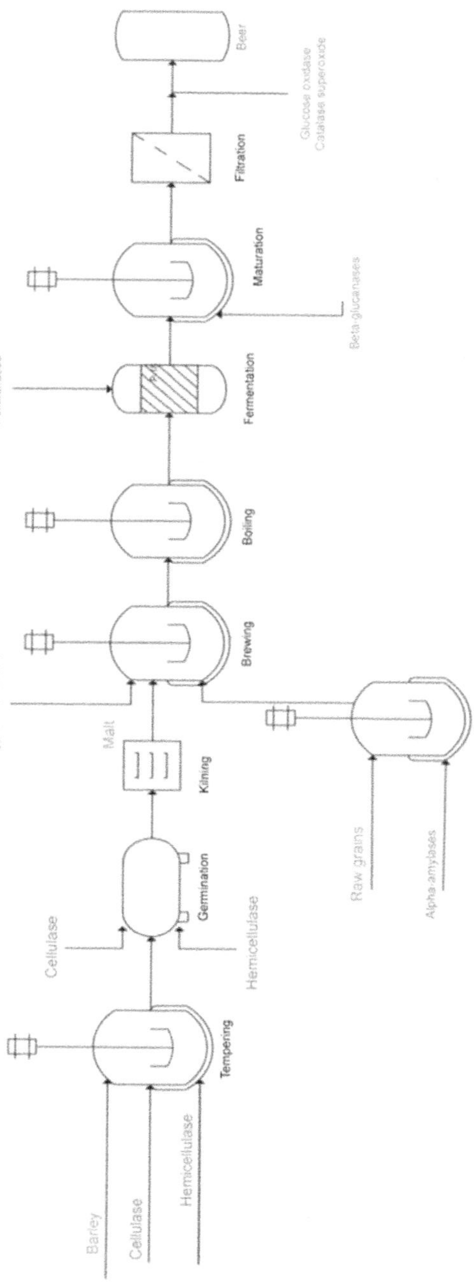

Figure 6.2. *Illustration of the addition of exogenous enzymes in malthouses and breweries*

When beer is stored, dissolved oxygen in the neck of the bottle can cause oxidation problems.

Glucose oxidase is used in brewing to remove oxygen by consuming it, producing gluconic acid, hydrogen peroxide and water.

The catalase present in the enzymatic extracts will, in turn, transform the hydrogen peroxide into water and oxygen.

6.3. Starch and its by-products

Starchy agricultural raw materials (wheat, corn, potato, etc.) are widely used around the world to produce starches using processes involving various enzymes (Figure 6.3).

The first starch processes were developed in the United States (Peckham 1979) to produce isoglucoses and glucose syrups, as well as bioethanol.

The hydrolytic enzymes used to transform amylopectin into D-glucose are liquefying enzymes, such as α-amylases, saccharifying enzymes, such as β-amylases and amyloglucosidases, and debranching enzymes, such as pullulanases and isoamylases.

Combined liquefaction and saccharification technology produces high hydrolysis yields and hydrolysates, with a degree of hydrolysis (DE) of between 97 and 98, or syrups with a high maltose content.

In Europe, this double enzymatic conversion technique (a combination of α-amylase from *Bacillus amyloliquefaciens* and amyloglucosidase from *Rhizopus delemar*, then *Aspergillus niger*) was introduced in 1964 by Roquette Frère.

The hydrolysates obtained made it possible to isolate dextrose by crystallization in higher yields than those obtained by acid or an acid-enzyme conversion.

The various hydrolysis products, along with their functionalities, are grouped together in Table 6.2 (Jiang et al. 2018).

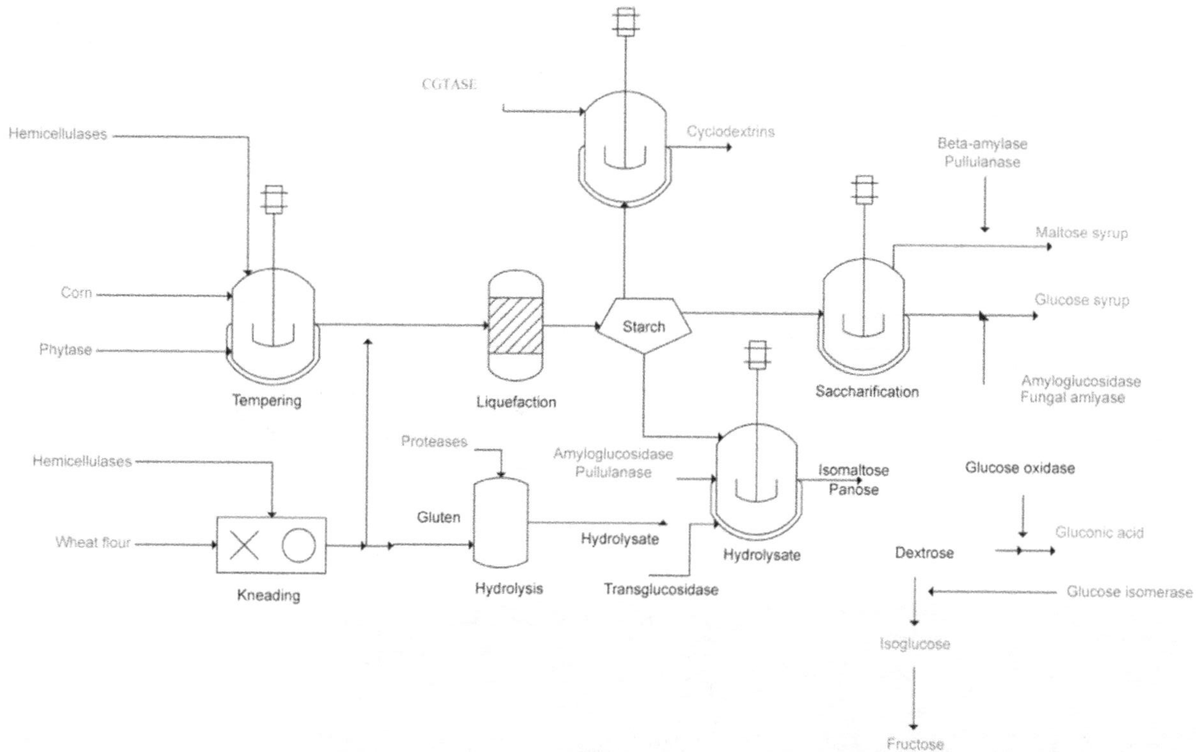

Figure 6.3. *Illustration of the use of enzymes in starch production*

Hydrolysats	Definition	Properties/applications
Maltodextrins	$DE \leq 20$ Obtained by acid or enzymatic hydrolysis or a combination of both	– Maltodextrins $2 \leq DE \leq 6$ have encapsulating, flavoring and gelling properties. – $DE \geq 6$ maltodextrins are used as drying or spray media, infant food, soups and sauces in sachets.
Glucose syrups	$20 \leq DE \leq 100$	– Syrups with a high DE have sweetening power. – Liquid form: confectionery, tinned fruit, candied fruit, jams, compotes, liqueurs, soft drinks and brewing. – Powder form: cold meats, cured meats, sauces, biscuits, ice cream, bakery aids and health foods. – High-maltose syrups (55–80%, saccharification by β-amylases or a mixture of β-amylases + pullulanase + maltogenase): non-cariogenic substitutes for sucrose in confectionery, chewing gum and chocolate.
Starch hydrolysates (Ho)	$96 \leq DE \leq 98$	– Raw fermentation materials. – Dextrose production.
Dextrose	Starch hydrolysate crystallization Use of pure amyloglucosidase	Use as a medium for spices or flavorings, bulking agents in instant products, fermentation substrate in cured meats, browning agent in pastries, sweetener, etc.

Table 6.2. *Different starch hydrolysis products: definition and properties*

6.4. Milk

Milk is rich in carbohydrates, lipids, proteins, minerals (calcium, potassium, sodium) and vitamins (D, B12). It also contains numerous enzymes and undergoes numerous transformation reactions that are initiated by these endogenous enzymes, or by the addition of exogenous enzymes. The purpose of these transformations is to improve nutritional, taste and aromatic qualities, so as to develop new by-products from milk with high added values, as well as increase the shelf life of milk and its by-products (Trani et al. 2017). Enzymes are also used in the cleaning stage of industrial plants involved in milk processing (pipes, vats, heat-exchangers and microfiltration). The most commonly used enzymes are lipases, proteases, lactase, catalases and transglutaminases. They are most frequently used in cheese production processes during coagulation, ripening and flavor development stages (Khan and Selamoglu 2020). The first uses of rennet to accelerate the coagulation process of cheese date back to 6,000 B.C. In contrast, the reduction of lactose through its enzymatic transformation into galactose and glucose, and its oligomerization and the engineered hydrolysis of fatty acids by lipases are more recent applications. The enzymes used have different origins (microbial, plant, animal and fungal). On an industrial scale, enzymes are often used in the form of cocktails, with a range of specific activities depending on the functionalities desired and the matrix used.

Table 6.3 summarizes the functionalities and technological effects of the most common enzymes used in various milk processing and functionalization processes.

Enzyme	Organoleptic/nutritional effects	Technological effects	References
Proteases (trypsin and chymotrypsin)	Improved solubility, emulsification and foaming power of milk proteins	Controlled hydrolysis of milk proteins	(Banach et al. 2013)
Proteases (peptidases)	Improved flavor and faster ripening of cheese	Protein hydrolysis and peptide production	(Haard and Patel 1987; Van Dijck et al. 2014)
Proteases (rennet)	Protein coagulation for cheese production	Production of paracasein and start of the gelation process in the presence of calcium	(Trani et al. 2017)
Proteases (endopeptidase)	Cheese debittering	Elimination of small peptides resulting from the action of endopeptidases	(Saha and Hayashi 2001)
Lipases	Improved flavor and faster ripening of cheese	Controlled triglyceride hydrolysis	(Hernandèz et al. 2005; Alkan et al. 2007)

Enzyme	Organoleptic/nutritional effects	Technological effects	References
Lipases (phospholipase A1)	Improved emulsifying properties	Increased cheese yield through fat and water retention	(Lilbæk et al. 2006)
Lipases (phospholipase)	Improved thermal stability and foaming and emulsifying properties of milk and milk by-products	Hydrolysis of phospholipids in milk	(Lilbæk et al. 2007)
Lactases (β-galactosidase)	Improved taste and nutrition through the elimination of lactose	Hydrolysis of lactose into galactose and glucose	(Wilkinson et al. 2003; Novalin et al. 2005)
Lactase (β-galactosidases)	Production of oligosaccharides with probiotic properties	Implementation of hydrolysis and transgalatoglycosulation reactions with lactose as an acceptor and donor substrate	(Foda and Lopez-Leiva 2000)
Peptidoglycan N-acetylmuramoylhydrolase (lysosyme)	Improved milk shelf life	Peptidoglycan hydrolysis of bacterial walls	(Masschalck and Michiels 2003)
Lactoperoxidase	Improved milk shelf life	Damage to bacterial walls by a combined action with hydrogen peroxide	(Kamau et al. 1991; Seifu et al. 2005)
Lactoferrin	Longer shelf life for milk and milk by-products	Essential iron chelation for bacterial growth	(Aguila La 2001)
Transglutaminase	Improved milk gel structure	Development of a cross-linking reaction between caseins and soluble proteins	(Jaros et al. 2010; Şanli et al. 2011)
Catalase	Improved quality and preservation of milk and milk by-products	Removal of hydrogen peroxide, indicator of heat treatment efficiency	(Singh et al. 2019)

Table 6.3. *Organoleptic, nutritional and technological effects of enzymes in milk processing and functionalization*

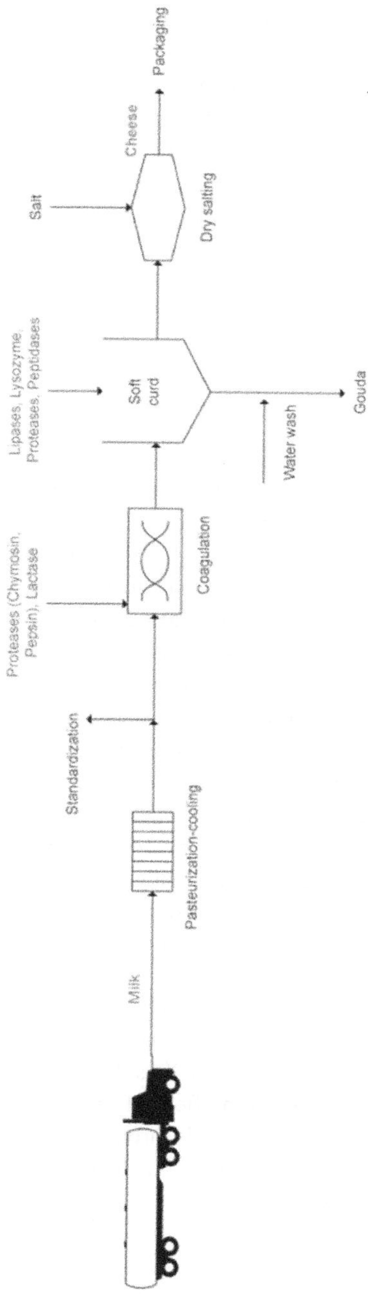

Figure 6.4. *How enzymes are used in the cheese production process*

Table 6.3 shows that many enzymes are used, or have the potential to be used, in the various stages of milk processing. The use of enzymes in the processing of milk's various constituents, notably fat and proteins, is growing, opening up new prospects for the development of healthy foods (oligosaccharides, peptide- or DHA-enriched protein hydrolysates, lactose-free formulations, etc.) and new products, with new flavor profiles, as well as improved, specific functionalities. These new ways of processing milk are enabled by recent advances in enzyme engineering, as well as significant improvements in enzyme extraction and purification production processes.

The main uses of enzymes in milk processing are illustrated in Figure 6.4.

6.5. Fat

The extraction of oils from the plant matrix and the biotransformation of fat, to obtain various by-products of interest in food and non-food applications, were essentially carried out by physicochemical processes (crushing, pressing, extraction, etc.) and reactions involving chemical catalysts. These processes and reactions have numerous drawbacks, due to their impact on the environment, the production of undesirable by-products and increased energy consumption. To overcome these disadvantages, enzymatic processes have been proposed as an alternative to chemical catalysts since 1985. These processes are carried out under mild temperature and pressure conditions, thus avoiding coloration of the reaction medium and initiation of secondary reactions. Similarly, enzymes are able to carry out hydrolysis and synthesis reactions in a selective and specific manner, facilitating post-reaction processing stages and obtaining highly specific products, depending on the application. The first industrial applications of lipases exploited their ability to hydrolyze triglycerides to produce free fatty acids, monoglycerides and diglycerides. These enzymes were later used in synthesis reactions (esterification) and acyl transfer by interesterification (transesterification, acidolysis and alcoholysis).

However, the first processes were unprofitable due to the high cost of enzymes and the difficulty of controlling the enzyme macro- and micro-environment under suitable conditions. It was not until the 2000s that the first efficient industrial applications of enzymes were found in the field of fat conversion and oil extraction. This increase in efficiency was achieved through the development of higher performance enzymes and their immobilization on innovative substrates, enabling them to be recycled over several cycles of use. Thanks to these improvements in enzyme engineering and processes, numerous applications have been developed. These include the production of structured lipids with new physicochemical, rheological and nutritional properties, which have been exploited in the fields of infant nutrition, the production of low-calorific compounds and the production of omega-3-rich extracts. The production of butter substitutes with melting temperatures controlled and adjusted according to the intended application is another outstanding application of enzymes. Enzymes are also used to synthesize specific flavors from milk fat.

The use of enzymes at the industrial stage has also been successfully applied to improve oil extraction processes from various plant matrices. The use of these enzymes has led to significant improvements in extraction yields, the energy balance and the environmental impact of these processes. Enzymes have also been used in the synthesis of specific esters for cosmetics, food emulsifiers and in the biodiesel production process.

Table 6.4 summarizes the organoleptic, nutritional and technological effects of some of the enzymes used in fat extraction and processing.

Enzyme	Organoleptic/nutritional effects	Technological effects	References
Phospholipases (PLA1, PLA2 and PLC)	– Removal of impurities – Improved taste – Increased extraction yield – Reduced water and energy consumption – Selective hydrolysis	Oil refining by phospholipid hydrolysis	(Guerrand 2017)
Lipases	– Improved emulsifying power and texture – Improved flavor and texture	In situ hydrolysis of wheat lipids	(Soe et al. 2017)
Lipases	Development of specific aromas and improved cheese texture	Hydrolysis of milk fat	(Aravindan et al. 2007)

Enzyme	Organoleptic/nutritional effects	Technological effects	References
Lipases	Improved textural and emulsifying properties of various food products	Oil esterification, mono and diester production	(Rathi 2017)
Cellulases	– Increased extraction yield – Improved oil quality	Hydrolysis of cell walls to release the oils contained in the plant matrix	(Ricochon and Muniglia 2010)
Hemicellulases	Increased oil extraction yields	Complementary action to cellulases in the hydrolysis of oilseed cellulases	(Kalia et al. 2001)
Pectinase	– Improved extraction – Color enhancement – Improved stability – Flavor enhancement	Cleavage of cell walls and release of plant matrix enzymes	Novozymes: "The use of enzymes in the extraction of olive oil. Documentation technique"
Triacylglycerol hydrolases	Improved rheological and nutritional properties: production of cocoa butter equivalent and Betapol for infant nutrition	Structured triglyceride synthesis	(Bornscheuer 2013; Holm et al. 2018)

Table 6.4. *Organoleptic, nutritional and technological effects of enzymes in fat extraction and processing*

The use of enzymes to improve food processes is growing at a steady pace and has led to the development of numerous new processes and products. These processes operate in a continuous mode, using less energy, eco-friendly solvents, and gentle temperature and pressure conditions. In addition to process improvements, enzyme activity and selectivity have also been enhanced by enzyme engineering and by the discovery of new enzyme production methods. Figure 6.5 illustrates the use of enzymes in a vegetable oil extraction process.

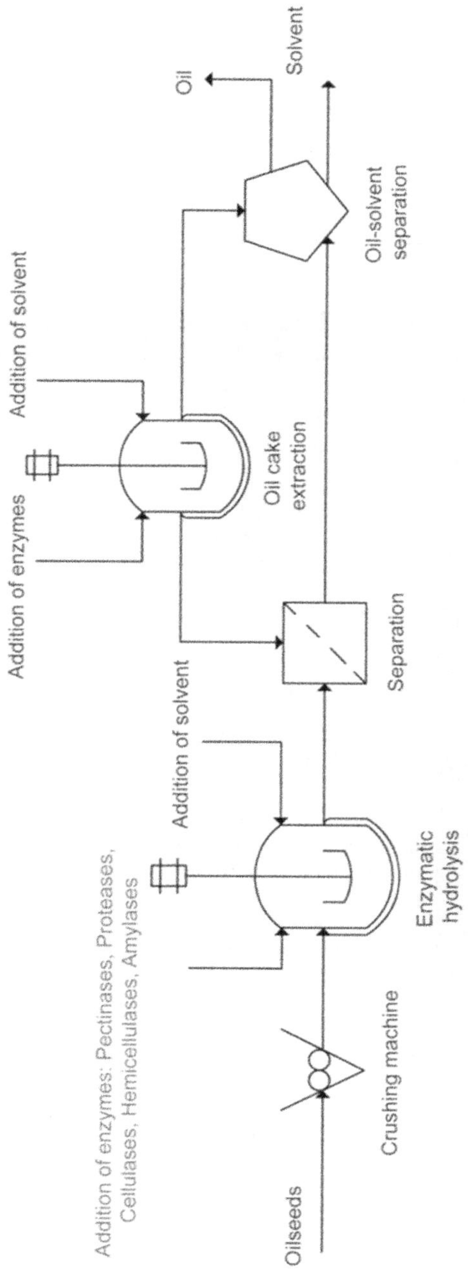

Figure 6.5. *Illustration of the use of enzymes in an oil extraction process*

6.6. Meat and seafood products

6.6.1. *Meat products*

Endogenous and exogenous enzymes are used in meat products to solve problems linked to meat tenderization or "meat staling", but also to reconstitute pieces of meat.

Table 6.5 lists the various enzymes used, as well as their technological contribution to meat products.

Enzymes	Technological effects	References
Exogenous enzymes: papain, ficin, collagenases	– Tenderization of muscles with lower collagen content – Uniform tenderness of grilled meats – Improved cuts that can then be eaten as steak or boiled	(Cunningham and Tiede 1981; Schenkova et al. 2007)
Calpains, cathepsins, proteasome	Modification of the myofibrillar structure of meat	(Penny 1980; Dransfield 1994)
Transglutaminase, TGase (EC 2.3.2.13)	– Restructured meat production – Increased firmness of heated meat products	(Nielsen et al. 1995; Tseng et al. 2000; Kuraishi et al. 2001; Dimitrakopoulou et al. 2005; Ahhmed et al. 2007)
Tyrosinase (EC 1.12.18.1)	Increased gelation of meat proteins, increased firmness of meat gels	(Lantto et al. 2006, 2007)
Laccase (EC 1.10.3.2)	Improved firmness of meat gels	(Yamaguchi 2000)

Table 6.5. *Technological effects of enzymes on meat products*

The endogenous enzymes involved in meat texture changes are cascaded. Calpains come first, with high ATP production. The degradation of ATP releases phosphoric acid, which acidifies the environment and activates cathepsins. Lastly, the proteins are eliminated by the proteasome.

These enzymes are effective in rapid maturation processes (10 days for beef and a few hours for chicken).

6.6.2. *Seafood products*

Proteases are used to hydrolyze fish flesh. Fish protein hydrolysate can be prepared with the addition of protease to increase endogenous hydrolytic activity and shorten hydrolysis time. Enzymatic hydrolysis has a greater advantage over chemical hydrolysis or autolysis caused by endogenous enzymes in fish. Hydrolysates can be obtained by using endogenous protease, but the process is time consuming.

To speed up the hydrolysis process, commercial proteases have been widely used to hydrolyze proteins, particularly those obtained as by-products or from under-utilized species. Commercial proteases used for fish protein hydrolysate production include alkalase, neutrase, pepsin, chymotrypsin, trypsin, papain and pronase E (Benjakul et al. 1997; Byun and Kim 2001; Je et al. 2004; Kim et al. 2007; Klompong et al. 2007, 2008; Thiansilakul et al. 2007a, 2007b; Cinq-Mars et al. 2008) (Figure 6.6).

Hydrolysis makes it possible to obtain fish hydrolysates with interesting antioxidant activities and angiotensin-converting enzyme (ACE) inhibitor activity.

In addition, neuroactive peptides and immunoactive peptides have been found in fish protein hydrolysates (Gildberg et al. 1996; Bernet et al. 2000). Bioactive peptides can be released from parent proteins by enzymes during gastrointestinal digestion or food processing (Meisel 1997).

Antioxidant peptides can act as natural antioxidants to prevent the modification of DNA by free radicals, proteins, lipids and small cellular molecules, which are associated with a number of pathological processes. These include atherosclerosis, arthritis, diabetes, and inflammatory and neurological disorders (Jun et al. 2004). ACE inhibitory peptides have significant antihypertensive potential (Vercruysse et al. 2005).

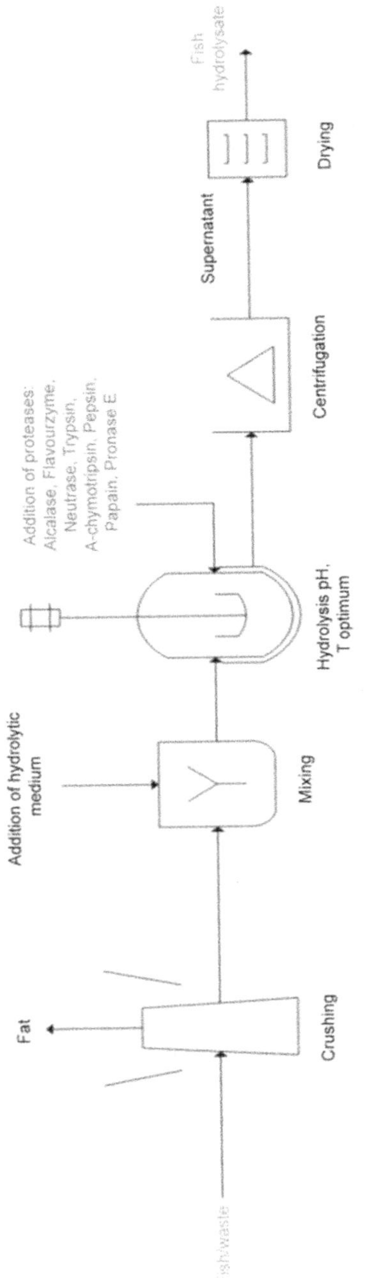

Figure 6.6. *Illustration of the process of converting fish into hydrolysate using enzymes*

TGase is a catalyst for the acyl transfer reaction between the γ-carboxyamide groups of glutamine residues in proteins or peptides and various primary amines. When the lysine amine group acts as an acyl acceptor, it leads to inter- or intramolecular polymerization and cross-linking of the protein through the formation of isopeptide bonds. This occurs through the exchange of the ε-amino group of the lysine residue for ammonia at the carboxyamine group of a protein glutamine residue. The formation of covalent cross-links between proteins underpins TGase's ability to modify the physical properties of food proteins (fish, shellfish, meat products, etc.).

Endogenous TGase in fish muscle varies from species to species and is most likely to be associated with gelation. Fish muscle naturally contains TGase, ranging from 0.10 to 2.41 U g^{-1} (Araki and Seki 1993). This enzyme induces cross-linking of muscle proteins (myosin) and affects the textural properties of fish protein gels. Benjakul et al. (2004) confirmed the role of TGase in the structuring of tropical fish surimi gel.

Modified TGase (MTGase) is an enzyme that can be produced by certain microorganisms, such as *Streptoverticillium mobaraense* (Gerber et al. 1994), *Streptoverticillium* sp. (Ando et al. 1989), *Bacillus subtilis* (Ramanujam and Hangeman 1990) and *Streptoverticillium ladakanum* (Tsai et al. 1996).

6.7. Flavors and additives

In recent years, enzymes have been at the heart of flavor synthesis. Whether of plant, animal or microbial origin, these enzymes have the advantage of being enantioselective.

Several processes have been developed to improve the performance of enzymatic reactions using immobilized enzymes to recycle the biocatalyst (Chen and Yang 1992).

Performance is also enhanced through the use of biphasic reaction media or in the presence of supercritical CO_2 (Baiao Dias et al. 2018). Enzymatically synthesized chemicals contribute organoleptic properties to foods, such as odor, taste and flavor. Table 6.6 summarizes the enzymes involved in the production of molecules that can be used in the food industry as flavorings.

Enzymes	Substrates/products	Effects	References
β-Glycosidases, mandelonitrile-lyase	Amygdalin/benzaldehyde	Production of fragrant terpene compounds in grape must	(Gunata et al. 1991)
Pectinase, cellulase and/or hemicellulose β-Glucosidase	Heterosides/simple sugars/vanillin	Degradation of cells, hydrolysis of heterosides, reduced production time and increased vanillin yield	(Brunerie 1993)
β-Glucosidase Alcohol dehydrogenase	Betuloside/Betuligenol Betuligenol/raspberry ketone	90–100% hydrolysis Natural raspberry flavor	(Dumont et al. 1995)
Rhamnosidases (commercial pectinases)	Naringin or hesperidin/ 2,5-dimethyl-4-hydroxy -2,3-dihydrofuran-3-one or furaneol	Hydrolysis, fruity odor production	(Cheetham 1993)
Acrylsulfatase or acid phosphatase	Sulfates or phosphates/phenols	Flavor in cow's or goat's milk	(Belin and Husson 1997)
β-Thioglycosidases (myrosinase)	Thioglycosides/sinalbine, sinigrin	Mustard odor and taste	(Richard 1992)
Lipase from *Mucor miehei* Lipase from *Candida antarctica*	Triglycerides (C4, C6 or C8)/butyric acid	Butter-like flavor	(Baiao et al. 2018)
Lipase from *Candida cylindracea*	Triglycerides/short-chain fatty acids	Natural cheese flavor concentrate	(Chen and Yang 1992)
Lipase from *Aspergillus niger*/lipase from *Mucor miehei*	Geranyl isobutyrate or citronellyl propionate/ terpene esters	Fruity flavor	(Gatfield 1988)
Lipase from *Candida rugosa*	L(-) Menthol/L(-) menthyl esters	Menthol flavor	(Vorlova et al. 2002)
Lipase from *Mucor miehei*	Hydroxylated fatty acids (4-hydroxy-decanoic acid, 4-hydroxy-butyric acid/lactones, ϒ-decalactone, ϒ-butyrolatone)	Fruity flavor	(Gatfield and Sommer 1993)
Immobilized protease from *Penicellium duponti* Pronase	Soy proteins/peptides and amino acids Casein/peptides, amino acids	Non-volatile aromatic fraction (flavor enhancers)	–

Enzymes	Substrates/products	Effects	References
Bacterial methioninase	Amino acids containing a sulfur/methanethiol atom	Cheddar flavor	(Lindsay and Rippe 1986)
Papain	Peptides/glutamic acid	Meat flavor	(Grosch and Zeiler-Hilgart 1992)
Glutaminase	Increases glutamic acid levels	Flavor enhancement in small pickled vegetables	(Daiwa-Kasei 1991)
Elastase from *Pseudomonas aeruginosa*	Formation of an amine function between the NH_2 of aspartic acid and the COOH group of L-phenylalanine	Aspartame precursor	(Lee et al. 1992)
Vegetable (bean or soy flour) and microbial lipoxygenases	Linolenic acid, linoleic acid, oxygen/monohydroperoxides	Crumb whitening, production of volatile compounds (hexanal, hexanol, 1-penten-3-ol, 1-pentanol and 2-heptanone)	(Marczy et al. 2002)
Hydroperoxide lyase (HPO lyase) in higher plants	Volatile hydroperoxides/aldehydes and oxoacids	Green-grassy notes (hexanal or cis-3-hexenal)	(Liu and Hua 2015)
Hydroperoxide lyase (HPO lyase) in fungi and algae	Hydroperoxides/alcohols and oxoacids	Raw mushroom flavor	(Gargouri et al. 2014)
Alcohol dehydrogenases (ADH), horse liver ADH, baker's yeast ADH **ADH from acetic bacteria**	Aldehydes C6/alcohol Ethanol/acetaldehyde Geraniol/geranial Trans-cinnamyl aldehyde/trans-cinnamyl alcohol Ketones/alcohols	Flavor precursor Flavor of certain fruits and foods	(Konkit et al. 2016)

Table 6.6. *Main enzymes involved in flavor production*

6.8. Fruit juices and wines

6.8.1. *Fruit juices*

Fruit juices are a source of sugars, vitamins, minerals and proteins, providing consumers with an important source of carbon, nitrogen, energy, as well as growth factors and inorganic salts. They are extracted from various plant matrices of varying complexity depending on the nature of the fruit. These matrices are made up of cellulose, hemicellulose, pectins, phenolic compounds, proteins and various other constituents. Juice extraction efficiency is affected by these constituents, and more specifically by the significant quantities of pectins and cellulose present in the cell walls. Pectin acts as a cement between the various constituents of the plant matrix and limits the release of juice during the pressing stage.

Early extraction processes use various operating combinations (crushing, pressing, decanting, filtering, centrifuging, heat treatment, etc.), depending on the type of fruit. However, these operations do not extract all the juice, making it difficult to clarify the juice because of the increase in viscosity, which can also affect the physicochemical (turbidity, color), rheological and nutritional properties of the juice obtained. They also promote the oxidation of phenolic compounds by polyphenol oxidases released into the medium in the presence of a large quantity of oxygen introduced during the crushing stage.

Several authors have recommended the use of enzymes to improve these mechanical processes (Quang et al. 2014). The first industrial implementations were made between 1960 and 1980 in a few applications (apple juice, orange juice), then generalized to all fruit and vegetable juices. This use takes place upstream of the pressing operation and at the clarification stage.

Enzymatic processing prior to the pressing process either involves total liquefaction for certain fruits through the combined action of pectinases, cellulases and hemicellulases (Baron et al. 2002), or as part of a maceration process for nectar production through the action of polygalacturonases (Sharma et al. 2017), or simply involves increasing the pressing yield through partial solubilization of cell walls under the combined effects of pectin methyl esterases, polygalacturonases and pectin lyase. In a second stage, enzymatic processing can be used for clarification by destabilizing the juice's colloidal equilibrium, and flocculating the insoluble particles formed (Pilnik and Voragen 1993). This destabilization is achieved through the combined effects of pectin methyl esterase, which degrades the ester bond, and polygalacturonases, which cause cleavage of the α (1-4) bonds between the galacturonic acids of pectin. The action of pectinases on the pectin chain results in the production of methanol. It is therefore important to maintain this concentration at a level that complies with current legislation.

For some juices, cloudiness is sought after and considered by consumers as an important quality. It is caused by the suspended presence of proteins, lipids, hemicellulose, cellulose and, above all, pectin. Loss of cloudiness is often linked to the action of endogenous pectin methyl esterase, which leads to the formation of calcium pectinate, forming aggregates with suspended particles that settle during storage (Klavons et al. 1991; Croak and Corredig 2006). It is therefore necessary to deactivate this enzyme through an appropriate physicochemical and/or thermal treatment.

The enzymes most commonly used in juice production processes are pectin methyl esterases, pectin lyases, cellulases and hemicellulases. Depending on the functionality required, these enzymes are used alone or in cocktails. It should be noted that the last two enzymes are only authorized in Europe for plant-derived juices (Council Directive 95/2/EC of February 20, 1995, on food additives other than colorants and sweeteners).

Other enzymes are also used, such as naranginase, limonoate dehydrogenase (Humanes et al. 1997; Ghosh et al. 2006) for juice debittering, laccases for phenolic compound reduction and taste enhancement (Giovanelli and Ravasini 1993), and amylases for starch degradation in certain fruits, such as bananas.

The quantity, duration of treatment, temperature and type of enzyme depend on the composition of the plant matrix used, the degree of methylation of the pectin and the desired functionality. These enzymes are often marketed as specific cocktails for each type of fruit or plant matrix. These enzymes are also capable of retaining their activity over a wide pH and temperature range, which enables almost all fruit to be processed.

The use of enzymes (pectinases, cellulases, hemicellulases and pectin lyases) has led to significant improvements in extraction yields, color, stability, shelf-life and the organoleptic and nutritional qualities of juices. These improvements have been amplified in recent years by the development of new enzymatic processes assisted by ultrasound, pulsed electric fields or microwaves.

Table 6.7 summarizes the functionalities and technological effects of using a number of enzymes in different stages of the fruit juice production process.

Enzyme	Organoleptic/ nutritional effects	Technological effects	References
Pectin methyl esterase, pectate lyases	Improved extraction yields	Pressing aid through partial de-esterification of pectin	(Rogerson et al. 2000; Kashyap et al. 2001; Dixit et al. 2003)
Polygalacturonases and polymethylestaerases	Nectar production by maceration	Breaking the bonds between non-esterified galacturonic acid residues in the pectin chain	(Yamaski et al. 1964; Zetelaki-Horvath and Vas 1981)
Pectin methyl esterases, pectate lyases, polygalacturonases	Improved stability and color	Juice clarification	(Massiot et al. 1997; Wicker et al. 2002; Kohli et al. 2015; Dal Magro et al. 2016)
Pectin methyl esterases	Increased fruit firmness through a combination of enzymes and the addition of NaCl and $CaCl_2$	Increased fruit shelf life and appearance	(Suutarinen et al. 2000; Degraeve et al. 2003; Kohli et al. 2015)
Pectin esterases	Controlling the gelation process to produce jams with a low sugar concentration	Modulation of pectin methylation levels	(Morris et al. 2000; Schmelter et al. 2002)
Pectinases	Clarification and higher juice extraction yields	Mango liquefaction	(Dzogbefia and Dajokoto 2006; Sakhale et al. 2016)
Limonate dehydrogenases	Debitterization of juice	Degradation of limonate into limonin	(Humanes et al. 1997; Ghosh et al. 2006)
Laccases	Improved juice color and taste	Partial elimination of phenolic compounds and reduction of oxidation by polyphenol oxidases	(Giovanelli and Ravacini 1993)

Table 6.7. *Organoleptic, nutritional and technological effects of enzymes in fruit juice production processes*

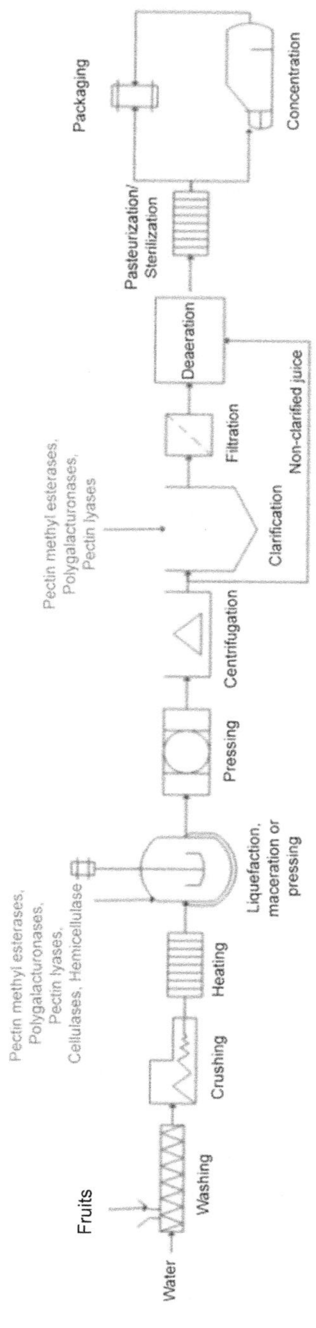

Figure 6.7. *Illustration of the use of enzymes in a fruit juice production process*

The use of enzymes in fruit juice extraction and stabilization processes not only increases yields and juice stability, but also helps reduce energy consumption and production costs.

Combining the use of enzymes with innovative technologies such as pulsed electric fields, ultrasound and microwaves contributes to a significant improvement in taste and color, as well as the emergence of new products that are difficult to access using conventional processes.

The discovery of new enzymes with greater thermal resistance and more selective catalytic mechanisms, and the development of continuous, high-performance processes have led to the widespread introduction of this technology on an industrial scale.

Figure 6.8 illustrates the use of a number of enzymes in the key stages of the extraction process for various fruit juices. The concentration and nature of the enzymes used, as well as the layout of the unit operations, vary according to the nature of the fruit processed and the functionalities sought after.

6.8.2. *Wines*

The use of pectinases in wine clarification was first reported over 25 years ago (Cruess and Besone 1941). Initially, pectic enzymes were simply added to wines that did not clarify spontaneously or did not respond to other clarification methods, such as the use of bentonite or gelatin.

In red wines, press wines are notoriously difficult to clarify. In French red wines, it proved beneficial to clarify press wines that had been made from continuously pressed must with pectic enzymes. This method resulted in a less astringent and less bitter press wine.

β-Glucosidases are also used in white winemaking to release certain aromas, such as terpenes, from their non-odorous glycosylated by-products (Jie et al. 2021). These same enzymes have also been shown to release flavanones, whose glycosylated by-products can bring about a certain bitterness to the wine (Riou et al. 1998).

Figure 6.8 illustrates the use of enzymes in a wine production process.

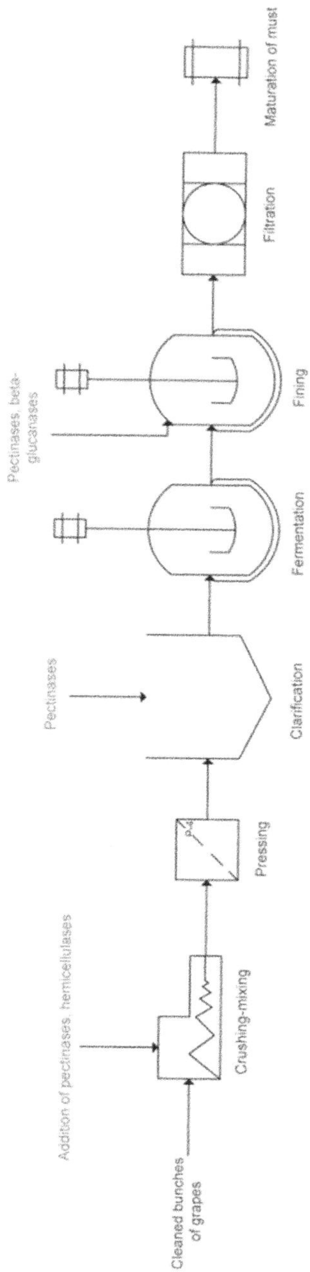

Figure 6.8. *Illustration of the use of enzymes in a wine production process*

Part 1 References

Adhikari, S. (2019). Application of immobilized enzymes in the food industry. In *Enzymes in Food Biotechnology*, Kuddus, M. (ed.). Academic Press, Amsterdam.

Aggelis, G. (2007). *Microbiology and Microbial Technology*, 1st edition. Stamoulis, Athens.

Aggelis, G. (2017). *Microbiology and Microbial Technology*, 2nd edition. Stamoulis, Athens.

Aguila La, O.A. (2001). Lactoferrin: Antimicrobial and diagnostic properties. *Biotechnol. Appl.*, 18(2), 76–83.

Ahhmed, A.M., Kawahara, S., Ohta, K., Nakade, K., Soeda, T., Mugumura, M. (2007). Differentiation in improvements of gel strength in chicken and beef sausages induced by transglutaminase. *Meat Sci.*, 76(3), 455–462.

Ahmad, R. and Sardar, M. (2015). Enzyme immobilization: An overview on nanoparticles as immobilization matrix. *Biochem. Anal. Biochem.*, 4(2), 1–8.

Ahmad, R., Mishra, A., Sardar, M. (2014). Simultaneous immobilization and refolding of heat treated enzymes on TiO_2 nanoparticles. *Adv. Sci. Eng. Med.*, 6(12), 1264–1268.

Albayati, S.H., Masomian, M., Ishak, S.N.H., Ali, M.S.M., Thean, A.L., Shariff, F.M., Noor, N.D.M., Rahman, R.N.Z.R.A. (2020). Main structural targets for engineering lipase substrate specificity. *Catalysts*, 10, 747.

Alkan, H., Baysal, Z., Uyar, F., Dogru, M. (2007). Production of lipase by a newly isolated *Bacillus coagulans* under solid-state fermentation using melon wastes. *Appl. Biochem. Biotechnol.*, 136(2), 183–192.

Alonso, F.O.M., Oliveira, E.B.L., Dellamora-Ortiz, G.M., Pereira-Meirelles, F.V. (2005). Improvement of lipase production at different stirring speeds and oxygen levels. *Braz. J. Chem. Eng.*, 22, 9–18.

Amaral, P.F.F., Rocha-Leão, M.H.M.M., Marrucho, I.M., Coutinho, J.A.P., Coelho, M.A.Z. (2006). Improving lipase production using a perfluorocarbon as oxygen carrier. *J. Chem. Technol. Biotechnol.*, 81, 1368–1374.

Implementation of Enzymatic Processes and Lactic Bacteria in the Food Industries, coordinated by Mohamed GHOUL. © ISTE Ltd 2024.

Amin, F., Bhatti, H.N., Bilal, M., Asgher, M. (2017a). Multiple parameter optimizations for enhanced biosynthesis of exopolygalacturonase enzyme and its application in fruit juice clarification. *Int. J. Food Eng.*, 13, 20160256.

Amin, F., Bhatti, H.N., Bilal, M., Asgher, M. (2017b). Purification, kinetic, and thermodynamic characteristics of an exopolygalacturonase from *Penicillium notatum* with industrial perspective. *Appl. Biochem. Biotechnol.*, 183, 426–443.

Amin, F., Bhatti, H.N., Bilal, M. (2019). Recent advances in the production strategies of microbial pectinases – A review. *Int. J. Biol. Macromol.*, 122, 1017–1026.

Amirbandeh, M. and Taheri-Kafrani, A. (2016). Immobilization of glucoamylase on tria-zinefunctionalized Fe_3O_4/graphene oxide nanocomposite: Improved stability and reusability. *Int. J. Biol. Macromol.*, 93, 1183–1191.

Ando, H., Adachi, M., Umead, K.K., Matsuura, A., Nonaka, M., Uchio, R., Tanaka, H., Motoki, M. (1989). Purification and characteristics of novel transglutaminase derived from microorganisms. *Agric. Biol. Chem.*, 53(10), 2613–2617.

Ansari, S.A. and Husain, Q. (2012). Potential applications of enzymes immobilized on/in nano materials: A review. *Biotechnol. Adv.*, 30(3), 512–523.

Araki, H. and Seki, N. (1993). Comparison of reactivity of transglutaminase to various fish actomyosins. *Nippon Suisan Gakkaishi*, 95, 711–716.

Aravindan, R., Anbumathi, P., Viruthagiri, T. (2007). Lipase applications in food industry. *Indian J. Biotechnol.*, 6(2), 141–158.

Arnold, F.H. (2018). Directed evolution: Bringing new chemistry to life. *Angew. Chem. Int. Ed.*, 57, 4143–4148.

Athenaki, M., Gardeli, C., Diamantopoulou, P., Tchakouteu, S.S., Sarris, D., Philippoussis, A., Papanikolaou, S. (2018). Lipids from yeasts and fungi: Physiology, production and analytical considerations. *J. Appl. Microbiol.*, 124, 336–367.

Baiao Dias, A.L., dos Santos, P., Martinez, J. (2018). Supercritical CO_2 technology applied to the production of flavor ester compounds through lipase-catalyzed reaction: A review. *J. CO_2 Util.*, 23, 159–178.

Bailey, M. and Pessa, E. (1990). Strain and process for production of polygalacturonase. *Enzyme Microb. Technol.*, 12(4), 266–271.

Balatsouras, G. (2006). *Food Microbiology*. Embryon, Athens.

Banach, J.C., Lin, Z., Lamsal, B.P. (2013). Enzymatic modification of milk protein concentrate and characterization of resulting functional properties. *LWT-Food Sci. and Technol.*, 54(2), 397–403.

Baron, A. (2002). Jus de fruits. In *Technologies de transformation des fruits*, Albagnac, G., Varoquaux, P., Montigaud, J.C. (eds). Lavoisier, Paris.

Bashari, M., Abdelhai, M.H., Abbas, S., Eibaid, A., Xu, X.M., Jin, Z.Y. (2016). Effect of ultrasound and high hydrostatic pressure (US/HHP) on the degradation of dextran catalyzed by dextranase. *Ultrason. Sonochem.*, 21(1), 76–83.

Basso, A. and Serban, S. (2019). Industrial applications of immobilized enzymes – A review. *Mol. Catal.*, 479, 110607.

Belafi-Bako, K., Eszterle, M., Kiss, K., Nemestothy, N., Gubicza, L. (2007). Hydrolysis of pectin by *Aspergillus niger* polygalacturonase in a membrane bioreactor. *J. Food Eng.*, 78(2), 438–442.

Belin, J.M., Husson, F., Lareta-Garde, V. (1997). *Enzymes en agroalimentaire*. Lavoisier, Paris.

Benjakul, S. and Morrissey, M.T. (1997). Protein hydrolysates from Pacific whiting solid wastes. *J. Agric. Food Chem.*, 45(9), 3423–3430.

Benjakul, S., Visessanguan, W., Pecharat, S. (2004). Suwari gel properties as affected by transglutaminase activator and inhibitors. *Food Chem.*, 85(1), 91–99.

Beopoulos, A., Cescut, J., Haddouche, R., Uribelarrea, J.L., Molina-Jouve, C., Nicaud, J.M. (2009). *Yarrowia lipolytica* as a model for bio-oil production. *Prog. Lipid Res.*, 48, 375–387.

Bernet, F., Montel, V., Noel, B., Dupouy, J.P. (2000). Diazepam-like effects of a fish protein hydrolysate (Gabolysat PC60) on stress responsiveness of the rat pituitary-adrenal system and sympa thoadrenal activity. *Psychopharmacology*, 149(1), 34–40.

Bezbradica, D., Mijin, D., Mihailović, M., Knežević-Jugović, Z. (2009). Microwave-assisted immobilization of lipase from Candida rugosa on Eupergit® supports. *J. Chem. Technol. Biotechnol.*, 84, 1642–1648.

Bhutada, G., Kavšček, M., Ledesma-Amaro, R., Thomas, S., Rechberger, G.N., Nicaud, J.M., Natter, K. (2017). Sugar versus fat: Elimination of glycogen storage improves lipid accumulation in *Yarrowia lipolytica*. *FEMS Yeast Res.*, 17(3).

Bilal, M. and Iqbal, H.M.N. (2020). State-of-the-art strategies and applied perspectives of enzyme biocatalysis in food sector – Current status and future trends. *Crit. Rev. Food Sci. Nutr.*, 60(12), 2052–2066.

Binod, P., Janu, K.U., Sindhu, R., Pandey, A. (2011). Hydrolysis of lignocellulosic biomass for bioethanol production. In *Biofuels*, Pandey, A., Larroche, C., Ricke, S.C., Dussap, C.G., Gnansounou, E. (eds). Academic Press, Amsterdam.

Bornscheuer, U.T. (2013). Enzymes in lipid modification: From classical biocatalysis with commercial enzymes to advanced protein engineering tools. *OCL*, 20(1), 45–49.

Bornscheuer, U.T. and Höhne, M. (2018). Methods in molecular biology. *YASARA: A Tool to Obtain Structural Guidance in Biocatalytic Investigations*, 1685, 43–67.

Brena, B.M. and Batista-Viera, F. (2006). Immobilization of enzymes. In *Immobilization of Enzymes and Cells*, 2nd edition, Guisan, J.M. (ed.). Humana Press Inc, Totowa.

Brunerie, P. (1993). Procédé d'obtention d'arôme naturel de vanille par traitement enzymatique des gousses de vanille verte et arôme obtenu. Patent, WO 93 04597.

Buchholz, K., Kasche, V., Bornscheuer, U.T. (2005). *Biocatalysts and Enzyme Technology*. Wiley VCH, Weinheim.

Byun, H.G. and Kim, S.K. (2001). Purification and characterization of angiotensin I converting enzyme (ACE) inhibitory peptides from *Alaska pollack* (Theragra chalcogramma) skin. *Process Biochem.*, 36, 1155–1162.

Calik, P., Calik, G., Ozdamar, T.H. (1998). Oxygen transfer effects in serine alkaline protease fermentation. *Bacillus licheniformis*: Use of citric acid as the carbon source. *Enz. Microb. Technol.*, 23, 451–461.

Calik, P., Calik, G., Ozdamar, T.H. (2000). Oxygen-transfer strategy and its regulation effects in serine alkaline protease production by *Bacillus licheniformis*. *Biotechnol. Bioeng.*, 96, 301–311.

Castilho, L.R., Mdronho, R.A., Alves, T.L.M. (2000a). Production and extraction of pectinases obtained by solid state fermentation of agroindustrial residues with *Aspergillus niger*. *Biores. Technol.*, 71, 45–50.

Castilho, L.R., Alves, T.L.M., Mdronho, R.A. (2000b). Recovery of pectolytic enzymes produced by solid state culture of *Aspergillus niger*. *Proc. Biochem.*, 34, 181–186.

Cazaban, D., Illanes, A., Wilson, L., Betancor, L. (2018). Bio-inspired silica lipase nanobiocatalysts for the synthesis of fatty acid methyl esters. *Process Biochem.*, 74, 86–93.

Celińska, E., Borkowska, M., Białas, W. (2017). Enhanced production of insect raw-starch-digesting alpha-amylase accompanied by high erythritol synthesis in recombinant *Yarrowia lipolytica* fed-batch cultures at high-cell-densities. *Proc. Biochem.*, 52, 78–85.

Celińska, E., Nicaud, J.M., Białas, W. (2021). Hydrolytic secretome engineering in *Yarrowia lipolytica* for consolidated bioprocessing on polysaccharide resources: Review on starch, cellulose, xylan, and inulin. *Appl. Microbiol. Biotechnol.*, 105, 975–989.

Chander, M. and Puri, P. (2019). Recent advances in microbial production of proteases. *Int. J. Res. Anal. Rev.*, 6, 64z–76z.

Chandrasekaran, M. (2016). *Enzymes in Food and Beverage Processing*, 1st edition. CRC Press, Boca Raton.

Chandrasekaran, S., Ramanathan, S., Basak, T. (2013). Microwave food processing – A review. *Food Res. Int.*, 52(1), 243–261.

Chatzifragkou, A., Papanikolaou, S., Kopsahelis, N., Kachrimanidou, V., Dorado, M.P., Koutinas, A.A. (2014). Biorefinery development through utilization of biodiesel industry by-products as sole fermentation feedstock for 1,3-propanediol production. *Biores. Technol.*, 159, 167–175.

Cheetham, P.S.J. (1993). The use of biotransformations for the production of flavours and fragrances. *Trends Biotechnol.*, 11(11), 478–488.

Chen, H.C. and Chang, C.C. (1996). Production of γ-linolenic acid by the fungus *Cunning-hamella echinulata* CCRC 31840. *Biotechnol. Prog.*, 12, 338–341.

Chen, H.C. and Liu, T.M. (1997). Inoculum effects on the production of γ-linolenic acid by the shake culture of *Cunninghamella echinulata* CCRC 31840. *Enz. Microb. Technol.*, 21, 137–142.

Chen, J.P. and Yang, B.K. (1992). Enhancement of release of short-chain fatty acids from milk fat with immobilized microbial lipase. *J. Food Sci.*, 57(3), 781–782.

Chica, R.A., Doucet, N., Pelletier, J.N. (2005). Semi-rational approaches to engineering enzyme activity: Combining the benefits of directed evolution and rational design. *Curr. Opin. Biotechnol.*, 16, 378–384.

Chowdhury, R. and Maranas, C.D. (2020). From directed evolution to computational enzyme engineering – A review. *AIChE J.*, 66(3), e16847.

Cinq-Mars, C.D., Hu, C., Kitts, D.D., Li-Chan, E.C.Y. (2008). Investigations into inhibitor type and mode, simulated gastrointestinal digestion, and cell transport of the angiotensin I-converting enzyme inhibitory peptides in Pacific hake (*Merluccius productus*) fillet hydrolysate. *J. Agric. Food Chem.*, 56(2), 410–419.

Conney, A.H., Lou, Y.R., Xie, J.G., Osawa, T., Newmark, H.L., Liu, Y., Chang, R.L., Huang, M.T. (1997). Some perspectives on dietary inhibition of carcinogenesis: Studies with curcumin and tea. *Proc. Soc. Exp. Bio. Med.*, 216(2), 234–245.

Jares Contesini, F., Davanço, M.G., Borin, G.P., Garcia Vanegas, K., Cirino, J.P.G., de Melo, R.R., Mortensen, U.H., Hildén, K., Campos, D.R., Carvalho, P.D.O. (2020). Advances in recombinant lipases: Production, engineering, immobilization and application in the pharmaceutical industry. *Catalysts*, 10(9), 1032.

Cornish-Bowden, A., Jamin, M., Saks, V. (2005). *Cinétique enzymatique*. EDP Sciences, Grenoble.

Couri, S., Terzi, S., Pinto, G.A.S., Freitas, S.P., da Costa, C.A. (2000). Hydrolytic enzyme production in solid-state fermentation by *Aspergillus niger* 3T5B8. *Proc. Biochem.*, 36, 255–261.

Croak, S. and Corredig, M. (2006). The role of pectin in orange juice stabilization: Effect of pectin methylesterase and pectinase activity on the size of cloud particles. *Food Hydrocoll.*, 20(7), 961–965.

Cruess, W.V. and Besone, J. (1941). Observations on the use of pectic enzymes in wine making. *Fruit Prod. J.*, 20, 365–367.

Cui, H., Stadtmüller, T.H.J., Jiang, Q., Jaeger, K.E., Schwaneberg, U., Davari, M.D. (2020). How to engineer organic solvent resistant enzymes: Insights from combined molecular dynamics and directed evolution study. *ChemCatChem.*, 12, 4073–4083.

Cunningham, F.E. and Tiede, L.M. (1981). Properties of selected poultry products treated with a tenderizing marinade. *Poult. Sci.*, 60, 2475–2479.

Curry, A. (2013). The milk revolution. *Nature*, 500, 20–22.

Dadwal, A., Sharma, S., Satyanarayana, T. (2020). Progress in ameliorating beneficial characteristics of microbial cellulases by genetic engineering approaches for cellulose saccharification. *Front. Microbiol.*, 11, 1387.

Dahiya, S., Bajaj, B.K., Kumar, A., Kumar, S., Singh, B.T. (2020). A review on biotechnological potential of multifarious enzymes in bread making. *Process Biochem.*, 99, 290–306.

Daiwa-Kasei, K.K. (1991). Method for improving flavor of processed food. Patent, JP 03240432 A2 011025.

Dal Magro, L., Dalagnol, L.M.G., Manfroi, V., Hertz, P.F., Klein, M.P., Rodrigues, R.C. (2016). Synergistic effects of pectinex ultra clear and lallzyme beta on yield and bioactive compounds extraction of Concord grape juice. *LWT-Food Sci. Technol.*, 72, 157–165.

Das, N., Kayastha, A.M., Malhotra, O.P. (1998). Immobilization of urease from pigeon pea (*Cajanus cajan* L.) in polyacrylamide gels and calcium alginate beads. *Biotechnol. Appl. Biochem.*, 27, 25–29.

Datta, S., Christena, L.R., Rajaram, Y.R.S. (2013). Enzyme immobilization: An overview on techniques and support materials. *3 Biotech*, 3(1), 1–9.

De Souza, J.R.C.G., Ledingham, W.M., De Morais, J.R.M.A. (2001). Utilisation of cheese whey as an alternative growth medium for recombinant strains of *Kluyveromyces marxianus*. *Biotechnol. Lett.*, 23, 1413–1416.

Declerck, N., Machius, M., Joyet, P., Wiegand, G., Huber, R., Gaillardin, C. (2002). Engineering the thermostability of *Bacillus licheniformis* alpha-amylase. *Biologia, Bratisl.*, 57(11), 203–211.

Degraeve, P., Saurel, R., Coutel, Y. (2003). Vacuum impregnation pretreatment with pectinmethylesterase to improve firmness of pasteurized fruits. *J. Food Sci.*, 68(2), 716–721.

Delgado-Povedano, M.M. and Luque de Castro, M.D. (2015). A review on enzyme and ultrasound: A controversial but fruitful relationship. *Anal. Chim. Acta*, 889, 1–21.

Diamantopoulou, P. and Philippoussis, A. (2001). Production attributes of *Agaricus bisporus* white and off-white strains and the effect of calcium chloride irrigation on productivity and quality. *Sci. Hortic.*, 91, 379–391.

Diamantopoulou, P., Stoforos, N.G., Xenopoulos, E., Sarris, D., Psarianos, D., Philippoussis, A., Papanikolaou, S. (2020). Lipid production by *Cryptococcus curvatus* growing on commercial xylose and subsequent valorization of fermentation waste-waters for the production of edible and medicinal mushrooms. *Biochem. Eng. J.*, 162, 107706.

Diamantopoulou, P., Gardeli, C., Papanikolaou, S. (2021). Impact of olive mill wastewaters on the physiological behavior of a wild-type new *Ganoderma resinaceum* isolate. *Environ. Sci. Poll. Res.*, 28, 20570–20585.

Dimitrakopoulou, M.A., Ambrosiadis, J.A., Zetou, F.K., Bloukas, J.G. (2005). Effect of salt and transglutaminase (TG) level and processing conditions on quality characteristics of phosphate-free, cooked, restructured pork shoulder. *Meat Sci.*, 70(4), 743–749.

Dixit, S., Upadhyay, S.K., Singh, H., Pandey, B., Chandrashekar, K., Verma, P.C. (2013). Pectin methylesterase of Datura species, purification, and characterization from Datura stramonium and its application. *Plant Signal Behav.*, 8(10), e25681.

Djekrif, D.S., Gillman, L., Bennamoun, L., Ait-Kaki, A., Labbani, K., Nouadri, T., Meraihi, Z. (2016). Amylolytic yeasts: Producers of α-amylase and pullulanase. *Int. J. Life Sci. Scient. Res.*, 2, 339–354.

Dombkowski, A.A. (2003). Disulfide by Design™: A computational method for the rational design of disulfide bonds in proteins. *Bioinformatics*, 19(14), 1852–1853.

Dominguez, A., Deive, F.J., Sanromán, M.A., Longo, M.A. (2003). Effect of lipids and surfactants on extracellular lipase production by *Yarrowia lipolytica*. *J. Chem. Technol. Biotechnol.*, 78, 1166–1170.

Dransfield, E. (1994). Modelling post-mortem tenderization-V: Inactivation of calpains. *Meat Sci.*, 37(3), 391–409.

Dumont, B., Hugueny, P., Belin, J.M. (1995). Production par bioconversion de cétone framboise. Patent, EP 0 707 072 A1.

Dwivedi, U.N., Singh, P., Pandey, V.P., Kumar, A. (2011). Structure-function relationship among bacterial, fungal and plant laccases. *J. Mol. Catal. B Enzym.*, 68, 117–128.

Dyal, S.D. and Narine, S.S. (2005). Implications for the use of *Mortierella* fungi in the industrial production of essential fatty acids. *Food Res. Int.*, 38, 445–467.

Dzogbefia, V.P. and Djokoto, D.K. (2006). Combined effects of enzyme dosage and reaction time on papaya juice extraction with aid of pectic enzymes-preliminary report. *J. Food Biochem.*, 30(1), 117–122.

Economou, C.N., Diamantopoulou, P.A., Philippoussis, A.N. (2017). Valorization of spent oyster mushroom substrate and laccase recovery through successive solid state cultivation of *Pleurotus, Ganoderma*, and *Lentinula* strains. *Appl. Microbiol. Biotechnol.*, 101, 5213–5222.

Economou, C.N., Philippoussis, A.N., Diamantopoulou, P.A. (2020). Spent mushroom substrate for a second cultivation cycle of *Pleurotus* mushrooms and dephenolization of agro-industrial wastewaters. *FEMS Microbiol. Lett.*, 367, fnaa060.

Elisashvili, V.I., Kachlishvili, E.T., Wasser, S.P. (2009). Carbon and nitrogen source effects on basidiomycetes exopolysaccharide production. *Appl. Biochem. Microbiol.*, 45, 531–535.

Ellouz, Y.T., Bayoudh, S., Kammoun, S., Gharsallah, N., Nasri, M. (2001). Production of protease by *Bacillus subtilis* grown on sardinelle heads and viscera flour. *Biores. Technol.*, 80, 49–51.

Ellouz, Y.T., Ghorbel, B., Souissi., N., Kammoun, S., Nasri, M. (2003). Biosynthesis of protease by *Pseudomonas aeruginosa* MN7 grown on fish substrate. *World J. Microbiol. Biotechnol.*, 19, 41–45.

Enari, T.M. (1983). Microbial cellulases. In *Microbial Enzymes and Biotechnology*, Fogarty, W.M. (ed.). Applied Sciences Publishers, London/New York.

European Parliament and Council of the European Union (1995). European Parliament and Council Directive No 95/2/EC of 20 February 1995 on food additives other than colours and sweeteners [Online]. Available at: https://leap.unep.org/countries/eu/national-legislation/european-parliament-and-council-directive-no-952ec-food-additives.

Fawole, O.B. and Odunfa, S.A. (2003). Some factors affecting production of pectic enzymes by *Aspergillus niger*. *Internat. Biodeter. Biodegr.*, 52, 223–227.

Ferrero, M.A., Castro, G.R., Abate, C.M., Baigorí, M.D., Siñeriz, F. (1996). Thermostable alkaline protease of *Bacillus licheniformis* MIR 29: Isolation, production and characterization. *Appl. Microbiol. Biotechnol.*, 45, 327–332.

Fickers, P., Nicaud, J.M., Gaillardin, C., Destain, J., Thonart, P. (2004). Carbon and nitrogen sources modulate lipase production in the yeast *Yarrowia lipolytica*. *J. Appl. Microbiol.*, 96, 742–749.

Fickers, P., Benetti, P.H., Waché, Y., Marty, A., Mauersberger, S., Smit, M.S., Nicaud, J.M. (2005). Hydrophobic substrate utilisation by the yeast *Yarrowia lipolytica*, and its potential applications. *FEMS Yeast Res.*, 5, 527–543.

Filippousi, R., Antoniou, D., Tryfinopoulou, P., Nisiotou, A.A., Nychas, G.J.E., Koutinas, A.A., Papanikolaou, S. (2019). Isolation, identification and screening of yeasts towards their ability to assimilate biodiesel-derived crude glycerol: Microbial production of polyols, endopolysaccharides and lipid. *J. Appl. Microbiol.*, 127, 1080–1100.

Foda, M. and Lopez-Leiva, M. (2000). Continuous production of oligosaccharides from whey using a membrane reactor. *Process Biochem.*, 35(6), 581–587.

Fogarty, W.M. and Kelly, C.T. (1983). Pectic enzymes. In *Microbial Enzymes and Biotechnology*, Fogarty, W.M. (ed.). Applied Sciences Publishers, London/New York.

Fogarty, W.M. and Kelly, C.T. (1990). *Microbial Enzymes and Biotechnology*, 2nd edition. Elsevier, New York.

Freeman, A., Woodley, J.M., Lilly, M.D. (1993). In situ product removal as a tool forbioprocessin. *Biotechnology*, 11, 1007–1012.

Friederich, J., Cinerman, A., Steiner, W. (1994). Concomitant biosynthesis of *Aspergillus niger* pectolytic enzymes and citric acid on sucrose. *Enz. Microb. Technol.*, 16, 703–707.

Galanie, S., Entwistle, D., Lalonde, J. (2020). Engineering biosynthetic enzymes for industrial natural product synthesis. *Nat. Prod. Rep.*, 37, 1122–1143.

Galiotou-Panayotou, M. (2007). *Food Enzymology*. Agricultural University of Athens, Athens.

Galiotou-Panayotou, M., Rodis, P., Kapantai, M. (1993). Enhanced polygalacturonase production by *Aspergillus niger* NRRL-364 grown on supplemented citrus pectin. *Lett. Appl. Microbiol.*, 17, 145–148.

Galiotou-Panayotou, M., Kapantai, M., Kalatzi, O. (1997). Growth conditions of *Aspergillus* sp. ATHUM-3482 for polygalacturonase production. *Appl. Microbiol. Biotechnol.*, 47, 425–429.

Galiotou-Panayotou, M., Kalantzi, O., Aggelis, G. (1998). Modelling of simultaneous production of polygalacturonase and exopolysaccharide by *Aureobasidium pullulans* ATHUM 2915. *Antonie van Leeuwenhoek*, 73, 155–162.

Gao, X., Dong, X., Li, X., Liu, Z., Liu, H. (2020). Prediction of disulfide bond engineering sites using a machine learning method. *Sci. Rep.*, 10(1), 10330.

Gargouri, M., Drouet, P., Legoy, M.D. (2004). Hydroperoxide-lyase activity in mint leaves volatile C6-aldehyde production from hydroperoxy-fatty acids. *J. Biotechnology*, 111(1), 59–65.

Gatfield, I.L. (1988). Production of flavor and aroma compounds by biotechnology. *Food Technology*, 42(10), 110–169.

Gatfield, I.L. and Sommer, H. (1993). The enzymatic and fermentative production of lactones and their use in natural flavors. In *Recent Developments in Flavor and Fragrance Chemistry*, Hopp, R. and Mori, K. (eds). Wiley VCH, Weinheim.

Geissinger, C., Gastl, M., Becker, T. (2021). Enzymes from cereal and fusarium metabolism involved in the malting process. *J. Am. Soc. Brew. Chem.*, 80(1), 1–16.

Gerber, U., Jucknischke, U., Putzein, S., Fuchsbauer, H.L. (1994). A rapid and simple method of purification of transglutaminase from *Streptovercillium mabaraense*. *J. Biochem.*, 299, 825–829.

Ghosh, M., Abhijit, G., Mallik, M. (2006). Evidence of indigenous NAH plasmid of naphthalene degrading *Pseudomonas putida* PpG7 strain implicated in limonin degradation. *J. Microbiol.*, 44(4), 473–479.

Gihaz, S., Kanteev, M., Pazy, Y., Fishman, A. (2018). Filling the void: Introducing aromatic interactions into solvent tunnels to enhane lipase stability in methanol. *Appl. Environ. Microbiol.*, 84(23), e02143-18.

Gildberg, A., Bogwald, J., Johansen, A., Stenberg, E. (1996). Isolation of acid peptide fractions from a fish protein hydrolysate with strong stimulatory effect on Atlantic salmon (*Salmon salar*) head kidney leucocytes. *Comp. Biochem. Physiol.*, 114B, 97–101.

Giovanelli, G. and Ravasini, G. (1993). Apple juice stabilization by combined enzyme-membrane filtration process. *LWT*, 26(1), 1–7.

Giuseppe, P.O., Souza, T.A.C.B., Souza, F.H.M., Zanphorlin, L.M., Machado, C.B., Ward, R.J., Jorge, J.A., Furriel, R.P.M., Murakami, M.T. (2014). Structural basis for glucose tolerance in GH1 beta-glucosidases. *Acta Cryst.*, D70, 1631–1639.

Gopinath, S.C.B., Anbu, P., Arshad M.K.M., Lakshmipriya, T., Voon, C.H., Hashim, U., Chinni, S.V. (2017). Biotechnological processes in microbial amylase production. *BioMed Res. Internat.*, 1272193.

Gorecka, E. and Jastrzebska, M. (2011). Immobilization techniques and biopolymer carriers. *Biotechnol. Food Sci.*, 75, 65–86.

Görgüç, A., Bircan, C., Yilmaz, F.M. (2019). Sesame bran as an unexploited by-product: Effect of enzyme and ultrasound-assisted extraction on the recovery of protein and antioxidant compounds. *Food Chem.*, 283, 637–645.

Goswami, S., Das, S., Datta, S. (2017). Understanding the role of residues around the active site tunnel towards generating a glucose-tolerant beta-glucosidase from *Agrobacterium tumefaciens*. *Protein Eng. Des. Sel.*, 30(7), 523–530.

Grosch, W. and Zeiler-Hilgart, G. (1992). Formation of meatlike flavor compounds. In *Flavor Precursors*, Teranishi, R., Takeoka, G., Gunthert, M. (eds). A.C.S., Washington, DC.

Grosová, Z., Rosenberg, M., Rebros, M., Sipocz, M., Sedlácková, S. (2008). Entrapment of β-galactosidase in polyvinylalcohol hydrogel. *Biotechnology Letters*, 30(4), 763–767.

Guerrand, D. (2017). Lipases industrial applications: Focus on food and agroindustries. *OLC*, 24(4), D403.

Gunata, Y.Z., Bayonove, C., Cordonnier, R.E., Arnaud, A., Galzy, P. (1990). Hydrolysis of grape monoterpenyl-β-D-glucosides by *Candida molischiana* and *Candida wickerhamii* β-glucosidases. *J. Sci. Food Chem.*, 38, 1232–1236.

Guo, B., Amano, Y., Nozaki, K. (2016). Improvements in glucose sensitivity and stability of *Trichoderma reesei* beta-glucosidase using site-directed mutagenesis. *PLoS ONE*, 11(1), e0147301.

Guo, S.W., Zhang, L., Zhang, L., Zhao, M.N., Meng, X.H. (2018). Inhibition kinetics of oligochitosan as an uncompetitive inhibitor on Fuji apple polyphenol oxidase. *J. Food Biochem.*, 42(5), 12585.

Gupta, R., Beg, Q.K., Khan, S., Chauhan, B. (2002). An overview on fermentation, downstream processing and properties of microbial alkaline proteases. *Appl. Microbiol. Biotechnol.*, 60, 381–395.

Gupta, S., Bhattacharya, A., Murthy, C.N. (2013). Tune to immobilize lipases on polymer membranes: Techniques, factors and prospects. *Biocatal. Agric. Biotechnol.*, 2, 171–190.

Haard, N.L. and Patel, T.R. (1987). Bacterial enzyme used as cheese ripening aid. Patent, US 4, 690, 826.

Hadeball, W. (1991). Production of lipase by *Yarrowia lipolytica*, I. Lipases from yeasts. *Acta Biotechnol.*, 11, 159–167.

Hadj-Taieb, N., Ayadi, M., Trigui, S., Bouabdallah, F., Gargouri, A. (2002). Hyperproduction of pectinase activities by a fully constitutive mutant (CT1) of *Penicillium occitanis*. *Enz. Microb. Technol.*, 30, 662–666.

Han, Z.L., Han, S.Y., Zheng, S.P., Lin, Y. (2009). Enhancing thermostability of a *Rhizomucor miehei* lipase by engineering a disulfide bond and displaying on the yeast cell surface. *Appl. Microbiol. Biotechnol.*, 85, 117–126.

Han, H., Ling, Z., Khan, A., Virk, A.K., Kulshrestha, S., Li, X. (2019). Improvements of thermophilic enzymes: From genetic modifications to applications. *Bioresour. Technol.*, 279, 350–361.

Hernandèz, I., de Renobales, M., Virto, M., Perez-Elortondo, F.J., Barron, L.J.R., Flanagan, C., Albisu, M. (2005). Assessment of industrial lipases for flavour development in commercial Idiazabal (ewe's raw milk) cheese. *Enzyme Microb. Technol.*, 36(7), 870–879.

Holm, H.C., Nielsen, P.M., Longin, F. (2018). Upscaling of enzymatic processes. In *Lipid Modification by Enzymes and Engineered Microbes*, Bornscheuer, U.T. (ed.). AOCS Press, Champaign.

Humanes, L., Lopez-Ruiz, A., Merino, M.T., Roldan, J.M., Diez, J. (1997). Purification and characterization of limonoate dehydrogenase from Rhodococcus fascians. *Appl. Environ. Microbiol.*, 63(9), 3385–3389.

Illanes, A. (2008). *Enzyme Biocatalysis: Principles and Applications*. Springer, Dordrecht.

Isaksen, G.V., Åqvist, J., Brandsdal, B.O. (2016). Enzyme surface rigidity tunes the temperature dependence of catalytic rates. *PNAS*, 113(28), 7822–7827.

Iyer, P.V. and Ananthanarayan, L. (2008). Enzyme stability and stabilization-aqueous and non-aqueous environment. *Process Biochemistry*, 43, 1019–1103.

Janssen, M.H.A., van Langen, L.M., Pereira, S.R.M., van Rantwijk, F., Sheldon, R.A. (2002). Evaluation of the performance of immobilized penicillin G acylase using active-site titration. *Biotech. Bioeng.*, 78, 425–432.

Jaros, D., Jacob, M., Jacob, M., Rohm, H. (2010). Excessive cross-linking of caseins by microbial transglutaminase and its impact on physical properties of acidified milk gels. *Int. Dairy J.*, 20(5), 321–327.

Jayani, R.S., Saxena, S., Gupta, R. (2005). Microbial pectinolytic enzymes: A review. *Proc. Biochem.*, 40, 2931–2944.

Je, J.Y., Park, P.J., Kwon, J.Y., Kim, S.K. (2004). A novel angiotensin I converting enzyme inhibitory peptide from Alaska pollack (*Theragra chalcogramma*) frame protein hydrolysate. *J. Agric. Food Chem.*, 52(26), 7842–7845.

Jie, Z., Tieru, W., Zhao, N., Xu, J., Qi, Y., Wei, X., Fan, M. (2021). Performance of a novel beta-glucosidase BGL0224 for aroma enhancement of Cabernet Sauvignon wines. *LWT-Food Sci. and Technol.*, 144, 111244.

Jin, J., Ma, H., Wang, W., Luo, M., Wang, B., Qu, W., He, R., Owusu, J., Li, Y. (2016). Effects and mechanism of ultrasound pretreatment on rapeseed protein enzymolysis. *J. Sci. Food Agric.*, 96(4), 1159–1166.

Joan, Q.S. and Drost-Lustenberger, C. (2002). Enzymes for bread, pasta and noodle products. In *Enzymes in Food Technology*, Whitehurst, R.J. and Law, B.A. (eds). Sheffield Academic Press, Sheffield/CRC Press, Boca Raton.

Jun, S.Y., Park, P.J., Jung, W.K. (2004). Purification and characterization of an antioxidative peptide from enzymatic hydrolysate of yellowfin sole (*Limanda aspera*) frame protein. *Eur. Food Res. Technol.*, 219, 20–26.

Kachrimanidou, V., Kopsahelis, N., Chatzifragkou, A., Papanikolaou, S., Yanniotis, S., Kookos, I.K., Koutinas, A.A. (2013). Utilisation of by-products from sunflower-based biodiesel production processes for the production of fermentation feedstock. *Waste Biomass Valor.*, 4, 529–537.

Kalia, V.C., Rashmi Lal, S., Gupta, N.M. (2001). Using enzymes for oil recovery from edibles seeds. *J. Scient. Ind. Res.*, 60(4), 298–310.

Kamau, D.N., Doores, S., Pruitt, K.M. (1991). Activation of the lactoperoxidase system prior to pasteurisation for shelf-life extension of milk. *Milchwissenschaft*, 46, 213–214.

Kamzolova, S., Morgunov, I., Aurich, A., Perevoznikova, S., Shiskanova, N., Stottmeister, U., Finogenova, T. (2005). Lipase secretion and citric acid production in *Yarrowia lipolytica* yeast grown on animal and vegetable fat. *Food Technol. Biotechnol.*, 43, 113–122.

Kapoor, S., Rafiq, A., Sharma, S. (2017). Protein engineering and its applications in food industry. *Crit. Rev. Food Sci. Nutr.*, 57(11), 2321–2329.

Kashyap, D.R., Vohra, P.K., Chopra, S., Tewari, R. (2001). Applications of pectinases in the commercial sector: A review. *Bioresour. Technol.*, 77(3), 215–227.

Kazlauskas, R.J. (2018). Engineering more stable proteins. *Chem. Soc. Rev.*, 47, 9026–9045.

Kazlauskas, R.J., Weissfloch, A.N.E., Rappaport, A.T., Cuccia, L.A. (1991). A rule to predict which enantiomer of a secondary alcohol reacts faster in reactions catalyzed by cholesterol esterase, lipase from *Pseudomonas cepacia*, and lipase from *Candida rugosa*. *J. Org. Chem.*, 56(8), 2656–2665.

Kefalogianni, I. and Aggelis, G. (2006). Modeling growth and biochemical activities of *Azospirillum* spp. *Appl. Microbiol. Biotechnol.*, 58, 352–357.

Keying, Q., Changzhong, R., Zaigui, L. (2009). An investigation on pretreatments for inactivation of lipase in naked oat kernels using microwave heating. *J. Food Eng.*, 95, 280–284.

Khan, A.A. and Alzohairy, M.A. (2010). Recent advances and applications of immobilized enzymes technologies: A review. *Res. J. Biol. Sci.*, 5(8), 565–575.

Khan, U. and Selamoglu, Z. (2020). Use of enzymes in dairy industry. A review of current progress. *Archives of Razi Institute*, 75(1), 131–136.

Khan, A.A., Akhtar, S., Hussain, Q. (2006). Direct immobilization of polyphenol oxidases on celite 545 from ammonium sulphate fractionated proteins of potato (*Solanum tuberosum*). *J. Mol. Catal., B Enzym.*, 40, 5863.

Kim, J.W., Park, T.J., Ryu, D.D., Kim, J.Y. (2000). High cell density culture of *Yarrowia lipolytica* using a one-step feeding process. *Biotechnol. Prog.*, 16, 657–660.

Kim, S.Y., Je, J.Y., Kim, S.K. (2007). Purification and characterization of antioxidative peptide from hoki (*Johnius belengerii*) frame protein by gastrointestinal digestion. *Biochem.*, 18(1), 31–38.

Klavons, J.A., Bennett, R.D., Vannier, S.H. (1991). Nature of the protein constituent of commercial orange juice cloud. *J. Agric. Food Chem.*, 39(9), 1545–1548.

Klompong, V., Benjakul, S., Kantachote, D., Shahidi, F. (2007). Antioxidative activity and functional properties of protein hydrolysate of yellow stripe trevally (*Selaroides leptolepis*) as influenced by the degree of hydrolysis and enzyme type. *Food Chem.*, 102(4), 1317–1327.

Klompong, V., Benjakul, S., Kantachote, D., Shahidi, F. (2008). Comparative study on antioxidative activity of yellow stripe trevally protein hydrolysate produced from Alcalase and Flavourzyme. *Int. J. Food Sci. Technol.*, 43(6), 1019–1026.

Knob, A., Fortkamp, D., Prolo, T., Izidoro, S.C., Almeida, J.M. (2014). Agro-residues as alternative for xylanase production by filamentous fungi. *BioRes.*, 9, 5738–5773.

Kohli, P., Kalia, M., Gupta, R. (2015). Pectin methylesterases: A review. *J. Bioprocess Biotech.*, 5, 227. doi: 10.4172/2155-9821.1000227.

Kojima, Y., Sakamoto, T., Kishida, M., Sakai, T., Kawasaki, H. (1999). Acidic condition-inducible polygalacturonase of *Aspergillus kawachii*. *J. Mol. Catal. B.: Enzym.*, 6, 351–357.

Konkit, M., Choi, W.J., Kim, W. (2016). Aldehyde dehydrogenase activity in Lactococcus chungangensis: Application in cream cheese to reduce aldehyde in alcohol metabolism. *J. Dairy Sci.*, 99(3), 1755–1761.

Korendovych, I.V. (2018). Rational and semirational protein design. *Methods Mol. Biol.*, 1685, 15–23.

Koutinas, A., Vlysidis, A., Pleissner, D., Kopsahelis, N., Garcia, I.L., Kookos, I.K., Papanikolaou, S., Kwan, T.H., Lin, C.S.K. (2014). Valorization of industrial waste and by-product streams via fermentation for the production of chemicals and biopolymers. *Chem. Soc. Rev.*, 43, 2587–2627.

Kuddus, M. (2018). *Enzymes in Food Technology – Improvements and Innovations*. Springer Verlag, Singapore.

Kuddus, M. (2019). *Enzymes in Food Biotechnology – Production, Applications, and Future Prospects*, 1st edition. Academic Press, Cambridge.

Kumar, S., Tsai, C.J., Nussinov, R. (2000). Factors enhancing protein thermostability. *Protein Eng.*, 13(3), 179–191.

Kumar, V., Sinha, A.K., Makkar, H.P.S., Becker, K. (2010). Dietary roles of phytate and phytase in human nutrition: A review. *Food Chem.*, 120(4), 945–959.

Kumari, N., Sushil., Rani, B., Malik, K., Avtar, R. (2019). Microbial amylases: An overview on recent advancement. *J. Entomol. Zool. Stud.*, 7, 198–205.

Kuraishi, C., Yamazaki, K., Susa, Y. (2001). Transglutaminase: Its utilization in the food industry. *Food Rev. Int.*, 17(2), 221–246.

Lantto, R., Plathin, P., Niemisto, M., Buchert, J., Autio, K. (2006). Effects of transglutaminase, tyrosinase and freeze-dried apple pomace powder on gel forming and structure of pork meat. *LWT-Food Sci. and Technol.*, 39(10), 1117–1124.

Lantto, R., Puolanne, E., Kruus, K., Buchert, J., Autio, K. (2007). Tyrosinase-aided protein cross-linking: Effects on gel formation of chicken breast myofibrils and texture and water-holding of chicken breast meat homogenate gels. *J. Agric. Food Chem.*, 55(4), 1248–1255.

Lareta-Garde, V. (1998). *Ensymes en agroalimentaires*. Lavoisier Tec et Doc, Paris.

Larsen, M.W., Zielinska, D.F., Martinelle, M., Hidalgo, A., Jensen, L.J., Bornscheuer, U.T., Hult, K. (2010). Suppression of water as a nucleophile in *Candida antarctica* lipase B catalysis. *Chem. Bio. Chem.*, 11, 796–801.

Ledesma-Amaro, R., Dulermo, T., Nicaud, J.M. (2015). Engineering *Yarrowia tlipolytica* to produce biodiesel from raw starch. *Biotechnol. Biofuel.*, 8, 148.

Lee, K.H., Lee, P.M., Siaw, Y.S., Morihara, K. (1992). Effects of methanol on the synthesis of aspartame precursor catalysed by *Pseudomonas aeruginosa* elastase. *Biotechnol. Lett.*, 14, 779–784.

Li, Y., Zhang, Y., Sui, X.N., Zhang, Y.N., Feng, H.X., Jiang, L.Z. (2014). Ultrasound-assisted aqueous enzymatic extraction of oil from perilla (*Perilla frutescens* L) seeds. *Cyta-J. Food*, 12(1), 16–21.

Li, Z., Duan, X., Chen, S., Wu, J. (2017). Improving the reversibility of thermal denaturation and catalytic efficiency of *Bacillus licheniformis* alpha-amylase through stabilizing a long loop in domain B. *PLoS ONE*, 12(3), e0173187.

Li, G., Fang, X., Su, F., Chen, Y., Xu, L., Yan, Y. (2018). Enhancing the thermostability of *Rhizomucor miehei* lipase with a limited screening library by rational-design point mutations and disulfide bonds. *Appl. Environ. Microbiol.*, 84(2), e02129–17.

Li, F.M., Fu, YF., Yang, H., Tang, YM. (2021). The inhibition mechanism of luteolin on peroxidase based on multispectroscopic techniques. *Int. J. Biol. Macromol.*, 166, 1072–1081.

Lilbæk, H.M., Broe, M.L., Høier, E., Fatum, T.M., Ipsen, R., Sørensen, N.K. (2006). Improving the yield of mozzarella cheese by phospholipase treatment of milk. *J. Dairy Sci.*, 89(11), 4114–4125.

Lilbæk, H.M., Fatum, T.M., Ipsen, R., Sørensen, N.K. (2007). Modification of milk and whey surface properties by enzymatic hydrolysis of milk phospholipids. *J. Agric. Food Chem.*, 55(8), 2970–2978.

Lin, C.S.K., Koutinas, A.A., Stamatelatou, K., Mubofu, E.B., Matharu, A.S., Kopsahelis, N., Pfaltzgraff, L.A., Clark, J.H., Papanikolaou, S., Kwan, T.H. et al. (2014). Current and future trends in food waste valorization for the production of chemicals, materials and fuels: A global perspective. *Biofuel. Bioprod. Bioref.*, 8, 686–715.

Lindsay, R.C. and Rippe, J.K. (1986). Enzymatic generation of methanathiol to assist in the flavor development of cheddar cheese and other foods. In *Biogeneration of Aromas*, Croteau, R. and Parliament, T. (eds). A.C.S., Washington, DC.

Liu, Q. and Hua, Y. (2015). Continuous synthesis of hexanal by immobilized hydroperoxide lyase in packed-bed reactor. *Bioprocess Biosyst. Eng.*, 38(12), 2439–2449.

Liu, L., Deng, Z., Yang, H., Li, J., Shin, H.D., Chen, R.R., Du, G., Chen, J. (2014). In silico rational design and systems engineering of disulfide bridges in the catalytic domain of an alkaline alpha-amylase from *Alkalimonas amylolytica* to improve thermostability. *Appl. Environ. Microbiol.*, 80(3), 798–807.

Lowe, D.A. (2001). Production of enzymes. In *Basic Biotechnology*, 2nd edition, Ratledge, C. and Kristiansen, B. (eds). Cambridge University Press, London.

Machius, M., Declerck, N., Huber, R., Wiegand, G. (1998). Activation of *Bacillus licheniformis* alpha-amylase through a disorder toorder transition of the substrate-binding site mediated by a calcium–sodium–calcium metal triad. *Structure*, 6, 281–292.

Macedo, G.A., Santana, A., Crawford, L.M., Wang, S.C., Gonçalves Dias, F.F., De Moura Bell, J.M.L.N. (2021). Integrated microwave- and enzyme-assisted extraction of phenolic compounds from olive pomace. *LWT*, 138(110621).

Macrae, A.R. (1983). Extracellular microbial lipases. In *Microbial Enzymes and Biotechnology*, Fogarty, W.M. (ed.). Applied Sciences Publishers, London/New York.

Madzak, C. (2021). *Yarrowia lipolytica* strains and their biotechnological applications: How natural biodiversity and metabolic engineering could contribute to cell factories improvement. *J. Fungi*, 7, 548.

Madzak, C., Gaillardin, C., Beckerich, J.M. (2004). Heterologous protein expression and secretion in the non-conventional yeast *Yarrowia lipolytica*: A review. *J. Biotechnol.*, 109, 63–81.

Mahmoudi, M., Arab, S.S., Zahiri, J., Parandian, Y. (2016). An overview of the protein thermostability prediction: Databases and tools. *J. Nanomed. Res.*, 3(6), 00072.

Maldonado, M.R., Alnoch, R.C., Almeida, J.M., Santos, L.A., Andretta, A.T., Ropaín, R.P.C., Souza, E.M., Mitchell, D.A., Krieger, N. (2021). Key mutation sites for improvement of the enantioselectivity of lipases through protein engineering. *Biochem. Eng. J.*, 172, 108047.

Manachini, P.L., Parini, C., Fortina, G. (1988). Pectic enzymes from *Aureobasidium pullulans* LV 10. *Enz. Microb. Technol.*, 10, 682–685.

Mantovani, C.F., Geimba, M.P., Brandelli, A. (2005). Enzymatic clarification of fruit juices by fungal pectin lyase. *Food Biotechnol.*, 19, 173–181.

Manzo, R.M., Ceruti, R.J., Bonazza, H.L., Adriano, W.S., Sihufe, G.A., Mammarella, E.J. (2018). Immobilization of carboxypeptidase A into modified chitosan matrixes by covalent attachment. *Appl. Biochem. Biotechnol.*, 185(4), 1029–43.

Marczy, J.S., Nemeth, A.S., Samu, Z., Hager-Veress, A., Szajani, B. (2002). Production of hexanal from hydrolyzed sunflower oil by lipoxygenase and hydroperoxid lyase enzymes. *Biotechnol. Lett.*, 24(20), 1673–1675.

Martin, C. and Christopher, B. (1990). *Enzyme Technology*. Cambridge University Press, London.

Masschalck, B. and Michiels, C.W. (2003). Antimicrobial properties of lysozyme in relation to foodborne vegetative bacteria. *Crit. Rev. Microbiol.*, 29(3), 191–214.

Massiot, P., Perron, V., Baron, A., Drilleau, J.F. (1997). Release of methanol and depolymerization of highly methyl esterifed apple pectin with an endopolygalacturonase from *Aspergillus niger* and pectin methylesterases from *Aspergillus niger* or from orange. *LWT-Food Sci. and Technol.*, 30(7), 697–702.

Matsui, K.N., Granado, L.M., Oliveira, P.V., Tadini, C.C. (2007). Peroxidase and polyphenol oxidase thermal inactivation by microwaves in green coconut water simulated solutions. *LWT*, 40, 852–859.

Matsumoto, M., Masuda, H., Hubacz, R., Horie, T., Iyota, H., Shimoyamada, M., Ohmura, N. (2021). Enzymatic starch hydrolysis performance of Taylor-Couette flow reactor with ribbed inner cylinder. *Chem. Eng. Sci.*, 231, 116270.

Mehta, D. and Satyanarayana, T. (2016). Bacterial and archaeal alpha-amylases: Diversity and amelioration of the desirable characteristics for industrial applications. *Front. Microbiol.*, 7, 1129.

Meisel, H. (1997). Biochemical properties of bioactive peptides derived from milk proteins: Potential nutraceuticals for food and pharmaceutical applications. *Livest. Prod. Sci.*, 50(2), 125–138.

Mejias, L., Cerda, A., Barrena, R., Gea, T., Sánchez, A. (2018). Microbial strategies for cellulase and xylanase production through solid-state fermentation of digestate from biowaste. *Sustainability*, 10, 2433.

Melais, N., Aribi-Zouioueche, L., Riant, O. (2016). The effect of the migrating group structure on enantioselectivity in lipase-catalyzed kinetic resolution of 1-phenylethanol. *C.R. Chimie*, 19, 971–977.

Melanouri, E.M., Dedousi, M., Diamantopoulou, P. (2022a). Cultivating *Pleurotus ostreatus* and *Pleurotus eryngii* mushroom strains on agro-industrial residues in solid-state fermentation. Part I: Screening for growth, endoglucanase, laccase and biomass production in the colonization phase. *Carbon Resour. Convers.*, 5, 61–70.

Melanouri, E.M., Dedousi, M., Diamantopoulou, P. (2022b). Cultivating *Pleurotus ostreatus* and *Pleurotus eryngii* mushroom strains on agro-industrial residues in solid-state fermentation. Part II: Effect on productivity and quality of carposomes. *Carbon Resour. Convers.*, 5, 52–60.

Meleiro, L.P., Salgado, J.C.S., Maldonado, R.F., Carli, S., Moraes, L.A.B., Ward, R.J., Jorge, J.A., Furriel, R.P.M. (2017). Engineering the GH1 beta-glucosidase from *Humicola insolens*: Insights on the stimulation of activity by glucose and xylose. *PLoS ONE*, 12(11), e0188254.

Metzger, J.O. and Bornscheuer, U. (2006). Lipids as renewable resources: Current state of chemical and biotechnological conversion and diversification. *Appl. Microbiol. Biotechnol.*, 71, 13–22.

Michaelis, M. and Menten, M. (1913). Die kinetic der invertinwirkung. *Biochem. Z.*, 49, 333– 369.

Ming, M., Bo, J., Zhengyu, J., BeMiller, James, N. (2018). Microbial starch-converting enzymes: Recent insights and perspectives. *Compr. Rev. Food Sci. Food Saf.*, 17(5), 1238–1260.

Miotto, M., Olimpieri, P.P., Rienzo, L.D., Ambrosetti, F., Corsi, P., Lepore, R., Tartaglia, G.G., Milanetti, E. (2019). Insights on protein thermal stability: A graph representation of molecular interactions. *Bioinformatics*, 35(15), 2569–2577.

Mishra, S.S., Ray, R.C., Rosell, C.M., Panda, D. (2017). Microbial enzymes in food applications: History of progress. In *Microbial Enzyme Technology in Food Applications*, Ray, R.C. and Rosell, C.M. (eds). CRC Press, Boca Raton.

Misra, N.N., Koubaa, M., Roohinejad, S., Juliano, P., Alpas, H., Inácio, R.S., Saraiva, J.A., Barba, F.J. (2017). Landmarks in the historical development of twenty first century food processing technologies. *Food Res. Intern.*, 97, 318–339.

Moharana, T.R. and Rao, N.M. (2020). Substrate structure and computation guided engineering of a lipase for omega-3 fatty acid selectivity. *PLoS ONE*, 15(4), e0231177.

Mojovic, J., Siler-Marinkovic, S., Kukic, G., Vunjak-Novakovic, G. (1993). Rhizopus arrhizus lipase-catalyzed interesterification of the midfraction of palm oil to a cocoa butter equivalent fat. *Enzyme Microb. Technol.*, 15(5), 438–443.

Montesinos, J.L., Dalmau, E., Casas, C. (2003). Lipase production in continuous culture of *Candida rugosa*. *J. Chem. Technol. Biotechnol.*, 78, 753–761.

Morris, G.A., Foster, T.J., Harding, S.E. (2000). The effect of the degree of esterification on the hydrodynamic properties of citrus pectin. *Food Hydrocoll.*, 14(3), 227–235.

Nadar, S.S., Rao, P., Rathod, V.K. (2018). Enzyme assisted extraction of biomolecules as an approach to novel extraction technology: A review. *Food Research International*, 108, 309–330.

Nagayasu, T., Miyanaga, M., Tanaka, T., Sakiyama, T., Nakanishi, K. (1994). Synthesis of aspartame precursor with an immobilized thermolysin in tert-amyl alcohol. *Biotech. Bioeng.*, 43(11), 1118–1123.

Naidu, G.S.N. and Panda, T. (1998). Production of pectolytic enzymes – A review. *Bioproc. Eng.*, 19, 355–361.

Narisetty, V., Cox, R., Bommaredy, R., Agrawal, D., Ahmad, E., Pant, K.K., Chandel, A.K., Bhatia, S.K., Kumar, D., Binod, P. et al. (2022a). Valorisation of xylose to renewable fuels and chemicals, an essential step in augmenting the commercial viability of lignocellulosic biorefineries. *Sustain. Energ. Fuels*, 6, 29–65.

Narisetty, V., Cox, R., Bommaredy, R., Agrawal, D., Ahmad, E., Pant, K.K., Chandel, A.K., Bhatia, S.K., Kumar, D., Binod, P., Gupta, V.K., Kumar, V. (2022b). Valorisation of xylose to renewable fuels and chemicals, an essential step in augmenting the commercial viability of lignocellulosic biorefineries. *Sustain. Energ. Fuels*, 6, 29–65.

Nestorson, A., Neoh, K.G., Kang, E.T., Järnström, L., Leufvén, A. (2008). Enzyme immobilization in latex dispersion coatings for active food packaging. *Packag. Technol. Sci.*, 21(4), 193–205.

Nicaud, J.M., Madzak, C., Van der Broek, P., Gysler, C., Duboc, P., Niederberger, P., Gaillardin, C. (2002). Protein expression and secretion in the yeast *Yarrowia lipolytica. FEMS Yeast Res.*, 2, 371–379.

Nielsen, G.S., Petersen, B.R., Møller, A.J. (1995) Impact of salt, phosphate and temperature on the effect of a transglutaminase (F XIIIa) on the texture of restructured meat. *Meat Sci.*, 41(3), 293–299.

Nighojkar, S., Phanse, Y., Sinha, D., Nighojkar, A., Kumar, A. (2006). Production of polygalacturonase by immobilized cells of using *Aspergillus niger* orange peel as inducer. *Proc. Biochem.*, 41, 1136–1140.

Nisha, S., Karthick, S.A., Gobi, N. (2012). Application and properties of immobilized. *Enzyme. Chem. Sci. Review Lett.*, 1(3), 148–155.

Niu, K., Liu, Z., Feng, Y., Gao, T., Wang, Z., Zhang, P., Du, Z., Gao, D., Fang, X. (2020). A novel strategy for efficient disaccharides synthesis from glucose by beta-glucosidase. *Bioresour. Bioprocess.*, 7, 45.

Novalin, S., Neuhaus, W., Kulbe, K.D. (2005). A new innovative process to produce lactose-reduced skim milk. *Journal of Biotechnol.*, 119(2), 212–218.

Novozymes (2022). The use of enzymes in extraction of olive oil [Online]. Available at: https://www.novozymes.com/en/solutions/juice-fruit-vegetables/olive-oil.

O'Donnell, C.P., Tiwari, B.K., Bourke, P., Cullen, P.J. (2010). Effect of ultrasonic processing on food enzymes of industrial importance. *Trends Food Sci. Technol.*, 21(7), 358–367.

Obeng, E.M., Adam, S.N.N., Budiman, C., Ongkudon, C.M., Maas, R., Jose, J. (2017). Lignocellulases: A review of emerging and developing enzymes, systems, and practices. *Bioresour. Bioprocess.*, 4(16), 22.

Olsson, L., Christensen, T.M.I.E., Hansen, K.P., Palmqvist, E.A. (2003). Influence of the carbon source on production of cellulases, hemicellulases and pectinases by *Trichoderma reesei* Rut C-30. *Enz. Microb. Technol.*, 33, 612–619.

Özrenk, E. (2006). The use of transglutaminase in dairy products. *Int. J. Dairy Technol.*, 59(1), 1–7.

Papanikolaou, S. and Aggelis, G. (2003). Modeling lipid accumulation and degradation in *Yarrowia lipolytica* cultivated on industrial fats. *Curr. Microbiol.*, 46, 398–402.

Papanikolaou, S. and Aggelis, G. (2010). *Yarrowia lipolytica*: A model microorganism used for the production of tailor-made lipids. *Eur. J. Lipid Sci. Technol.*, 112, 639–654.

Papanikolaou, S. and Aggelis, G. (2011). Lipids of oleaginous yeasts. Part I: Biochemistry of single cell oil production. *Eur. J. Lipid Sci. Technol.*, 113, 1031–1051.

Papanikolaou, S. and Aggelis, G. (2019). Sources of microbial oils with emphasis to Mortierella (Umbelopsis) isabellina fungus. *World J. Microbiol. Biotechnol.*, 35(4), 63.

Papanikolaou, S., Chevalot, I., Komaitis, M., Aggelis, G., Marc, I. (2001). Kinetic profile of the cellular lipid composition in an oleaginous *Yarrowia lipolytica* capable of producing a cocoa-butter substitute from industrial fats. *Antonie van Leeuwenhoek*, 80, 215–224.

Papanikolaou, S., Muniglia, L., Chevalot, I., Aggelis, G., Marc, I. (2003). Accumulation of a cocoa-butter-like lipid by *Yarrowia lipolytica* cultivated on agro-industrial residues. *Curr. Microbiol.*, 46, 124–130.

Papanikolaou, S., Galiotou-Panayotou, M., Fakas, S., Komaitis, M., Aggelis, G. (2007a). Lipid production by oleaginous Mucorales cultivated on renewable carbon sources. *Eur. J. Lipid Sci. Technol.*, 109, 1060–1070.

Papanikolaou, S., Chevalot, I., Galiotou-Panayotou, M., Komaitis, M., Marc, I., Aggelis, G. (2007b). Industrial derivative of tallow: A promising renewable substrate for microbial lipid, single-cell protein and lipase production by *Yarrowia lipolytica. Electron. J. Biotechnol.*, 10, 425–435.

Papanikolaou, S., Dimou, A., Fakas, S., Diamantopoulou, P., Philippoussis, A., Galiotou-Panayotou, M., Aggelis, G. (2011). Biotechnological conversion of waste cooking olive oil into lipid-rich biomass using *Aspergillus* and *Penicillium* strains. *J. Appl. Microbiol.*, 110, 1138–1150.

Parashar, D. and Satyanarayana, T. (2018). An insight into ameliorating production, catalytic efficiency, thermostability and starch saccharification of acid-stable alpha-amylases from acidophiles. *Front. Bioeng. Biotechnol.*, 6, 125.

Park, D., Haam, S., Jang, K., Ahn, I.S., Kim, W.S. (2005). Immobilization of starch-converting enzymes on surface-modified carriers using single and co-immobilized systems: Properties and application to starch hydrolysis. *Process Biochem*, 40, 53–61.

Payne, C.M., Knott, B.C., Mayes, H.B., Hansson, H., Himmel, M.E., Sandgren, M., Ståhlberg, J., Beckham, G.T. (2015). Fungal cellulases. *Chem. Rev.*, 115, 1308–1448.

Peckham, B.W. (1979). Economics and invention: A technological history of the corn refining history of the United States. PhD Thesis, University of Wisconsin-Madison, Madison.

Peijun, J., Huishan, T., Xin, X., Wei, F. (2010). Lipase covalently attached to multi-walled carbon nanotubes as an efficient catalyst in organic solvent. *AIChE J.*, 56, 3005–3011.

Pellequer, J.L. and Chen S.W. (2006). Multi-template approach to modeling engineered disulfide bonds. Proteins: Structure, function, and bioinformatics, 65(1), 192–202.

Pelmont, J. (1997). *Enzymes. Catalyseurs du monde vivant.* Presses Universitaires de Grenoble, Grenoble.

Penny, I.F. (1980). The enzymology of conditioning. In *Developments in Meat Science, Vol. 1*, Lawrie, R.A. (ed.). Elsevier, London/New York.

Pereira-Meirelles, F.V., Rocha-Leão, M.H.M., Sant'Anna, J.G.L. (1997). A stable lipase from Candida lipolytica: Cultivation conditions and crude enzyme characteristics. *Appl. Biochem. Biotechnol.*, 63–65, 73–85.

Philippoussis, A. (2009). Production of mushrooms using agro-industrial residues as substrates. In *Biotechnology for Agro-industrial Residues Utilization*, Singh-Nigam, P. and Pandey, A. (eds). Springer, New York.

Philippoussis, A. and Diamantopoulou, P. (2012). Exploitation of the biotechnological potential of agro-industrial by-products through mushroom cultivation. In *Mushroom Biotechnology and Bioengineering*, Petre, M. and Berovic, M. (eds). University of Pitesti/CD Press, Bucharest.

Philippoussis, A., Zervakis, G., Diamantopoulou, P. (2001). Bioconversion of agricultural lignocellulosic wastes through the cultivation of the edible mushrooms *Agrocybe aegerita, Volvariella volvacea* and *Pleurotus* spp. *World J. Microbiol. Biotechnol.*, 17, 191–200.

Philippoussis, A., Diamantopoulou, P., Papadopoulou, K., Lakhtar, H., Roussos, S., Parissopoulos, G., Papanikolaou, S. (2011). Biomass, laccase and endoglucanase production by *Lentinula edodes* during solid state fermentation of reed grass, bean stalks and wheat straw residues. *World J. Microbiol. Biotechnol.*, 27, 285–297.

Pignède, G., Wang, H., Fudalej, F., Gaillardin, C., Seman, M., Nicaud, J.M. (2000). Characterization of an extracellular lipase encoded by LIP2 in *Yarrowia lipolytica. J. Bacteriol.*, 182, 2802–2810.

Pilnik, W. and Voragen, A.G.J. (1993). Pectic enzymes in fruit and vegetable juice manufacture. In *Enzyme in Food Processing*, 3rd edition, Nagodawithan, T. and Reeds, G. (eds). Academic Press, Amsterdam.

Potus, J. and Drapon, R. (1997). Les enzymes dans les industries de cuisson des céréales. In *Enzymes en agroalimentaire ?*, Larreta-Garde, V. (ed.). Lavoisier, Paris.

Prakash, D., Nawani, N., Prakash, M., Bodas, M., Mandal, A., Khetmalas, M., Kapadnis, B. (2013). Actinomycetes: A repertory of green catalysts with a potential revenue resource. *Biomed. Res. Int.*, 21, 264020.

Quang, P.S., Le, T.T.T., Le, V.V.M. (2014). Optimisation of ultrasonic treatment of apple (*Malus domestica*) mash in the extraction of juice with high antioxidant content. *IOSR J. Eng.*, 4(12), 18–21.

Ramanujam, M.V. and Hangeman, J.H. (1990). Intracellular transglutaminase (EC 2.3.2.13) in a prokaryote: Evidence from vegetative and sporulating cells of *Bacillus subtilis*. *FASEB J.*, 4, 2321–2328.

Rao, G.V. and Sobha, K. (2020). Lipases with preferred thermo-tolerance in food industry. *Res. J. Biotechnol.*, 15(7), 141–150.

Rao, M.B., Tanksale, A.M., Ghatge, M.S., Desphande, V.V. (1998). Molecular and biotechnological aspects of microbial proteases. *Microbiol. Mol. Biol. Rev.*, 62, 597–625.

Rathi, C.L. (2017). Enzymes for improving oil yield in palm oil extraction. In *World Congress on Industrial Biotechnology*, Montreal.

Ratledge, C. (1994). Yeasts, moulds, algae and bacteria as sources of lipids. In *Technological Advances in Improved and Alternative Sources of Lipids*, Kamel, B.S. and Kakuda, Y. (eds). Blackie Academic and Professional, London.

Raveendran, S., Parameswaran, B., Ummalyma, S.B., Abraham, A., Mathew, A.K., Madhavan, A., Rebello, S., Pandey, A. (2018). Applications of microbial enzymes in food industry. *Food Technol. Biotechnol.*, 56, 16–30.

Richard, H. (1992). *Épices et aromates*. Lavoisier, Paris.

Ricochon, G. and Muniglia, L. (2010). Influence of enzymes on the oil extraction processes in aqueous media. *OLC*, 17(6), 356–359.

Rigoldi, F., Donini, S., Redaelli, A., Parisini, E., Gautieri, A. (2018). Review: Engineering of thermostable enzymes for industrial applications. *APL Bioeng.*, 2, 011501.

Riou, C., Salmon, J.M., Vallier, M.J., Günata, Z., Barre, P. (1998). Purification, characterization, and substrate specificity of a novel highly glucose-tolerant beta-glucosidase from *Aspergillus oryzae*. *Appl. Environ. Microbiol.*, 64(10), 3607–3614.

Rogerson, F.S.S., Vale, E., Grande, H., Silva, M.C.M. (2000). Alternative processing of port-wine using pectolytic enzyme. *Cienc. Technol. Aliment*, 2(5), 222–227.

Rouau, X. (1996). Les hémicellulases en panification (The hemicellulases in bread-making technology). *Ind. Cereal.*, 1, 13–19.

Saha, B.C. and Hayashi, K. (2001). Debittering of protein hydrolyzates. *Biotechnol. Adv.*, 19(5), 355–370.

Saini, R., Saini, H.S., Dahiya, A. (2017). Amylases: Characteristics and industrial applications. *J. Pharmacogn. Phytochem.*, 6, 1865–1871.

Sakhale, B.K., Pawar, V.N., Gaikwad, S.S. (2016). Studies on effect of enzymatic liquefaction on quality characteristics of Kesar mango pulp. *Int. Food Res. J.*, 23(2), 860–865.

Şanli, T., Sezgin, E., Deveci, O., Şenel, E., Benli, M. (2011). Effect of using transglutaminase on physical, chemical and sensory properties of set-type yoghurt. *Food Hydrocoll.*, 25(6), 1477–1481.

Santos, C.A., Morais, M.A.B., Terrett, O.M., Lyczakowski, J.J., Zanphorlin, L.M., Ferreira-Filho, J.A., Tonoli, C.C.C., Murakami, M.T., Dupree, P., Souza, A.P. (2019). An engineered GH1 beta-glucosidase displays enhanced glucose tolerance and increased sugar release from lignocellulosic materials. *Sci. Rep.*, 9, 4903.

Sarris, D. and Papanikolaou, S. (2016). Biotechnological production of ethanol: Biochemistry, processes and technologies. *Eng. Life Sci.*, 16, 307–329.

Schenková, N., Šikulová, M., Jeleníková, J., Pipek, P., Houška, M., Marek, M. (2007). Influence of high isostatic pressure and papain treatment on the quality of beef meat. *High Press. Res.*, 27(1), 163–168.

Schmelter, T., Wientjes, R., Vreeker, R., Klaffke, W. (2002). Enzymatic modifications of pectins and the impact on their rheological properties. *Carbohydr. Poly.*, 47(2), 99–108.

Seifu, E., Buys, E.M., Donkin, E.F. (2005). Significance of the lactoperoxidase system in the dairy industry and its potential applications: A review. *Trends Food Sci. Technol.*, 16(4), 137–154.

Semenova, M.V., Sinitsyna, O.A., Morozova, V.V., Fedorova, E.A., Gusakov, A.V., Okunev, O.N., Sokolova, L.M., Koshelev, A.V., Bubnova, T.V., Vinetskii, Y.P. et al. (2006). Use of a preparation from fungal pectin lyase in the food industry. *Appl. Biochem. Microbiol.*, 42, 598–602.

Sethi, B.K., Nanda, P.K., Sahoo, S. (2016). Enhanced production of pectinase by *Aspergillus terreus* NCFT 4269.10 using banana peels as substrate. *3 Biotechol.*, 6, 36.

Shah, N., Ul-Islam, M., Khattak, W.A., Park, J.K. (2013). Overview of bacterial cellulose composites: A multipurpose advanced material. *Carbohydr. Polym.*, 98(2), 1585–1598.

Shamel, M.M., Ramachandran, K.B., Hasan, M., Al-Zuhair, S. (2007). Hydrolysis of palm and olive oils by immobilised lipase using hollow-fibre reactor. *Biochem. Eng. J.*, 34(3), 228–235.

Sharma, S.K. and Leblanc, R.M. (2017). Biosensors based on β-galactosidase enzyme: Recent advances and perspectives. *Anal. Biochem.*, 535, 1–11.

Sharma, N. and Sharma, N. (2017). Microbial xylanases and their industrial applications as well as future perspectives: A review. *G.J.B.A.H.S.*, 6, 5–12.

Sharma, R., Chisti, Y., Banerjee, U.C. (2001). Production, purification, characterization and application of lipases. *Biotechnol. Adv.*, 19, 627–662.

Sharma, H.P., Patel, H., Sugandha. (2017). Enzymatic added extraction and clarification of fruit juices – A review. *Crit. Rev. Food Sci. Nutr.*, 57(6), 1215–1227.

Sharma, A., Gupta, G., Ahmad, T., Mansoor, S., Kaur, B. (2021). Enzyme engineering: Current trends and future perspectives. *Food Rev. Int.*, 37(2), 121–154.

Sheldon, R.A. (2007). Cross-linked enzyme aggregates (CLEAs): Stable and recyclable biocatalysts. *Biochem Soc.*, 35(6), 1583–1587.

Sheldon, R.A. and van Pelt, S. (2013). Enzyme immobilisation in biocatalysis: Why, what and how. *Chem. Soc. Rev.*, 42, 6223–6235.

Silva, D., Tokuioshi, K., Martins, E., Da Silva, R., Gomes, E. (2005). Production of pectinase by solid-state fermentation *with Penicillium viridicatum* RFC3. *Proc. Biochem.*, 40, 2885–2889.

Singh, R., Singh, A., Sachan, S. (2019). Enzymes used in the food industry: Friends or foes? In *Enzymes in Food Biotechnology: Production, Applications, and Future Prospects*, Kuddus, M. (ed.). Academic Press, Cambridge.

Soe, J.B., Mikkelsen, R., Jorgensen, T.L. (2015). Improved enzymatic modification of wheat phospholipids in bakery applications. Patent, WO 2017/112734 Al.

Sonderegger, M., Schümperli, M., Sauer, U. (2004). Metabolic engineering of a phosphoketolase pathway for pentose catabolism in *Saccharomyces cerevisiae. Appl. Environ. Microbiol.*, 70, 2892–2897.

Soni, S. (2021). Trends in lipase engineering for enhanced biocatalysis. *Biotechnol. Appl. Biochem.*, 187, 127–142.

Sowdhamini, R., Srinivasan, N., Shoichet, B., Santi, D.V., Ramakrishnan, C., Balaram, P. (1989). Stereochemical modeling of disulfide bridges. Criteria for introduction into proteins by site-directed mutagenesis. *Protein Eng.*, 3(2), 95–103.

Stepankova, V., Bidmanova, S., Koudelakova, T., Prokop, Z., Chaloupkova, R., Damborsky, J. (2013). Strategies for stabilization of enzymes in organic solvents. *ACS Catal.*, 3, 2823–2836.

Subin, R.S. and Bhat, S.G. (2015). Enzymes: Concepts, nomenclature, mechanism of action and kinetics, characteristics and sources of food-grade enzymes. In *Enzymes in Food and Beverage Processing*, Chandrasekaran, M. (ed.). CRC Press, Boca Raton.

Suganth, S.H., Swathi, K.V., Biswas, R., Basker, S., Ramani, K. (2019). Co-immobilization of multiple-enzymes onto surface-functionalized magnetic nanoparticle for the simultaneous hydrolysis of multiple substrates containing industrial wastes. *Appl. Nanosci.*, 9(7), 439–1457.

Sujoy, B. and Aparna, A. (2013). Enzymology, immobilization and applications of urease. *Int. Res. J. Biol. Sci.*, 2(6), 51–56.

Suutarinen, J., Honkapaa, K., Heinio, L., Autio, K., Mokkila, M. (2000). The effect of different prefreezing treatments on the structure of strawberries before and after jam making. *LWT-Food Sci. and Technol.*, 33, 188–201.

Szczęsna-Antczak, M., Antczak, T., Piotrowicz-Wasiak, M., Rzyska, M., Binkowska, N., Bielecki, S. (2006). Relationships between lipases and lipids in mycelia of two. *Mucor* strains. *Enz. Microb. Technol.*, 39, 1214–1222.

Taheri-Kafrani, A., Kharazmi, S., Nasrollahzadeh, M., Soozanipour, A., Ejeian, F., Etedali, P., Mansouri-Tehrani, H.A., Razmjou, A., Yek, S.M.G., Varma, R.S. (2020). Recent developments in enzyme immobilization technology for high-throughput processing in food industries. *Crit. Rev. Food Sci. Nutr.*, 61(19), 3160–3196.

Thiansilakul, Y., Benjakul, S., Shahidi, F. (2007a). Compositions, functional properties and antioxidative activity of protein hydrolysates prepared from round scad (*Decapterus maruadsi*). *Food Chem.*, 103(4), 1385–1394.

Thiansilakul, Y., Benjakul, S., Shahidi, F. (2007b). Antioxidative activity of protein hydorlysate from round scad muscle using Alcalase and Flavourzyme. *J. Food Biochem.*, 31(2), 266–287.

Trani, A., Loizzo, P., Cassone, A., Faccia, M. (2017). Enzymes applications for the dairy industry. In *Advances in Dairy Products*, Conto, F. (ed.). John Wiley & Sons Ltd, New York.

Tsagou, V. and Aggelis, G. (2006). Growth dynamics of *Azospirillum lipoferum* at steady and transitory states in the presence of NH^{+}. *J. Appl. Microbiol.*, 100, 286–295.

Tsai, G.J., Lin, S.M., Jiang, S.T. (1996). Transglutaminase from *Streptoverticillum ladakanum* and application to minced fish product. *J. Food Sci.*, 61(6), 1234–1240.

Tseng, T.F., Liu, D.C., Chen, M.T. (2000). Evaluation of transglutaminase on the quality of low-salt chicken meat-balls. *Meat Sci.*, 55(4), 427–431.

Tsouko, E., Kourmentza, C., Ladakis, D., Kopsahelis, N., Mandala, I., Papanikolaou, S., Paloukis, F., Alves, V., Koutinas, A.A. (2015). Bacterial cellulose production from industrial waste and by-product streams. *Int. J. Mol. Sci.*, 16, 14832–14849.

Tzirita, M., Papanikolaou, S., Chatzifragkou, A., Quilty, B. (2018). Waste fat biodegradation and biomodification by *Yarrowia lipolytica* and a bacterial consortium composed of *Bacillus* spp. and *Pseudomonas putida*. *Eng. Life Sci.*, 18, 932–942.

Vaghari, H., Eskandari, M., Sobhani, V., Berenjian, A., Song, Y., Jafarizadeh-Malmiri, H. (2015). Process intensification for production and recovery of biological products. *Am. J. Biochem. Biotechnol.*, 11, 37–43.

Valdés, G., Mendonça, R.T., Aggelis, G. (2020). Lignocellulosic biomass as a substrate for oleaginous microorganisms: A review. *Appl. Sci.*, 10, 7698.

Van Dam, H.W. and Hille, J.D.R. (1993). Yeast and enzymes in bread-making. *Cereal Food World*, 37(3), 246–252.

Van Dijck, A.A., Folkertsma, B., Jacobus, P., Dekker, T. (2014). Carboxypeptidase for cheese ripening. Patent, US20140205717 A1.

Vasiliadou, I.A., Bellou, S., Daskalaki, A., Tomaszewska-Hetman, L., Chatzikotoula, C., Kompoti, B., Papanikolaou, S., Vayenas, D., Pavlou, S., Aggelis, G. (2018). Biomodification of fats and oils and scenarios of adding value on renewable fatty materials through microbial fermentations: Modelling and trials with *Yarrowia lipolytica*. *J. Clean. Prod.*, 200, 1111–1129.

Veno, J., Rahman, R.N.Z.R.A, Masomian, M., Ali, M.S.M., Kamarudin, N.H.A. (2019). Insight into improved thermostability of cold-adapted staphylococcal lipase by glycine to cysteine mutation. *Molecules*, 24, 3169.

Vorlova, S., Bornscheuer, U.T., Gatfield, I., Hilmer, J.M., Bertram, H.J., Schmid, R.D. (2002). Enantioselective hydrolysis of D, L-menthyl benzoate to L-(-)-menthol by recombinant *Candida rugosa* lipase LIP1. *Adv. Synth. Catal.*, 344(10), 1152–1155.

Wang, Y., Xu, J., Luo, G., Dai, Y. (2008). Immobilization of lipase by ultrafiltration and cross-linking onto the polysulfone membrane surface. *Bioresour. Technol.*, 99, 2299–2303.

Wang, C.H., Liu, X.L., Huang, R.B., He, B.F., Zhao, M.M. (2018a). Enhanced acidic adaptation of *Bacillus subtilis* Ca-independent alpha-amylase by rational engineering of pKa values. *Biochem. Eng. J.*, 139, 146–153.

Wang, D., Yan, L., Ma, X., Wang, W., Zou, M., Zhong, J., Ding, T., Ye, X., Liu, D. (2018b). Ultrasound promotes enzymatic reactions by acting on different targets: Enzymes, substrates and enzymatic reaction systems. *Int. J. Biol. Macromol.*, 119, 453–461.

Ward, O.P. (1983). Proteinases. In *Microbial Enzymes and Biotechnology*, Fogarty, W.M. (ed.). Applied Sciences Publishers, London/New York.

Wicker, L., Ackerley, J.L., Corredig, M. (2002). Clarification of juice by thermolabile valencia pectinmethylesterase is accelerated by cations. *J. Agric. Food Chem.*, 50(14), 4091–4095.

Wilkinson, A.P., Gee, J.M., Dupont, M.S., Needs, P.W., Mellon, F.A., Williamson, G., Johnson, I.T. (2003). Hydrolysis by lactase phlorizin hydrolase is the first step in the uptake of daidzein glucosides by rat small intestine in vitro. *Xenobiotica*, 33(3), 255–264.

Williams, J.A. (2002). Keys to bioreactor selections. *Chem. Eng. Prog.*, 98, 34–41.

Wohlgemuth, R., Plazl, I., Žnidaršič-Plazl, P., Gernaey, K.V., Woodley, J.M. (2015). Microscale technology and biocatalytic processes: Opportunities and challenges for synthesis. *Trends Biotechnol.*, 33, 302–314.

Wong, L.Y., Saad, W.Z., Mohamad, R., Tahir, M.D. (2017). Optimization of cultural conditions for polygalacturonase production by a newly isolated *Aspergillus fumigatus* R6 capable of retting kenaf. *Ind. Crop. Prod.*, 97, 175–183.

Wu, C., Zhou, G., Jiang, X., Ma, J., Zhang, H., Song, H. (2012). Active biocatalysts based on *Candida rugosa* lipase immobilized in vesicular silica. *Process Biochem.*, 47, 953–959.

Xu, F., Oruna-Concha, M.J., Elmore, J.S. (2016). The use of asparaginase to reduce acrylamide levels in cooked food. *Food Chem.*, 210, 163−171.

Yagoub, A.A., Ma, H., Zhou, C. (2017). Ultrasonic-assisted extraction of protein from rapeseed (*Brassica napus* L.) meal: Optimization of extraction conditions and structural characteristics of the protein. *Int. Food Res. J.*, 24(2), 621–629.

Yamaguchi, S. (2000). Method for cross-linking protein by using enzyme. Patent, 6121013.

Yamaski, M., Yasui, T., Arima, K. (1964). Pectic enzymes in the clarification of apple juice. *Agri. Chem.*, 28(11), 779–787.

Yang, Y., Zhang, X., Yin, Q., Fang, W., Fang, Z., Wang, X., Zhang, X., Xiao, Y. (2015). A mechanism of glucose tolerance and stimulation of GH1 beta-glucosi-dases. *Sci. Rep.*, 5, 17296.

Yang, G., Yao, H., Mozzicafreddo, M., Ballarini, P., Pucciarelli, S., Micelia, C. (2017). Rational engineering of a cold-adapted alpha-amylase from the antarctic ciliate *Euplotes focardii* for simultaneous improvement of thermostability and catalytic ability. *Appl. Environ. Microbiol.*, 83(13), e00449–17.

Zawawi, F.S.M., Karim, L., Omar, S.R., Ali, A. (2020). Enzyme activity and stability of lactase immobilized on two different supports: Calcium alginate and magnetic chitosan. *Mal. J. Fund. Appl. Sci.*, 16(4), 413–417.

Zetelaki-Horváth, K., Vas, K. (1981). Effect of oxygen transfer rate on the composition of the pectolytic enzyme complex of Aspergillus niger. *Biotechnol. Bioing.*, 23(10), 2231–2241.

Zhang, X., Cao, T., Tian, X., Gai, D. (2012). Effect of microwave irradiation on the structure of glucoamylase. *Process Biochem.*, 47(12), 2323–2328.

Zhang, C., Chen, S., Alvarez, P.J., Chen, W. (2015). Reduced graphene oxide enhances horseradish peroxidase stability by serving as radical scavenger and redox mediator. *Carbon*, 94, 531–538.

Zhang, Y., Geary, T., Simpson, B.K. (2019). Genetically modified food enzymes: A review. *Curr. Opin. Food Sci.*, 25, 14–18.

Zhu, F., He, B., Gu, F., Deng, H., Chen, C., Wang, W., Chen, N. (2020). Improvement in organic solvent resistance and activity of metalloprotease by directed evolution. *J. Biotechnol.*, 309, 68–74.

PART 2

Lactic Acid Bacteria and Bifidobacteria: A Fermented Food Source

Introduction to Part 2

Marie-Bénédicte Romond
Laboratoire de Virologie, Université de Lille, France

Food fermentation processes have been known for centuries and are the result of the proliferation of microorganisms.

Acidifying microorganisms improve food preservation by suppressing potentially pathogenic bacteria. They also modify the organoleptic characteristics and texture of food. The microorganisms involved are usually lactic acid bacteria and bifidobacteria, but yeasts are also present.

In addition to the technological aspects, lactic acid bacteria can have a beneficial effect on human health.

Scientific advances in microbiology and human physiology, combined with technological innovations, have made it possible (and will continue to do so) to better exploit the potential of these microorganisms through the formulation of functional foods.

7

Lactic Acid Bacteria and Bifidobacteria

Jennifer BURGAIN[1] and Marie-Bénédicte ROMOND[2]

[1] *LIBio, Université de Lorraine, Nancy, France*

[2] *Laboratoire de Virologie, Université de Lille, France*

7.1. Taxonomic notions and the definition of "probiotic"

The initial interest in bacteria stemmed from Louis Pasteur's discoveries, which demystified the theory of spontaneous generation. In particular, this work laid the scientific foundations for fermentation, vinification and beer brewing (Berche 2012).

Subsequent work by teams at the Institut Pasteur in Paris highlighted the special role of lactic acid bacteria, ultimately leading to a deeper understanding of these microorganisms (Figure 7.1).

In his thesis in 1900, a pediatrician at the Institut Pasteur by the name of Henri Tissier described bacilli as having a bifid appearance, isolated from stool samples of infants who were breast-fed by their mothers, and thought to confer the resistance observed in gastroenteritis (Tissier 1900). In his 1907 pamphlet *Essais optimistes*, Élie Metchnikoff, deputy director of the Pasteur Institute, reported on several experiments (including those carried out on mice by Bélonowsky at the Pasteur Institute) and observations on the beneficial effects of daily consumption of "Bulgarian bacillus" isolated from curdled milk (Metchnikoff 1907). He concluded that lactic acid bacteria play an undeniable role in combating intestinal putrefaction.

Implementation of Enzymatic Processes and Lactic Bacteria in the Food Industries,
coordinated by Mohamed GHOUL.

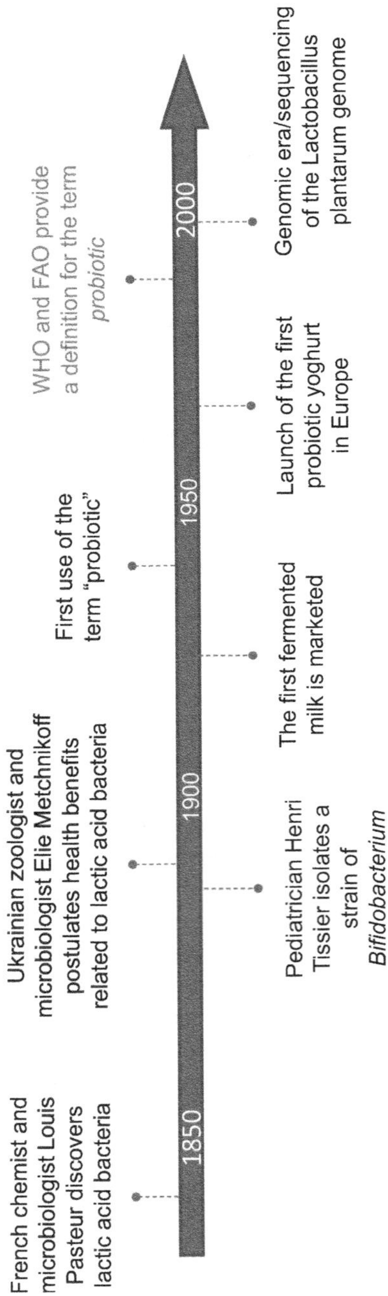

Figure 7.1. *Timeline of the history of probiotics (from O'Toole et al. (2017)).*
For a color version of this figure, see www.iste.co.uk/ghoul/implementation.zip

In reality, the notion of lactic acid bacteria, initially associated with lactobacilli alone, turns out to be much broader, since it is based on the production of lactic acid from a fermentable substrate (usually a sugar compound). The *Lactobacillales*, classified in the order *Firmicutes*, include the main families that produce lactic acid, almost exclusively, as short-chain fatty acids: *Carnobacteriaceae*, *Enterococcaceae*, *Lactobacillaceae* (including *Lacticaseibacillus*, *Lactiplantibacillus*, *Lactobacillus*, *Latilactobacillus*, *Lentilactobacillus*, *Levilactobacillus*, *Ligilactobacillus*, *Loigolactobacillus*, *Pediococcus*, *Secundilactobacillus*), *Leuconostocaceae* (including *Leuconostoc*, *Oenoccocus*), *Streptococcaceae* (including *Lactococcus*, *Streptococcus*). Some genera include pathogenic species (such as *Streptococcus* pyogenes) and commensal species of interest in dairy fermentation (such as *Streptococcus thermophilus*).

In contrast to exclusively lactic bacteria, *Bifidobacteriales* and *Propionibacteriales* are classified in the order *Actinobacteriae*. In addition to lactic acid, the former produce the predominant acetic acid. The latter produce propionic and acetic acids from lactic acid. While bifidobacteria are not known to be opportunistic, propionibacteria are divided into health-promoting (*Propionibacterium freundenreichii*) and potentially pathogenic bacteria (*Cutibacterium acnes*).

The production of short-chain fatty acids was sought to protect the hygienic quality of food products. However, research in human health came later. The term "probiotic" was first used in 1953, according to Hamilton-Miller et al. (2003), and was taken up again in 1965 by Lilly and Stillwell (1965). This definition was refined in 1989, when it was stated that "probiotics are microorganisms that are thought to improve the health of the host by modulating the properties of its gut microbiota" (Fuller 1989, 1992).

Following the Cordoba meeting in 2001, the World Health Organization (WHO) and the Food and Agriculture Organization of the United Nations (FAO) provided a consensus-based definition that describes probiotic bacteria as "live microorganisms that, when administered in adequate amounts, confer a health benefit on the host" (Gilliland et al. 2001; Hill et al. 2014). This definition has two fundamental elements:

– an adequate amount of live bacteria;

– a health benefit on the host.

The first point is an important one, as the requirement for "bacterial vitality" has long been the subject of debate. While France has always favored yoghurts based on live bacteria, the English-speaking countries and Germany are reluctant to use them for two reasons: the risk linked to the potentially harmful effects of live bacteria (some of which can act as opportunists, such as enterococci) and the cost of preserving this vitality (cold chain). Yoghurts are therefore heat-treated in countries where food hygiene does not tolerate live bacteria, with the addition of gelling agents providing a curd substitute.

In order for the principle of live bacteria to be accepted, bacterial vitality therefore had to be associated with a health effect. Between 1980 and 2000, there were many attempts to prove that live bacteria were necessary for any health effect. However, the situation remains unresolved. In fact, it all depends on the desired effect. Initially, the aim was to compensate for a deficit (sometimes assumed, in the case of human intestinal lactobacilli) of microbiota bacteria, to which certain physiological functions were attributed. Particular attention was paid to the capacities of bacterial strains, such as bifidobacteria, to increase populations using strains of *Bifidobacterium*, or strains that do not belong to the genus but show the same types of microbiota modulation.

Outside of any pathological situation, it is essential that the health benefits of such an increase be demonstrated. The impact of bifidobacteria on the immune system and bacterial translocation is a promising lead.

The creation of the European Food Safety Authority (EFSA) in 2004 made it possible to clarify "health" objectives in accordance with Articles 13 and 14 of Regulation (EC) 1924/2006 on health claims made on foods.

7.2. The probiotics market and health claims

At present, probiotics are found in food in two main forms: dietary supplements (capsules, dispersible powder), or ferments inoculated into a food matrix, as in the case of fermented products (including yoghurts). The global market for probiotics is growing steadily and is all the more significant as people in good health want to stay that way through a healthy diet. The claimed health benefits of probiotics, coupled with the technological advances that have been observed in recent years, are all factors boosting the probiotics market. In Asia, and more particularly in Japan, the aging of the population explains the increase in consumption of dietary supplements, which include probiotics.

Nevertheless, there are major hurdles to overcome, the main one probably being the regulatory requirements for health claims. However, there is no shortage of publications in the field, and research is very active in identifying new probiotic strains, even if research costs remain high.

In recent years, the European probiotic market has been forced to restructure itself following the publication by EFSA of Regulation (EC) 1924/2006 on nutrition and health claims made on foods (European Commission 2006). A health claim is defined as "any statement that suggests that a relationship exists between a food and health" (European Commission 2006). In order for the beneficial effects of probiotics on human health to be accepted as health claims, they must be assessed by regulatory institutions like the EFSA. From 2005 to 2013, the EFSA's Scientific Panel on Dietetic Products, Nutrition and Allergies analyzed numerous health claim applications. Note that 78% of these applications were considered likely to be beneficial to human health, particularly intestinal health, but the lack of scientific data meant that these health claims were almost all rejected (Binnendijk and Rijkers 2013).

The exception is the probiotic effect of *Lactobacillus delbrueckii subsp. bulgaricus* and *Streptococcus thermophilus*, the two bacteria used to make yoghurt, on lactose digestion (EFSA Scientific Panel on Dietetic Products, Nutrition and Allergies 2010). In this case, bacterial vitality must be preserved, as it guarantees effective galactosidase activity.

Dietary supplements sold in pharmacies (50% of the dietary supplements market) therefore do not meet any claims associated with the presence of probiotics. A quick analysis of labeling shows that authorized claims are linked to other compounds, in particular vitamins and minerals (Regulation (EU) No. 432/2012, establishing a list of authorized health claims made on foods, other than those referring to the reduction of disease risk and to children's development and health). As a result, companies such as Pilege and Procter & Gamble no longer market products that exclusively contain probiotics, since no claims can be made. On the other hand, the presence of vitamins and/or minerals, in authorized concentrations, does enable health claims to be made. According to the order of June 24, 2014 (published in the Journal Officiel JORF No. 0163 of July 17, 2014), establishing the list of plants authorized in food supplements and their conditions of use, associations with plants are an alternative solution.

As previously mentioned, part of probiotic activity is linked to bacterial vitality. However, bacteria lose vitality during product storage (Gomand et al. 2019) and during digestion. It is therefore necessary to find a solution to these survival issues, so that probiotic bacteria can exert their beneficial effects in the intestine. One way of doing this is through microencapsulation, which creates a favorable local environment that enables the bacteria to remain alive and functional until they reach their site of action.

7.3. Prebiotics

7.3.1. *Definition*

Prebiotics are carbohydrates, specifically non-digestible dietary oligosaccharides (Gibson and Roberfroid 1995), similar to human milk oligosaccharides (HMOs) (Donovan and Comstock 2016). HMOs exert prebiotic activity (mainly bifidogenic activity) by stimulating the growth of protective commensal intestinal bacteria (bifidobacteria) and are fermented into short-chain fatty acids (SCFAs). In turn, SCFAs play a key role in stimulating immune cells and intestinal epithelial cells, thereby improving intestinal integrity (Donovan and Comstock 2016). More than 200 distinct HMO structures have been identified, all with a lactose group on their reducing end (Wu et al. 2010; Urashima et al. 2020). Apart from 2-fucosyl-lactose (2'FL), the major oligosaccharide chains are lacto-N-tetraose (LNT) and lacto-N-neotetraose (LNnT), which associate a lactosamine with a lactose unit. Derivations with fucose or sialic acid allow oligosaccharides to be sub-classified as fucosylated or sialylated (El-Hawiet et al. 2015).

The term *prebiotic* was defined by Gibson and Roberfroid (1995) as "non-digestible food ingredients that beneficially affect the host by selectively stimulating the growth and/or activity of one or a limited number of bacteria in the colon, and thus improves host health". In 2017, the definition was refined by the International Scientific Association of Probiotics and Prebiotics (Gibson et al. 2017), and prebiotics are now described as "a substrate that is selectively utilized by host microorganisms conferring a health benefit".

Currently, the main prebiotics used are inulin and its by-products, fructo-oligosaccharides (FOS), galacto-oligosaccharides (GOS) and xylo-oligosaccharides (XOS). They can be obtained by extraction from naturally rich foods (such as inulin in chicory), fermentation or chemical synthesis.

Prebiotics must meet three criteria: be resistant to digestion in the stomach and upper intestine (1), be fermentable by the gut microbiota (2) and be able to specifically stimulate the growth and/or activity of gut bacteria beneficial to the health of the host (3) (Gibson et al. 2017).

FOS, GOS and inulin are the most frequently studied prebiotics. The effects of prebiotics and dietary fiber intake are often confused. While dietary fibers are poorly digestible and influence the gut microbiota overall, prebiotics are thought to favor only certain species, in particular bifidobacteria. However, the FOS/GOS ratio must be controlled to achieve this result. In fact, FOS are readily used by clostridia, leading to bloating in subjects with low dietary fiber in their diet. Bifidobacteria do not benefit from efficient fructosidases and are therefore penalized compared to clostridia. GOS, on the other hand, do not suffer from this undesirable effect.

7.3.2. *Prebiotics, synbiotics and health claims*

In terms of health claims, FOS isolated from inulin and GOS obtained after elimination of fructose units (AlphaGOS®-Olygose) have obtained approval for the reduction of postprandial glycemia (EFSA Scientific Panel on Dietetic Products, Nutrition and Allergies 2014). This post-prandial reduction is accompanied, when GOS is administered, by a reduction in clostridia.

A health claim concerning inulin extracted from chicory has also been shown to regulate transit, provided that minimum daily doses of 12 g are taken (EFSA Scientific Panel on Dietetic Products, Nutrition and Allergies 2015).

A synbiotic is defined as a product containing at least one prebiotic and at least one probiotic. Prebiotics thus have a beneficial effect on the stability, survival and implantation of the probiotic(s) incorporated into the product. The effect of synbiotics on health is probably attributable to the individual combination of a probiotic and a prebiotic (de Vrese and Schrezenmeir 2008). Given a very large number of possible combinations, the application of synbiotics for the modulation of gut microbiota in humans seems promising (Scavuzzi et al. 2014; Markowiak and Śliżewska 2017). For example, synbiotics can have anti-bacterial, anti-carcinogenic and anti-allergic effects. They can also help prevent transit disorders such as constipation and diarrhea. Ultimately, synbiotics can be highly effective in preventing osteoporosis, reducing lipids and carbohydrates in blood (Pandey et al. 2015; Markowiak and Śliżewska 2017). Numerous clinical trials have assessed the impact of synbiotics on human health, particularly with regard to obesity (Safavi et al. 2013; Sanchez et al. 2014; Ipar et al. 2015), type 2 diabetes (Moroti et al. 2012; Asemi et al. 2013; Shakeri et al. 2014; Tajadadi-Ebrahimi et al. 2014), irritable bowel syndrome (Šmid et al. 2016) or on symptoms associated with lactose intolerance (Post 2013).

8

Bifidobacteria: From Commensal Bacteria to Probiotics and Metabiotics/Postbiotics

Frank PIVA and Marie-Bénédicte ROMOND
Laboratoire de Virologie, Université de Lille, France

8.1. Taxonomy and ecological aspects of commensal bifidobacteria: discovery and identification criteria

Bifidobacteria were described at the dawn of the 20th century by Dr. Henri Tissier at the Pasteur Institute in Paris (Tissier 1900). In the stools of infants who were breastfed by their mothers, he observed the presence of Gram-positive bacilli, in a peculiar Y-shaped bifidobacterial form (from the Latin *bifidus*, "split in two"), with low oxygen tolerance. He named this bacterium *Bacillus bifidus* (Figure 8.1).

In 1924, Orla-Jensen (1924) discovered the genus *Bifidobacterium*. However, this proposal was debated until the 1960s, when Sebald et al. (1965) separated the genera *Bifidobacterium* and *Lactobacillus* on the basis of the GC content. It was not until 1980, with the validation by the ad hoc commission of the International Committee on Systematic Bacteriology (Skerman et al. 1980), that the name *Bifidobacterium* was accepted, the type species (now called *Bifidobacterium*

For a color version of all the figures in this chapter, see www.iste.co.uk/ghoul/implementation.zip.

Implementation of Enzymatic Processes and Lactic Bacteria in the Food Industries, coordinated by Mohamed GHOUL. © ISTE Ltd 2024.

bifidum) being that described by Tissier. In the meantime, the name *Lactobacillus bifidus* was used by some authors.

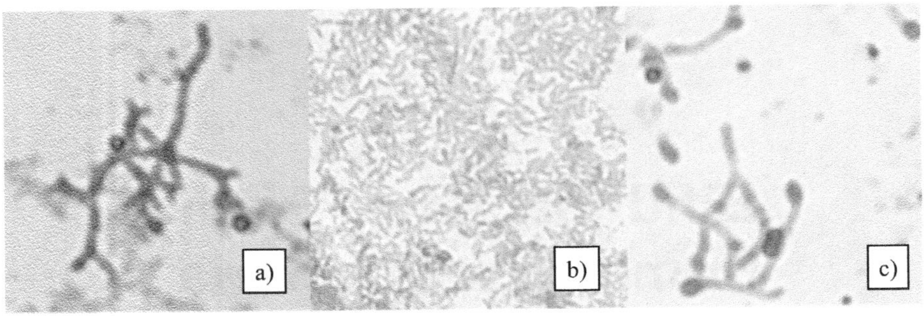

Figure 8.1. *Details of light microscopy images of different Bifidobacterium strains: (a) B. bifidum; (b) B. breve; (c) B. longum, spatula shapes. Gram staining, initial magnification × 1,000 (personal data)*

Closely related genera have been created on the basis of GC content of less than 50%, but it is close, if we consider an additional criterion such as the key enzyme in *Bifidobacterium* glycolysis, fructose-6-phosphate phosphoketolase (F6PPK). This is the *scardovia* group, which includes the genera *Aeriscardovia*, *Alloscardovia*, *Bombiscardovia*, *Galliscardovia*, *Gardnerella*, *Neoscardovia*, *Parascardovia*, *Pseudoscardovia* and *Scardovia*, all with a GC content of less than 50% (Mekadim et al. 2019). Together (including the genus *Bifidobacterium*), they form the *Bifidobacteriaceae* family.

Alongside GC content, F6PPK was rapidly proposed as an orientation criterion (Sgorbati et al. 1976). This key enzyme in glycolysis through the pentose pathway (de Vries and Stouthamer 1967) exists in two forms: D-xylulose 5-phosphate and D-fructose 6-phosphate phosphoketolases (Sgorbati et al. 1976; Grill et al. 1995). The activity is shared by the genera making up the *Bifidobacteriaceae* family, but the sequences differ (Cleusix et al. 2010).

A third criterion, the sequence of DNA coding for 16S ribosomal RNA, helps to classify an F6PPK-positive strain in the genus *Bifidobacterium* (Stackebrandt and Goebel 1994). Alignment between sequences of wild strains and type strains also makes it possible to define phylogenetic taxa (groups of living organisms descended from the same ancestor with certain common characters) that group together several bifid species, between the "genus" and "species" levels. For example, when defining the new species *Bifidobacterium aquikefiri*, the isolated kefir strain could be incorporated into the genus *Bifidobacterium*, following 97.1% and 97.4% sequence alignment of the 16S RNA-encoding gene with those of the type strains of *Bifidobacterium psychraerophilum* and *Bifidobacterium crudilactis*, respectively. These three species make up a phylogenetic taxon (Laureys et al. 2016).

The methods used to define a new species are extremely varied, including techniques combining phenospecies description by searching for enzymatic and/or fermentative profiles (such as the use of API galleries), genospecies analysis by hybridization between chromosomal DNAs (Gavini et al. 1991), molecular typing by ERIC-PCR (Duranti et al. 2019), multilocus sequence analysis (MLSA) of household gene sequences (clpC, dnaB, dnaG, dnaJ, hsp60 and rpoB) (Jian et al. 2001; Martens et al. 2008; Kim et al. 2010), ARDRA, RAPD-PCR and rep-PCR (Jarocki et al. 2016). Lastly, average nucleotide identity by Blast (ANIb), calculated from whole-genome sequencing, provides additional selective information (Modesto et al. 2018c; Duranti et al. 2019; Eckel et al. 2020).

Wild strains identification can be based on molecular methods targeting variations in different bifidobacterial genes (Ventura et al. 2001; Mullié et al. 2003; Baffoni et al. 2013). However, it is recommended to further characterize bifidobacterial strains for technological purposes; on the one hand, when intellectual protection is sought and, on the other, if specific effects are alleged. In this context, the characterization of the strain can either be apprehended based on its physiological traits or its genetic traits. In the latter case, the methods already mentioned (Jarocki et al. 2016) can be used, but genome sequencing is also recommended.

Lastly, the number of new bifidobacteria species has been growing exponentially since researchers began analyzing bifidobacteria from primates (mostly living in captivity) (Lugli et al. 2021; Neuzil-Bunesova et al. 2021). The species are validated by the ad hoc Subcommittee on Systematics of Prokaryotes and published in the *International Journal of Systematic and Evolutionary Microbiology* (Mattarelli et al. 2020; Oren and Garrity 2022).

Table 8.1 lists the species according to the origins of the first isolates. Bifidobacteria are found in the intestines of mammals (humans, non-human primates, cattle, sheep, rodents, leporids, bats, etc.), birds (geese, hens) and insects (bees, bumblebees) (Table 8.1)[1].

---

1 The list, which was updated in 2022, can be viewed at: https://lpsn.dsmz.de/genus/bifidobacterium.

Eleven species of bifidobacteria have been described in humans, most of them of intestinal origin. Only *B. dentium* (formerly known as *Actinomyces eriksonii*) and *B. scardovii* have been isolated from other biotopes, usually in pathological situations.

Within the hominids, three new species have recently been described. Genomic analysis of *B. moukalabense*, a species isolated from wild gorillas, reveals an adaptation to the gorilla's diet of fruits and leaves (Segawa et al. 2019).

	Host/environment	Species	Description
Hominids	Human (*Homo sapiens*)	*B. adolescentis*	(Reuter 1963)
		B. angulatum	(Scardovi and Crociani 1974)
		B. bifidum	(Tissier 1900; Orla-Jensen 1924)
		B. breve	(Reuter 1963)
		B. catenulatum	(Scardovi and Crociani 1974)
		B. dentium	(Scardovi and Crociani 1974)
		B. faecale	(Choi et al. 2014)
		B. gallicum	(Lauer 1990)
		B. longum	(Reuter 1963)
		B. pseudocatenulatum	(Scardovi et al. 1979)
		B. scardovii	(Hoyles et al. 2002)
	Chimpanzee (*Pan troglodytes*)	*B. panos*	(Neuzil-Bunesova et al. 2021)
	Western lowland gorilla (*Gorilla gorilla gorilla*)	*B. moukalabense*	(Tsuchida et al. 2014)
	Sumatran orangutan (*Pongo abelii*)	*B. pongonis*	(Lugli et al. 2021)

Table 8.1. *Source of the Bifidobacterium species. Hominids: human beings and anthropomorphic monkeys*

Primates also host specific species of bifidobacteria. As with chimpanzees, gorillas and orangutans, only one species is described in colobus, patas, lemurs and makis, all of which are classified as Old World primates.

Old World primates	Host/environment	Species	Description
Monkeys	Mantled guereza (*Colobus guereza*)	*B. colobi*	(Lugli et al. 2021)
	Common patas monkey (*Erythrocebus patas*)	*B. erythrocebi*	(Neuzil-Bunesova et al. 2021)
Lemurs	Black lemur (*Eulemur macaco*)	*B. eulemuris*	(Michelini et al. 2016b)
	Ring-tailed lemur (*Lemur catta*)	*B. lemurum*	(Modesto et al. 2015)

Table 8.2. *Source of the Bifidobacterium species and other Old World primates*

In contrast, among New World primates, the common marmoset alone harbors up to eight different species of bifidobacteria. The silvery marmoset also carries a substantial number of bifidobacterial species. This impression is tempered by the bifidobacterial biodiversity of tamarins. Contrasting situations make it impossible to draw a general conclusion on bifidobacteria carriage in primates.

Furthermore, they have only recently been described. In insects, new species have been added to the small panel of bifidobacteria described over 50 years ago. It is therefore too early to define the degree of bifidobacteria biodiversity among carriers until more literature is available.

However, these new species represent a reservoir of potentially probiotic bifidobacteria. An analysis of the protein sequences of their lipoproteins reveals homologies with those from species isolated from humans. As the latter lipoproteins are known for their promising biological activities, the wider choice of sequences and complexity in fatty acids and oligosaccharides offer new prospects for technological applications in food and health (see Figures 8.5 and 8.6).

	Host/environment	Species	Description
New World primates (marmosets)	Common marmoset (*Callithrix jacchus*)	*B. aesculapii*	(Modesto et al. 2014)
		B. callitrichos	(Endo et al. 2012)
		B. catulorum	(Modesto et al. 2018b)
		B. hapali	(Michelini et al. 2016a)
		B. jacchi	(Modesto et al. 2019a)
		B. myosotis	(Michelini et al. 2016a)
		B. reuteri	(Endo et al. 2012)
		B. tissieri	(Michelini et al. 2016)
	Silvery marmoset (*Mico argentatus*)	*B. amazonense*	(Lugli et al. 2021)
		B. miconis	(Lugli et al. 2021)
		B. miconisargentati	(Lugli et al. 2021)
		B. pluvialisilvae	(Lugli et al. 2021)
		B. santillanense	(Lugli et al. 2021)
	Pygmy marmoset (*Cebuella pygmaea*)	*B. margollesii*	(Lugli et al. 2018)
		B. parmae	(Lugli et al. 2018)
	Goeldi's marmoset (*Callimico goeldii*)	*B. callimiconis*	(Duranti et al. 2019)
		B. cebidarum	(Duranti et al. 2020)
		B. goeldii	(Duranti et al. 2019)
		B. moraviense	(Neuzil-Bunesova et al. 2021)

Table 8.3. *Source of the Bifidobacterium species and New World primates*

In the case of livestock and similar animals, the description of species is older, given the accessibility of dejecta, intestinal contents, etc., at the time of slaughter. Advances in knowledge demonstrate the multiplicity of hosts for certain bifidobacteria. *B. boum*, isolated from bovine rumen, frequently colonizes the intestines of pigs (especially piglets) and wild boar (Splichalova et al. 2020). It has also been found in Asian elephants (Bunesova et al. 2013).

	Host/environment	Species	Description
New World primates (tamarins)	Golden-headed lion tamarin (*Leontopithecus chrysomelas*)	*B. leontopitheci*	(Duranti et al. 2020)
	Golden lion tamarin (*Leontopithecus rosalia*)	*B. simiiventris*	(Lugli et al. 2021)
	Two-tone tamarin (*Saguinus bicolor*)	*B. saguinibicoloris*	(Lugli et al. 2021)
	Emperor tamarin (*Saguinus imperator*)	*B. callitrichidarum*	(Modesto et al. 2018a)
		B. imperatoris	(Lugli et al. 2018)
		B. simiarum	(Modesto et al. 2018c)
		B. vansinderenii	(Duranti et al. 2017)
	Golden-handed tamarin (*Saguinus midas*)	*B. biavatii*	(Endo et al. 2012)
		B. saguini	(Endo et al. 2012)
		B. stellenbochensee	(Endo et al. 2012)
	Moustached tamarin (*Saguinus mystax*)	*B. olomucense*	(Neuzil-Bunesova et al. 2021)
	Cotton-top tamarin (*Saguinus oedipus*)	*B. aerophilum*	(Michelini et al. 2016c)
		B. avesanii	(Michelini et al. 2016c)
		B. felsineum	(Modesto et al. 2018c)
		B. oedipodis	(Neuzil-Bunesova et al. 2021)
		B. primatium	(Modesto et al. 2018c)
		B. ramosum	(Michelini et al. 2016c)
		B. scaligerum	(Modesto et al. 2018c)

Table 8.4. *Source of the Bifidobacterium species and New World primates*

Similarly, *B. pseudolongum*, which was first isolated in pigs, is commonly found in rodents (Korgan et al. 2022; Takei et al. 2022). This colonization raises the question of using mouse models to assess probiotic strains of *B. pseudolongum* and extrapolating the results to humans (Guo et al. 2022). Indeed, this natural establishment could lead to beneficial effects by compensating for a deficiency. On the other hand, it is difficult to transpose the mechanism to a host that is poorly endowed with the species under consideration, as advanced genomic analysis of human stools confirms the rarity of *B. pseudolongum* (Xiao et al. 2021). The same concern applies to *B. animalis*, which was isolated for the first time in rats and widely used in food in Europe.

	Host/environment	Species	Description
Bovids (rumen)	Domestic cattle (*Bos taurus*)	*B. boum*	(Scardovi et al. 1979)
		B. merycicum	(Biavati and Mattarelli 1991)
		B. ruminantium	(Biavati and Mattarelli 1991)
Suidae	Boar (*Sus scrofa*)	*B. apri*	(Pechar et al. 2017)
	Pig (*Sus scrofa domesticus*)	*B. choerinum*	(Scardovi et al. 1979)
		B. pseudolongum	(Mitsuoka 1969)
		B. psychraerophilum	(Simpson et al. 2004)
		B. thermophilum	(Mitsuoka 1969)
Leporidae	Rabbit (*Oryctolagus cuniculus*)	*B. cuniculi*	(Scardovi et al. 1979)
		B. italicum	(Lugli et al. 2018)
		B. magnum	(Scardovi and Zani 1974)
Anatidae	Goose (*Anser anser domesticus*)	*B. anseris*	(Lugli et al. 2018)
Gallinaceous	Pheasant (*Phasianus colchicus*)	*B. phasiani*	(Lugli et al. 2021)
	Chicken (*Gallus gallus domesticus*)	*B. pullorum*	(Trovatelli et al. 1974)
By-products	Food (kefir, raw milk, traditional fermented milk)	*B. aquikefiri*	(Laureys et al. 2016)
		B. crudilactis	(Delcenserie et al. 2007)
		B. minimum	(Biavati et al. 1982)
		B. mongoliense	(Watanabe et al. 2009)
		B. tibiigranuli	(Eckel et al. 2020)
Effluents	Wastewater	*B. subtile*	(Biavati et al. 1982)
		B. thermacidophilum	(Dong et al. 2000)

Table 8.5. *Source of the Bifidobacterium species. Livestock, game and by-products*

Lastly, the detection of bifidobacteria in the environment (wastewater) raises the issue of the contaminating source, as conditions are not conducive to the proliferation of bifidobacteria. The B. *thermoacidophilum* species, which is initially isolated from wastewater, is traditionally carried by wild boar, indicating possible fecal contamination by these mammals. The link between *B. subtile* and fecal origin is more difficult to establish, since, at present, excluding wastewater, it is only found in dental infections (Mantzourani et al. 2009). The possibility of it having hospital origins is a hypothesis that remains to be explored.

The description of specific bifidobacteria in fermented products also raises questions regarding the source of contamination, although in the case of traditional fermented milks, there may be a continuum between productions. Research into the ecology of these new species would nonetheless be welcome, so as to support their use in the food industry.

	Host/environment	Species	Description
Rodents	European beaver (*Castor fiber*)	*B. castoris*	(Duranti et al. 2019)
	European hamster (*Cricetus cricetus*)	*B. criceti*	(Lugli et al. 2018)
	Golden hamster (*Mesocricetus auratus*)	*B. tsurumiense*	(Okamoto et al. 2008)
	Patagonian mara (*Dolichotis patagonum*)	*B. dolichotidis*	(Duranti et al. 2019)
	Rat (*Rattus norvegicus*)	*B. animalis*	(Mitsuoka 1969)
Canidae	Dog (*Canis lupus familiaris*)	*B. canis*	(Neuzil-Bunesova et al. 2020)
Chiroptera	Roussette (*Rousettus aegyptiacus*)	*B. rousetti*	(Modesto et al. 2019b)
		B. vespertilionis	(Modesto et al. 2019b)
Unaus	Two-toed sloth (*Choloepus didactylus*)	*B. choloepi*	(Modesto et al. 2020a)

Table 8.6. *Source of the Bifidobacterium species. Rodents and other mammals*

Furthermore, following the massive use of *B. animalis* in the food industry (particularly fermented milks), its detection in human feces is now very common, unlike in the past. It is not uncommon to detect its DNA in premature infants, even though they do not receive any products containing it. However, its culture from human feces remains uncertain. This example illustrates the limitations of genomic analysis alone, as well as the use of culture to confirm host colonization.

		Host/environment	Species	Description
Insects	Bees	Carpenter bee (*Xylocopa violacea*)	*B. aemilianum*	(Alberoni et al. 2019)
			B. xylocopae	(Alberoni et al. 2019)
		European bee (*Apis mellifera*)	*B. apousia*	(Chen et al. 2021)
			B. asteroides	(Scardovi and Trovatelli 1969)
			B. choladohabitans	(Chen et al. 2021)
			B. mizhiense	(Li et al. 2022)
			B. polysaccharolyticum	(Chen et al. 2021)
		Indian bee (*Apis cerana indica*)	*B. indicum*	(Scardovi and Trovatelli 1969)
	Bumblebees	Various bumblebees (*Bombus lapidarius*, *B. terrestris*, *B. hypnorum*)	*B. commune*	(Praet et al. 2015)
		White-tailed bumblebee (*Bombus lucorum*)	*B. actinocoloniiforme*	(Killer et al. 2011)
			B. bohemicum	(Killer et al. 2011)
			B. bombi	(Killer et al. 2009)

Table 8.7. *Source of the Bifidobacterium species. Insects*

Lastly, even species of bifidobacteria found in insects can be found in mammals. For example, *B. indicum* is found in the rumen of alfalfa-fed sheep. The absence of *B. indicum* in alfalfa-fed llamas suggests that this bacterium is part of the biotope, and not simply in transit (Pei et al. 2010).

8.1.1. *Distribution of bifid species in humans and animals*

With the exception of recently described bifidobacteria, which have not yet been subjected to an epidemiological study extended to new biotopes, several hosts have been found to harbor the same bifid species.

Species found in humans can thus be isolated from other mammals (Table 8.8).

	Human		Animals		Wastewater
	Infant	Adult	Mammals	Other	
B. adolescentis	+	+	+		+
B. angulatum	+	+	+		+
*B. animalis**	+	+	+	§P	+
B. bifidum	+	+	+		
B. breve	+	+	+		
B. catenulatum	+	+	+		+
B. dentium	+	+	+		
B. gallicum		+	+		
*B. globosum***	+	–	+	§P	+
B. longum-infantis	+	+	+		
*B. mongoliense***	–	+	+		
B. pseudocatenulatum	+	+	+		
*B. pullorum***	–	+	–	§P	
B. scardovii		+	+		
*B. thermophilum***	+	+	+		

Table 8.8. *Distribution of bifidobacteria species in mammals according to (Scardovi 1986; Crociani et al. 1994; Haarman and Knol 2005; Gavini et al. 2006; Milani et al. 2017; Milani et al. 2019)*

COMMENT ON TABLE 8.8.– *: *Recent observation corresponding to industrial use (Kato et al. 2017; Yang et al. 2019; Tsukimi et al. 2020);* **: *recent description in human secretions (for B. globosum, name B. pseudolongum subsp globosum) (Mathys et al. 2008; Kato et al. 2017);* §*P: chicken.*

The metagenomic approach shows that the bifdobacteria species most commonly found in the mammals under study are *B. longum-infantis* and *B. adolescentis*, followed by *B. bifidum* (Milani et al. 2017). However, the fact that bifidobacterial DNA is detected in the feces of mammals, either farmed or living in zoos, may only reflect a transmission of these bacterial residues through direct contact with humans and/or indirectly through industrial food.

However, the isolation of *B. bifidum*, *B. breve*, *B. pseudocatenulatum* and *B. pseudolongum subsp. globosum* from suckling calves, and of *B. adolescentis* from cows, pigs and rabbits, suggests that some species of bifidobacteria adapt to biotopes other than the human gut (Scardovi 1986; Crociani et al. 1994; Gavini et al. 2006; Wickramasinghe et al. 2020).

The case of *B. longum* is somewhat different, insofar as the sub-species of this continuum adapt to different biotopes. In pigs and calves, for example, *B. longum subsp. suis* will predominate (Scardovi 1986; Kelly et al. 2016).

The metagenomic study also showed that the proportion of bifidobacteria in infant stools is higher than in the young of other mammalian mothers, probably reflecting the particularly high concentration of oligosaccharides in women's milk (Albrecht et al. 2014). Ultimately, as described for women, mammalian milk contains bifidobacterial DNA. Transmission of bifidobacteria could therefore continue after delivery by breastfeeding in several mammalian species (Milani et al. 2017). Cow's milk very commonly contains DNA from *B. bifidum*, *B. longum subsp. longum*, *B. pseudolongum subsp. globosum* and *B.* mongoliense (Milani et al. 2019). Bifidobacteria isolation is also frequent in milk intended for cheese production, particularly *B. mongoliense*, which would explain their detection in consumer stools.

8.1.2. *Topographical and kinetic distribution in humans*

The fecal microbiota is said to only reflect a partial picture of the colonic microbiome. Ileal biotopes subject to intense peristalsis are even further apart in their composition. However, access to these sites is complex. Most data can only be acquired from stoma effluents or biopsies obtained during surgery, collected from patients that are often subjected to long-term antibiotic therapy. In infants and children, several species, including *B. bifidum*, *B. breve* and *B. dentium*, can be found in stoma effluents (Romond et al. 2011). Levels vary from 5 to 7 log cfu/mL of stoma effluent. However, these figures probably overestimate the reality of the situation, as fluxes are very low compared to normal peristalsis.

In adults, endoscopic brush collection detects similar quantities of bifidobacteria by metagenomics between ileum, cecum and sigmoid colon (around 1% of the collected microbiome) (Nishino et al. 2018). Culture techniques also give similar results between proximal colon and rectum from biopsies obtained under colonoscopy (Poxton et al. 1997). Under these sampling conditions, the presence of cell-adherent bifidobacteria is rare in the ileal area, which differentiates them from lactobacilli (Fakhry et al. 2009).

A study based on emergency cases of bowel resections – that is to say, without intestinal preparation by antibiotic therapy or barium lavage – shows that bifidobacteria are significantly less abundant on ileal mucosa than colonic mucosa (Ahmed et al. 2007). The low colonization by bifidobacteria of digesters mimicking ileal peristalsis, compared with those mimicking colonic peristalsis, corroborates these findings in intestinal modeling systems (Meddah et al. 2001). There would, therefore, be a gradual colonization by bifidobacteria between the poorly colonized ileum and the invaded colon, even at an older age. On the contrary, stools contain fewer and fewer bifidobacteria over time, at less than 0.1% of the microbiota in centenarians (Kato et al. 2017). Species also vary, with *B. dentium* becoming established late in life, while *B. breve* is rare in the elderly. Only *B. longum* is commonly found throughout life (Figure 8.2).

Apart from the intestine, other biotopes can be colonized by bifidobacteria, such as saliva or vaginal secretions. Saliva colonization is highest if teeth are decayed (Beighton et al. 2010; Kaur et al. 2013). Species-wise, we have *B. longum*, *B. dentium* and *B. adolescentis* (Hojo et al. 2007; Barroso et al. 2015; Eriksson et al. 2017). As for the vagina, we find the species are isolated in the intestine, in particular, *B. longum* and *B. breve* (Freitas and Hill 2017, 2018). Moreover, it is not possible to differentiate strains according to their vaginal or intestinal origin. On the other hand, the origin of intestinal bifidobacteria would not be linked to contamination by vaginal bifidobacteria, but by those hosted in the mother's intestine.

Lastly, bifidobacteria can be isolated from breast milk. Their origin gives rise to several hypotheses, from simple contamination by contact with the infant's soiled hands to translocation by catching intestinal bifidobacteria by maternal dendritic cells located in the intestinal wall, and release into the mammary gland (Rodríguez 2014; Kordy et al. 2020). The latter hypothesis is supported by the identity of strains isolated from the mother's milk and her child's intestine (Eshaghi et al. 2017). However, metagenomic analyses indicate that, in addition to species considered to be of human origin (*B. adolescentis*, *B. bifidum*, *B. breve*, *B. dentium* and *B. longum*), species associated with mammalian biotopes, such as

B. pseudolongum and *B. crudilactis*, can also be found (Lugli et al. 2020), indicating that catching dietary bifidobacteria is possible.

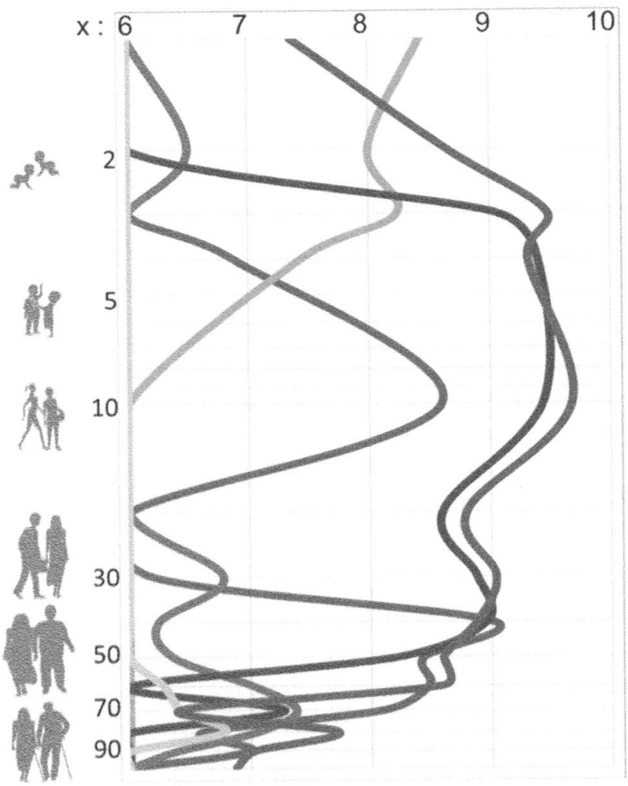

Figure 8.2. *Age-dependent changes in quantities of bifidobacteria from different groups and species. Red: B. adolescentis group; blue: B. bifidum; light green: B. breve; burgundy: B. catenulatum group; yellow: B. dentium; dark green: B. longum group. X: number of bacteria found per gram of dry stool, expressed in log10, according to (Arboleya et al. 2016; Kato et al. 2017)*

8.1.3. *Safety and health activity*

Bifidobacteria are renowned for their harmlessness. However, this assertion needs to be clarified. Their anaerobic nature restricts their isolation, and few medical analysis laboratories in towns or hospitals have the skills to cultivate them. In addition, the mastery of molecular techniques is not enough to make a diagnosis, as the mere detection of bacterial DNA cannot justify the introduction of antibiotic therapy.

In fact, *B. dentium*, formerly known as *Actinomyces eriksonii*, has been found in cases of actinomycosis of the sinuses, with possible spread to the brain and development of abscesses (Pulverer et al. 2003). Other bifidobacteria (*B. adolescentis* and *B. breve*) have also been isolated from cerebrospinal fluids in meningitis (Schwarz et al. 1993; Nakazawa et al. 1996).

These infections should be viewed in the context of oral carriage of bifidobacteria. Bifidobacteria proliferation is associated with the presence of cavities (Nakajo et al. 2010; Valdez et al. 2016; Manome et al. 2019). Tooth loss, on the other hand, is accompanied by a change in the carriage of bifidobacteria species, in particular, the reduced prevalence of *B. dentium* (Mantzourani et al. 2010).

In addition to diffusions from the oral sphere, it is possible that dietary bifidobacteria may enter the bloodstream from the gastrointestinal tract and induce symptoms (Ohishi et al. 2010; Bertelli et al. 2015; Zbinden et al. 2015; Esaiassen et al. 2016; Pruccoli et al. 2019). Bacteremias in infants are mainly caused by probiotic strains. Even if cases are rare in relation to use in pediatrics and neonatal intensive care, it is necessary to question the relevance of considerable intake of live bifidobacterial, compared with the doses delivered by breastfeeding. Bifidobacteria concentrations are highly variable throughout the day, when the levels of each species are measured. For example, it can be estimated that the highest quantity of *B. longum* that can be ingested by a 3 kg child is 1 million per day, distributed in variable doses over the day, some of which are dead (personal data). However, full-term or premature infants receive between 50 and 100 million probiotic bifidobacteria in one or two doses. Live bacteria pressure is probably exaggerated and may explain the bacteremia observed.

Furthermore, bifidobacterial bacteriuria are reported in patients in intensive care units, with urinary catheters or in elderly patients with recurrent urinary tract infections (Barberis et al. 2012; Guevara Nuñez et al. 2017; Rozenfeld et al. 2018).

Bifidobacteria can also cause infections in the gastrointestinal tract (Tena et al. 2014; Suwantarat et al. 2014). They can also colonize the stomach in cases of gastritis, but so far without any proven pathogenic effect (Mattarelli et al. 2014). On the other hand, their role in chronic inflammatory diseases is controversial. Several works highlight their over-representation by metagenomic analysis in patients with ulcerative colitis (Forbes et al. 2018). However, opposite results attest to an under-representation of bifidobacteria, more specifically *B. bifidum* (Duranti et al. 2016). Quantitative analyses (culture, qPCR) corroborate their decrease in the stools of patients with inflammatory bowel disease (Distrutti et al. 2016; Salem et al. 2019). This being said, the enumeration of localized bifidobacteria in biopsies from patients with colitis, during active and inactive phases, does not provide a clear-cut answer, as the quantities enumerated are variable and independent of the patient's condition (Poxton et al. 1997). Ultimately, in patients suffering from irritable bowel syndrome, the relative proportions of bifidobacteria and their role have yet to be defined, with metagenomic analyses yielding contradictory results, particularly in the small intestine (Kerckhoffs et al. 2009; Dlugosz et al. 2015). While there is consensus regarding the existence of an abnormal biotope in chronic inflammatory bowel disease, the role of bifidobacteria remains to be substantiated.

The healthy activity of bifidobacteria is also difficult to pin down in terms of those colonizing the host. Bacterial populations vary quantitatively as a function of nutrients and environmental stress. These fluctuations can serve as markers of imbalance, but demonstrating their impact is complex. The use of gnotobiotic animals has, however, enabled us to identify a regulatory function for bacterial translocation, particularly of enterobacteria, from the intestine to the inner environment (Yamazaki et al. 1985; Romond et al. 1997a, 2008). This function is only shared with a mucin-degrading strain of *Peptostreptococcus* (Carlstedt-Duke et al. 1986; Herías et al. 1998). This specific control depends on T-cell stimulation (Yamazaki et al. 1985, 1991). Moreover, the implantation of certain strains of *Bifidobacterium* protects animals from toxin production by *E. coli* (Yoshimura et al. 2010).

On the other hand, the human health activity of orally produced bifidobacteria is easier to determine.

There are several categories of effects, depending on the pathologies considered:

– longer periods of remission in inflammatory bowel disease (Saez-Lara et al. 2015);

– improved quality of life (including reduced intestinal pain) in irritable bowel syndrome (Guglielmetti et al. 2011; Andresen et al. 2020; Pratt and Campbell 2020);

– prevention of colds during periods of stress (Langkamp-Henken et al. 2015);

– weight reduction in obesity (Michael et al. 2020).

However, these effects cannot be generalized to the genus as a whole, nor to individual species. In most cases, these are strain-specific effects. Few mechanisms are known. Only the *B. bifidum* strain MIMBb75, currently on the market as a medical device (Kijimea), has a proven mechanism, based on a BopA lipoprotein present on the surface of the bifid wall, which has a high capacity to adhere to intestinal cells (Gleinser et al. 2012). However, even if the strain's clinical effects are proven, it is impossible to conclude that efficacy is linked to this lipoprotein alone, exposed on the surface of bifid cells.

8.2. Production of probiotics and metabiotics/postbiotics

8.2.1. *Nutritional requirements of bifidobacteria: use of complex oligosaccharides and production of exopolysaccharides*

Bifidobacteria are prototrophic microorganisms that require carbon in organic form. They can also use mineral sources of nitrogen. Several synthetic media have been developed, containing either ammonium acetate (Hassinen et al. 1951) or ammonium sulfate (Gyllenberg and Niemelä 1959) as the sole source of nitrogen. However, *B. bifidum* strains belonging to variant b cannot increase in the absence of growth factors containing N-acetylglucosamine (bifidogenic factor I), the latter being incorporated into the peptidoglycan (György et al. 1954). Oligosaccharides containing N-acetylglucosamine are found in breast milk as free oligosaccharides (gynolactose described by Polonovski and Lespagnol (1930)). More complex sources are mucins (O-glycosylated proteins) and N-glycosylated glycoproteins, some of which are present in the intestinal lumen (Romond et al. 1990a, 1990b). Only a few *B. bifidum* strains can utilize N-acetylglucosamine units without prior release.

The question of the use of oligosaccharides from human milk by other species of bifidobacteria has long been debated. Research has focused on the dominant species in the infant's intestine, *B. bifidum*, *B. breve* and *B. longum*, while considering the two sub-species in the latter, *B. longum subsp. longum* and *subsp. infantis*. Genomic and transcriptomic approaches have clarified the subject. They confirm that the *B. bifidum* species is not homogeneous in terms of oligosaccharide utilization, with some strains not using fucosyl-lactose or sialyl-lactose (Figure 8.3) (Garrido et al. 2015). In contrast, *B. longum subsp. infantis*, which uses ABC transporters to internalize oligosaccharides before degrading them in the cytosol using glycosyl hydrolases, shows a homogeneous transcriptomic response within the species. In particular, its dual transport system enables it to catch a wide range of fucosylated oligosaccharides (Sakanaka et al. 2019).

Comparative genomic analysis, focusing on the use of basic oligosaccharide structures (lactosamine, lactotetraose and lactoneotetraose), allows us to construct a similar degradation scheme in *B. breve*: internalization by ABC transporters, degradation by hydrolases, with the released monomers integrated into the pentose (glucose), Leloir (galactose) or osamine (N-acetylglucosamine) cycle.

Lastly, *B. longum subsp. longum* has a variable ability to utilize oligosaccharides from women's milk, as the gene encoding the enzyme lacto-N-biosidase (LnbX) is not present in all genome strains (Gotoh et al. 2018). It seems that, like *B. bifidum*, *B. longum subsp. longum* LnbX+ strains employ extracellular hydrolases to degrade oligosaccharides in women's milk; whereas, *B. breve*, *B. longum subsp. infantis* and *B. longum subsp. longum* LnbX– strains depend on transporters that are capable of catching oligosaccharides prior to intracellular use (James et al. 2016).

Furthermore, the release of sub-units from free oligosaccharides or mucins by *B. bifidum* is beneficial to *B. breve* and *B. longum* species (Turroni et al. 2011; Egan et al. 2014; Gotoh et al. 2018). *B. bifidum* is thought to play a pivotal role in the nutritional chain, providing small saccharide units that the membrane transporters of other species can internalize. Thus, it is mainly *B. bifidum* that degrades O-glycosylated glycoproteins such as mucins (Killer and Marounek 2011; Kim et al. 2013; Katoh et al. 2017). In contrast, N-glycosylated glycoproteins (including lactoferrin and kappa-casein) are accessible substrates for *B. breve* and *B. longum* species, whose oligosaccharide structures are released by specific endoglycosidases (Oda et al. 2014; Karav et al. 2016, 2017; Katoh et al. 2020).

In addition to their catabolic capacities, bifidobacteria are capable of producing exopolysaccharides (EPS) and polysaccharides closely linked to the cell wall that have properties of interest in human health. Currently, *B. adolescentis*, *B. animalis*, *B. breve*, *B. dentium*, *B. longum*, *B. pseudocatenulatum* and *B. thermophilum* display genes in their genome encoding glycansucrases and glycosyltransferases required for EPS synthesis (Pyclik et al. 2020). Nine gene clusters have been identified within bifidobacterial species, representing 44 genes per strain involved in EPS synthesis. However, bifidobacteria generally only produce one type of EPS, composed of three major units (D-glucose, D-galactose and L-rhamnose), with a molecular mass ranging from 10 to 1,000 kDa. This being said, chain length and the molar ratio between units depend on the different bifidobacterial-producing species. EPS-producing bifidobacteria are distinguished by the mucous appearance of their colonies. EPS production is influenced by the composition of the medium, milk being an unfavorable choice.

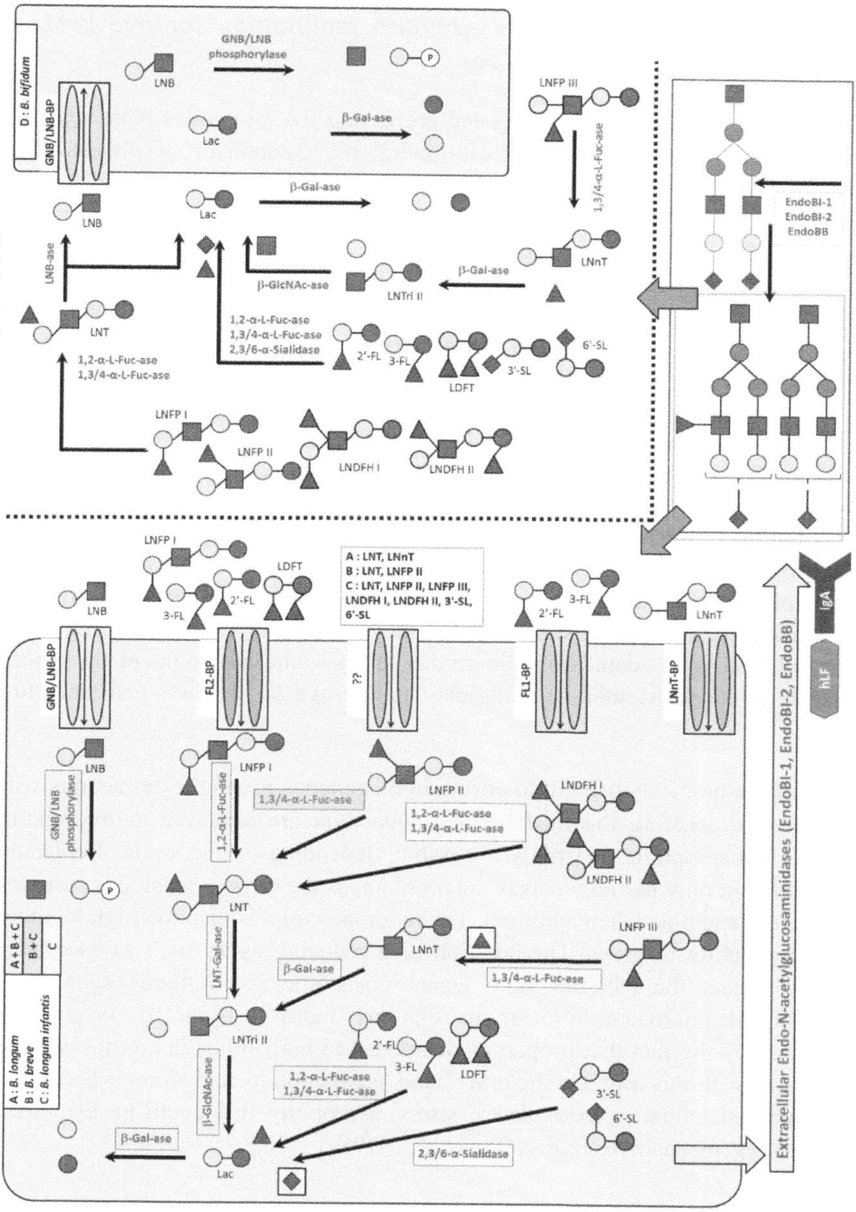

Figure 8.3. *Bifid strategies of oligosaccharide and glycoprotein catabolism in women's milk (according to Garrido et al. (2015), James et al. (2016), Karav et al. (2016, 2017) and Gotoh et al. (2018))*

8.2.2. *Concentration and preservation technology for live probiotics (strain selection, encapsulation)*

Technologies for concentrating and preserving live probiotics must adapt to the acidification of the environment, following the catabolism of oligosaccharides (Figure 8.4); the most important of these are the pentose shunt via F6PPK (glucose catabolism) and the Leloir cycle (galactose catabolism). These two metabolic pathways involved in lactose degradation are complemented by the amino sugar catabolic pathway (galacto-N-biose/lacto-N-biose cycle or GNB/LNB) (De Bruyn et al. 2013). This pathway, which is an alternative to galactose degradation, was first highlighted in *B. longum* and *B. bifidum* (Fushinobu 2010), then in several bifid species, including *B. breve* (James et al. 2016).

In order to optimize biomass, pH control is crucial. However, controlling pH consequentially increases salt concentration and osmotic pressure, which ultimately bends the growth curve. In the case of *B. bifidum*, *B. longum* and *B. breve* species, the addition of whey proteins or hydrolysates (controlled hydrolysis) is a solution that provides greater buffering capacity than ammonium salts and stimulates growth by supplying bifidogenic factors (N-glycosylated proteins). Other growth factors can be considered, such as yeast extract, but they have no effect on buffering capacity. One solution is to adapt probiotic strains to an acidic environment by subculture (Wei et al. 2019). Membrane modifications improve *B. longum*'s resistance to acid stress.

Furthermore, it is difficult to ensure homogeneous mixing in fermenters without partially oxygenating the medium. Bifidobacteria are sensitive to oxygen, being either microaerophilic or strictly anaerobic, depending on the strain. Replacing air with nitrogen only partially solves the problem, as the oxygen residue is sufficient to turn methylene blue when bubbling. The redox potential is then too high to allow the bifid population to grow. The addition of a reducing agent such as ascorbic acid cannot protect the medium from anaerobiosis under continuous agitation. The selection of microaerophilic strains that are more resistant to oxygenation is hampered by the fact that properties of interest to human health are not necessarily associated with this trait. On the other hand, adaptation to acid stress is accompanied by improved tolerance to oxidative stress, a property that could be exploited for highly oxygen-sensitive strains (Wei et al. 2019).

These media, dedicated to the final concentration of bacteria, must also be easily removed during preparation of the frozen bacterial concentrate. The classical method is to sediment the bacteria by centrifugation. If washing is necessary, the buffer should be reduced beforehand and a reducing agent (such as ascorbic acid) added to prevent oxidation, which is often fatal to the survival of bifidobacteria. Regardless of the precautions taken, frozen concentrates rarely exceed 10^{11} revivable bacteria per gram of concentrate, requiring pre-culture prior to food inoculation.

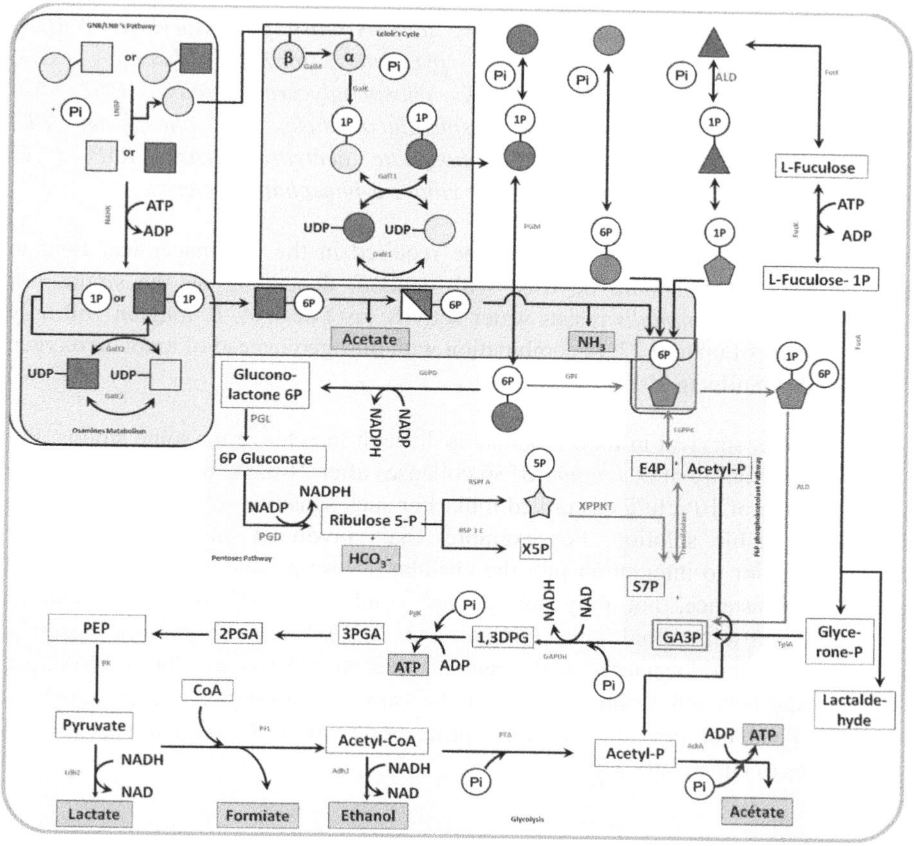

Figure 8.4. *Oligosaccharide catabolism by bifidobacteria, metabolic pathways and consequences (according to Pokusaeva et al. (2011), De Bruyn et al. (2013) and James et al. (2016))*

COMMENT ON FIGURE 8.4.– *Abbreviations: E4P: erythrose 4-P; S7P: sedoheptulose 7-P; GA3P: glyceraldehyde 3-P; X5P: xylulose 5-P; PEP: phosphoenol-pyruvate; 2PGA: glycerate 2-P; 3PGA: glycerate 3-P; 1,3DPG: 1,3 diphosphoglycerate; AckA: acetate kinase; Adh2: aldehyde-alcohol dehydrogenase 2; ALD: fructose-bisphosphate aldolase; F6PPK: fructose-6-phosphoketolase; FucA: L-fuculose-1P aldolase; FucI: L-fucose isomerase; FucK: L-fuculose kinase; G6PD: glucose-6-phosphate dehydrogenase; GalE1: UDP-glucose 4-epimerase; GalK: galactokinase; GalM: galactose mutarotase; GalT: galactose-1-phosphate uridylyl-transferase; GAPDH: glyceraldehyde-3-phosphate dehydrogenase C; Ldh2: lactate dehydrogenase 2; LNBP: lacto-N-biose phosphorylase; NAHK: N-acetyl-hexosamine1-kinase; PFL: pyruvate formate lyase; PGD: phosphogluconate dehydrogenase; PgK: phosphoglycerate kinase; PGL: 6-P phosphogluconolactonase; PGM: phosphoglucomutase; Pi: phosphate; PK: pyruvate kinase; PTA: acetyl-CoA phosphate acetyltransferase; R5P 3 E: ribulose-5-Phosphate 3-epimerase; R5PI: ribose-5-phosphate isomerase.*

Freeze-drying, a solution that may be required in the pharmaceutical field to avoid genetic drift, should be used with caution, depending on the strain. For example, while *B. animalis* resists water activity (aw) of 0.44, *B. longum* will only tolerate an aw of up to 0.22, in combination with nitrogen instead of ambient oxygen (Celik and O'Sullivan 2013).

Ultimately, survival in food products is difficult to achieve for some strains. For example, the vitality of *B. longum* B536 collapses after 21 days in fermented milk at a residual rate of 10^4 cfu/g of curdled milk. For such sensitive strains, encapsulation may be a viable solution. For example, trials involving the encapsulation of *B. bifidum* prior to integration into the cheddar cheese production process showed improved resistance, not only for storage conditions, but also for conditions mimicking gastrointestinal transit (Afzaal et al. 2020). Encapsulation also preserves the vitality of bifidobacteria in the human intestine (Cui et al. 2007). However, comparisons between strains show that the same encapsulation system (such as alginate) does not provide the same protection within *B. longum* species and sub-species (Yeung et al. 2016).

8.3. Industrial applications

8.3.1. *Fermentation manufacturers and characteristics of marketed products*

Two types of bifidobacterial probiotic strains are used in foods and dietary supplements in Europe: on the one hand, strains for which there is abundant literature, such as *Bifidobacterium animalis subsp. lactis* BB12 (Chr. Hansen), HN019 and Bl-04 (DuPont Nutritional and Health), *B. longum* B536 and *B. breve* M-16V (Morinaga), *B. bifidum* MIMBb75 (Synformulas GmbH) or *B. bifidum* GBN4 (BIFIDO Ltd); on the other hand, strains with qualified presumption of safety (QPS) status (Sarkar and Mandal 2016; Wong et al. 2019). Strains are produced under contract by general fermenters (Bioprox, Genibio, Lallemand, Lesaffre and Chr. Hansen) for formulas developed by their customers (fermented infant milks, yoghurts and dietary supplements).

In Europe, bifidobacteria have been granted QPS status by EFSA since 2007. This status, granted at the genus level, was confirmed in 2017 (Ricci et al. 2018). Non-exemption from GRAS (generally recognized as safe) status is a procedure that a company can apply for from the U.S. Food and Drug Administration. This approach has been implemented by Morinaga for its BB536 and MV16 strains, for use in infant nutrition (GRAS Notification No. GRN 000268, 000453 and 000454).

By contrast, to-date, no health claims have been issued under European regulation (EC) No. 1924/2006 for probiotic bifidobacteria or food products fermented with these bacteria. One of the main reasons for this is that the effects of probiotics have been demonstrated in patients suffering from pathologies, a condition which de facto excludes the use of studies to demonstrate benefits in healthy people.

The current solution for food supplements is to supplement probiotic formulas with vitamins and minerals, with authorized health claims (Regulation (EU) No. 1047/2012). This approach not only improves the stability of bifidobacterial strains (for example, ascorbic acid or vitamin C), but also authorizes benefit claims linked to the final concentration of these compounds.

Whilst the aim of the regulations was to clarify labeling, it has to be said that there is still a degree of vagueness about the characteristics of products on the market. The data available are obtained from heterogeneous products. For example, in vitro tests are carried out on media widely used in laboratories (Man-Rogosa-Sharpe or Columbia media), which are not adapted to human food and production conditions. However, bacterial wall composition is influenced by culture conditions. Moreover, while phenotypical expression may vary according to cultivation conditions, genetic drift must also be considered.

The *Bifidobacterium animalis subsp. lactis* strain, which has widely been used for around thirty years, is an interesting example. The first uses of *B. animalis* can be traced back to the late 1980s, when milks fermented with *Bifidobacterium* strains were launched in France. The first was BA fermented milk, the result of a collaboration between Morinaga and Saint-Hubert. This product initially included the *B. longum* BB536 strain. Success followed, and the BSN group (Danone) soon launched a competing fermented milk product using a *B. animalis* strain, first identified as *B. longum* (the carbohydrate fermentation profiles are similar). The first samples also included a strain of *B. bifidum*. This was the product that flooded the market under the Bio brand, then Activia. Equivalent products fermented with the *B. animalis* BB12 strain were marketed in various European countries. Isolation of bifidobacteria from fermented milks on the European market led to reclassification as a sub-species (Masco et al. 2004). Little variability has been found in the pangenome, with only three open reading frames differentiating strains of *B. animalis subsp. lactis* (Milani et al. 2013). The study of nucleotide polymorphism, however, shows divergence between genomic groups (Lomonaco et al. 2015). Variability in the number of transposases suggests that the genome may yet evolve, as may its phenotypic expression.

8.3.2. *Live biotechnology products and European Pharmacopoeia constraints*

Preventing genetic drift is a strategy implemented by the pharmaceutical vaccine industry. It involves conserving strains in primary banks (long-term conservation without replanting) and secondary banks (or working banks), from which productions are prepared.

While it is understandable that a medicinal approach is needed for virulent strains, the extra cost involved in using a specific workshop for production, inherent in pharmaceutical processes, is difficult to conceive in the food industry.

Moreover, developing a pharmaceutical record for a probiotic strain is a lengthy and costly procedure. The record must follow the principles of the International Council for Harmonisation (ICH), based on the Common Technical Document (CTD). Based on data defining the active substance (the bacteria) and the final formula (excipients), the CTD must provide evidence of the formula's safety and efficacy in given pathologies.

Up until 2017, a drug-status probiotic (Lyo Bifidus) was still available. However, the Haute Autorité de Santé (the French Health Authority) dealt a fatal blow to this type of product, pointing out the low level of benefit provided, which was attributed to the absence of recent clinical studies.

In addition, the removal of live microorganisms from the medical device record, when the legislation was recast (Regulation (EU) 2017/745), now prevents this regulatory basis from being changed, which was nevertheless appropriate for probiotics supplied in pharmaceutical form (for example, capsules). The Kijimea medical device must therefore change its regulatory framework.

8.4. Prospects for development and innovation: metabiotics/postbiotics (bifidobacterial lipoproteins), therapeutic targets

The supernatants from probiotic cultures contain compounds that are derived from bacteria (metabiotics or postbiotics, depending on the nomenclature proposed), whose potential activity in human health is still under-exploited. In the case of bifidobacteria, the first indications that abiotic compounds ("privative and bio: living or being alive") were effective antidiarrheals came about in the 1990s. Analysis of the microflora of healthy volunteers treated with erythromycin showed that protection, induced by taking milk fermented with *B. longum* BB536, was associated with the collapse of intestinal clostridia (Colombel et al. 1987). However, not all whey derived from the fermentation of milk by several strains of bifidobacteria are clostridia repressors in vivo (Romond et al. 1997b, 1998). Furthermore, clostridia population regulation is not achieved if strains are cultured on conventional media (SRM, heart-brain or Columbia). Other molecules modulating intestinal permeability are then isolated (Heuvelin et al. 2010).

Although the biological efficacy of whey appears to be linked to high-molecular-weight molecules (>100 kDa), initial attempts to isolate and characterize the related molecules of activity have not been successful. A first hypothesis is that the efficiency-boosting products are proteins, given the inactivation by heat (75°C, 5 min) and sensitivity to oxidation of anti-clostridia activities (Mullié et al. 2002). The presence of a 43–45 kDa protein with a CHAP domain (cysteine, histidine-dependent amidohydrolases/peptidases; amino acid position 333–459), whose known function is the hydrolysis of bacterial wall peptidoglycan, was not confirmed until the genomes were available (Scuotto et al. 2014). The CHAP domain is found in the *B. longum* peptide chain from position 324 to 451. A second feature of the protein is a lipobox structure (Scuotto et al. 2016).

The current question is how many species of bifidobacteria synthesize a same type of lipoprotein, potentially modulating the microbiota for further exploitation. The availability of bifidobacterial genome databases makes it possible to check in silico whether the sequence is universal in the genus *Bifidobacterium*, and whether it is systematically equipped with lipobox and CHAP functions. After listing validated species of bifidobacteria (research on typical species strains), the method used consists of extracting proteins of similar structure, organization and functionality to the lipoproteins of *B. longum* CBi0307 and *B. breve* C50, using the BLAST (Basic Local Alignment Search Tool). The databases explored are EMBL-EBI, Uniprot, NCBI and SIB Prosite. Protein selection criteria include belonging to the group of proteins containing a CHAP domain, as well as the names "amidase" and "secreted antigen", a protein chain length in the 420–520 amino acid range, and the existence of a signal peptide that may in some cases contain the predictive structure leading to a lipoprotein (or lipobox) (LipoP and Prosite software). Two lipoprotein sub-groups were identified, one including *B. bifidum* and characterized by the absence of a lipobox, the other including *B. longum* and *B. breve*. Multiple sequence alignment programs (Clustal Omega, COBALT), used to compare the proteins identified, revealed phylogenetic similarities, generally with species in hosts other than humans.

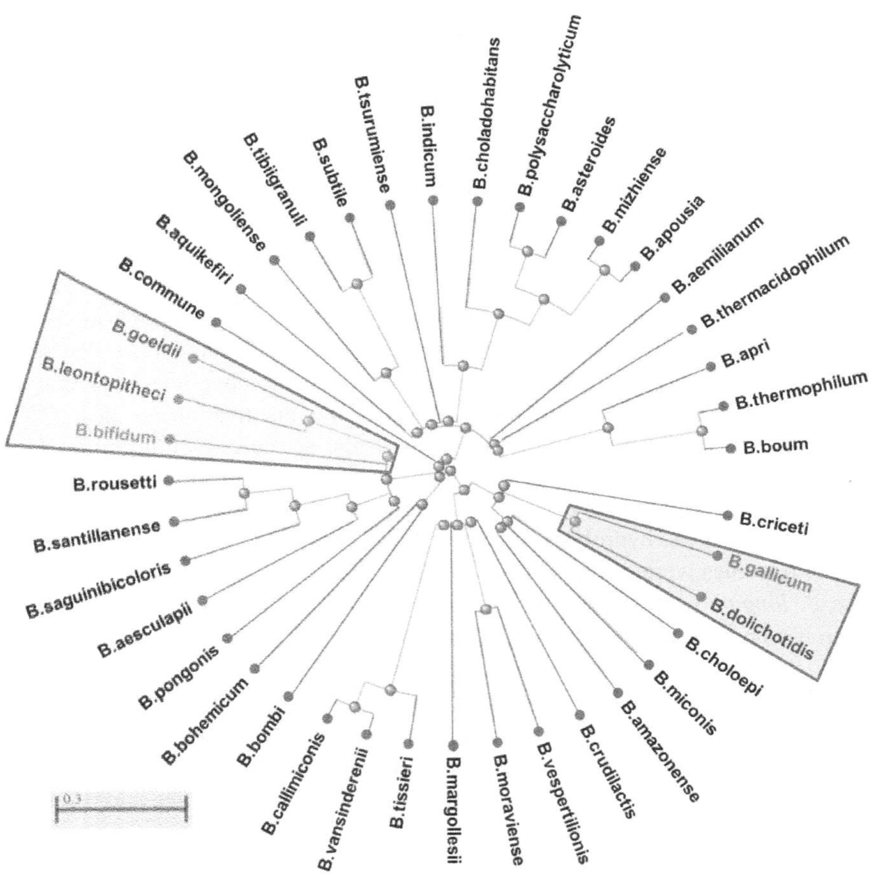

Figure 8.5. *Lipoprotein dendrogram of lipobox-free Bifidobacterium (sequences from UniprotKB or NCBI correspond to those of type strains)*

Thus, the *B. bifidum* protein is closer to those of *B. goeldii* and *B. leonpitheci*, the bifidobacterial species described in the Goeldi's marmoset and the golden-headed lion tamarin, than to those of *B. gallicum*, another human-hosted species (Figure 8.5).

The exceptions to this rule are the lipoproteins of *B. adolescentis* and *B. faecale*, on the one hand, and *B. catenulatum* and *B. pseudocatenulatum*, on the other hand (Figure 8.6).

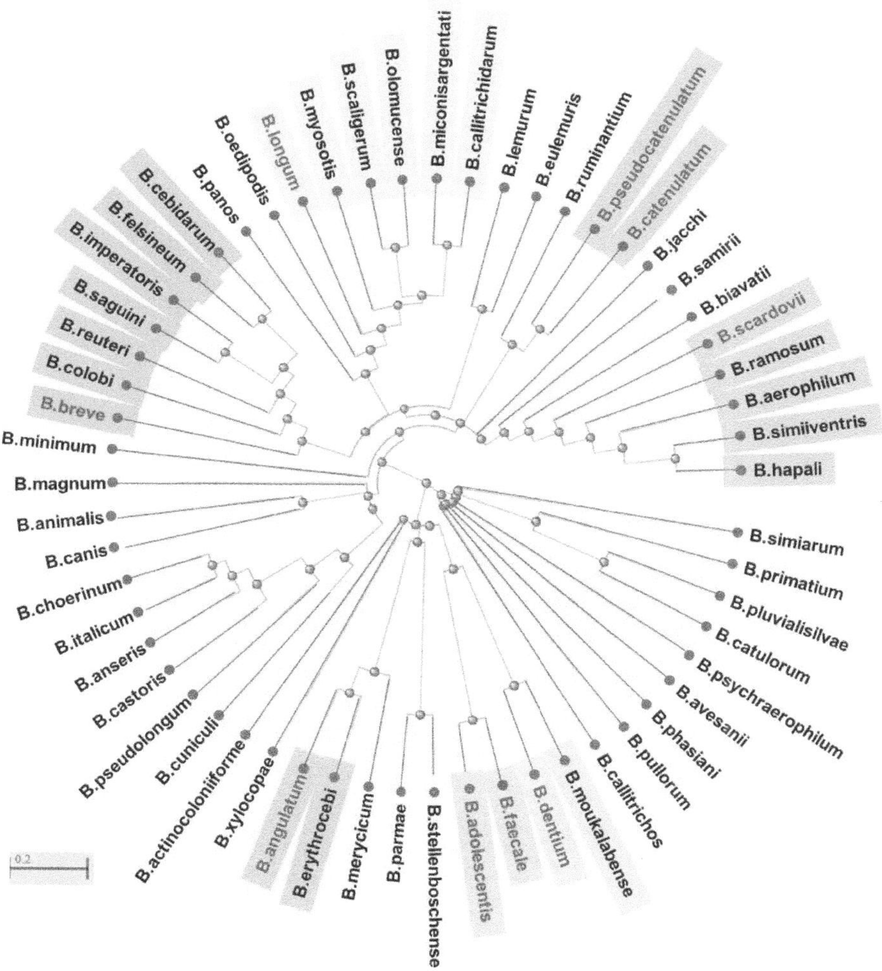

Figure 8.6. *Bifidobacterium lipoprotein dendrogram, including a lipobox zone (sequences from UniprotKB or NCBI correspond to those of type strains)*

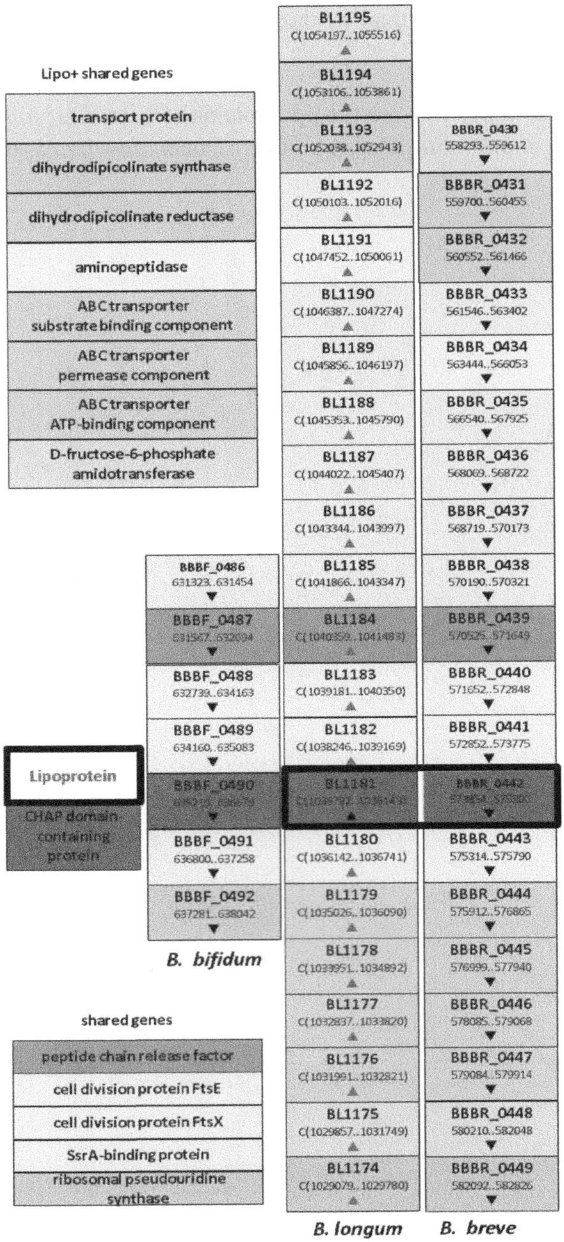

Figure 8.7. *Environment of the BBF_0490, BL1181 and BBBR_0442 genes encoding the protein sequence in* B. bifidum, B. longum *and* B. breve, *respectively*

As lipobox detection is limited by the number of sequences around an NH2-terminal cysteine described in the databases, an additional study focuses on analyzing the gene environment of the gene of interest, using genome explorers such as KEGG, *Genes Genome Browser* and *EnsemblBacteria*.

Proteins from typical strains of *B. bifidum*, *B. breve* and *B. longum* are selected. Figure 8.7 illustrates the differences observed between the gene environments.

The open reading frames do not cover the same number of genes. Several genes encoding ABC transporters and ribosomal pseudouridine synthase are missing from the *B. bifidum* genome.

The direct environment of the gene indicates that the secretion capacity in the culture medium is reduced compared to *B. longum* and *B. breve*, which may explain why *B. bifidum* culture supernatants do not induce any disruption of the microbiota when administered to animals (Romond et al. 1997b).

B. bifidum also has the BopA wall lipoprotein, whose sequence is far removed from those of *B. longum* and *B. breve* (Guglielmetti et al. 2008; Gleinser et al. 2012). It is only localized in the wall and therefore cannot participate in the effects of culture supernatants.

In reality, *B. bifidum* possesses several non-secreted CHAP-functional proteins, such as the TgaA enzyme, which has two areas of activity: lytic murein transglycosylase (LT) and CHAP activity (Guglielmetti et al. 2014).

The non-lipobox sub-group is therefore currently less accessible for the exploitation of lipoprotein-like proteins from *B. longum* and *B. breve*.

However, while the structures are common to the lipoproteins of these two species, their sequences are far apart, as shown in the dendrogram (Figure 8.8).

Proteins from *B. breve* and *B. longum* can thus be distinguished. On the other hand, protein sequences from *B. longum* are not grouped according to sub-species.

Sequence alignment analysis also shows distinct patterns in the signal peptide (Figure 8.9) and in the sequence preceding the CHAP domain (Figure 8.10).

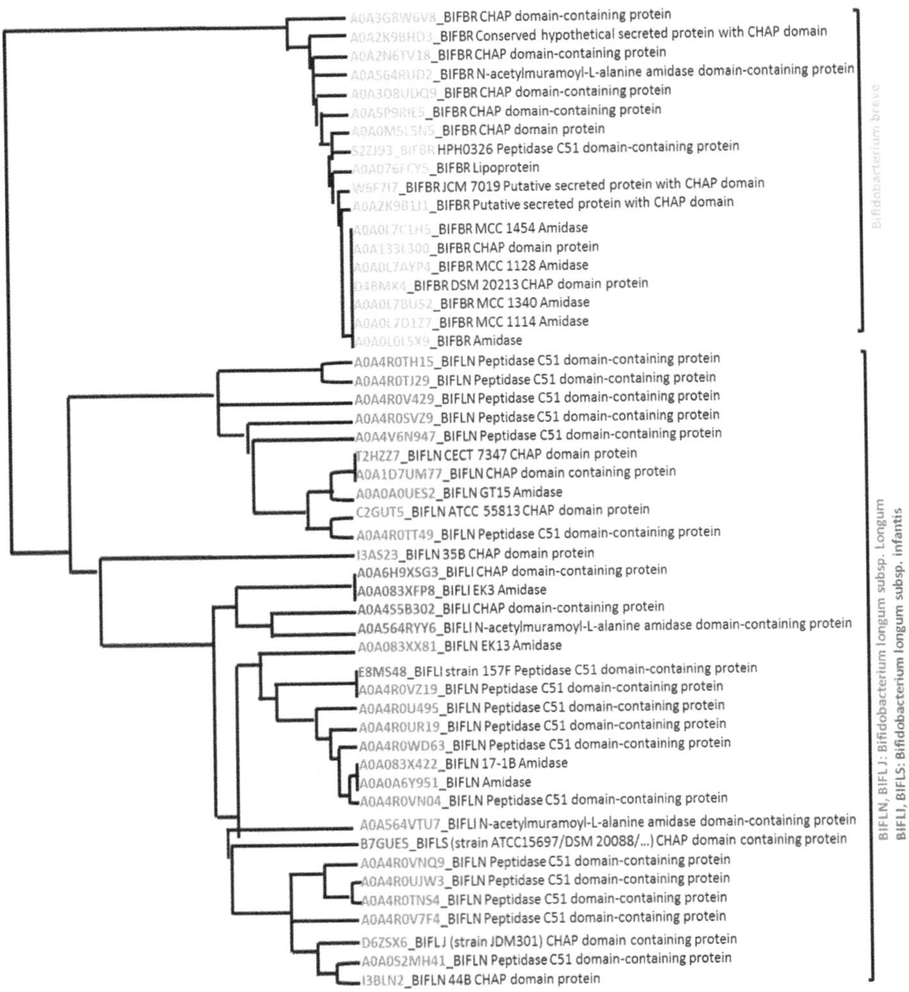

Figure 8.8. *Lipoprotein dendrogram of* B. breve *(yellow),* B. longum supsp. longum *(light green) and* B. longum supsp. infantis *(dark green) (sequences from UniprotKB)*

Gas chromatography–mass spectrometry analysis confirms the presence of acylated chains (*B. breve*: C20 and C22; *B. longum*: C14, C16, C22) in the protein extracts (Scuotto et al. 2014, 2016). The acylated chains are most likely grafted onto the cysteine of the AIC structure of the LAACAASAIC lipobox from *B. breve* and the AAC structure of the LAALVAAAAC lipobox from *B. longum* (Figure 8.9).

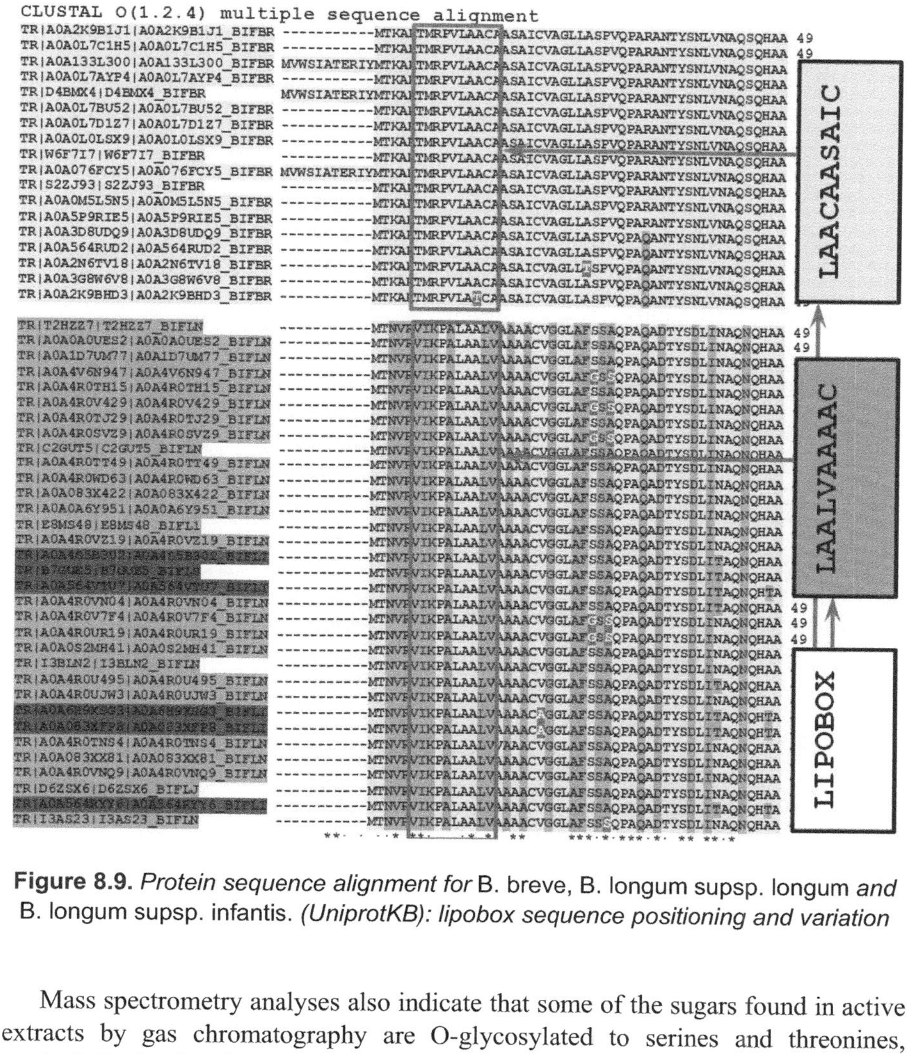

Figure 8.9. *Protein sequence alignment for* B. breve, B. longum supsp. longum *and* B. longum supsp. infantis. *(UniprotKB): lipobox sequence positioning and variation*

Mass spectrometry analyses also indicate that some of the sugars found in active extracts by gas chromatography are O-glycosylated to serines and threonines, particularly in the CHAP area (personal data). The presence of oligosaccharide chains would explain the reactivity of lipoproteins with galectin-1 (Scuotto et al. 2014, 2016). We can thus reconstruct the structural model of O-glycosylated lipoprotein in monomeric form (Figure 8.11). At pH 4 (unregulated lacto-fermentation), these amphiphilic O-glycosylated lipoproteins combine, giving rise to aggregates of over 600 kDa.

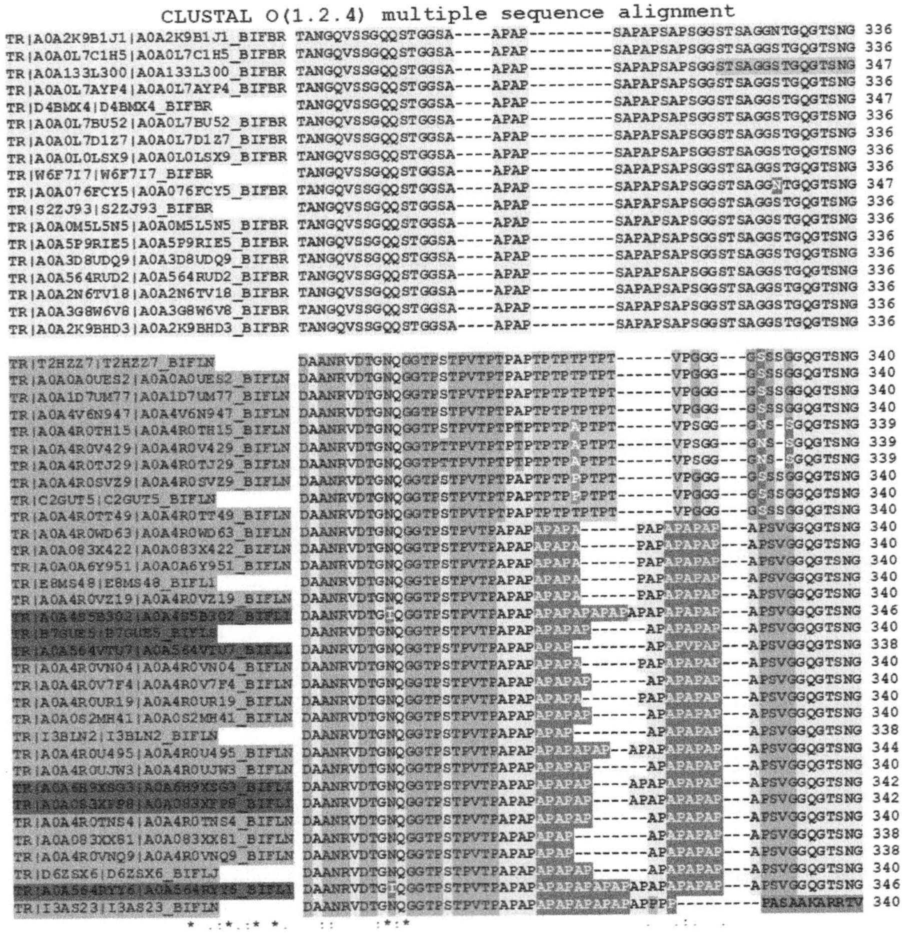

Figure 8.10. *Protein sequence alignment for* B. breve, B. longum supsp. longum *and* B. longum supsp. infantis. *(UniprotKB): high variability zone preceding CHAP domain*

The biological activity spectrum of lipoproteins is based on aggregates or monomeric forms, depending on the fermentation pH. Apart from clostridia, in simulated intestinal biotope or animal models (conventional or gnotobiotic), we observe a repression of *Bacteroides* of the *fragilis* group and a stimulation of bifidobacteria (Romond et al. 1997b; Meddah et al. 2001; Mullié et al. 2002; Scuotto et al. 2016). Aggregates are also capable of regulating bacterial translocation, particularly that of *Clostridium perfringens* (Blareau et al. 2008; Petay et al. 2008).

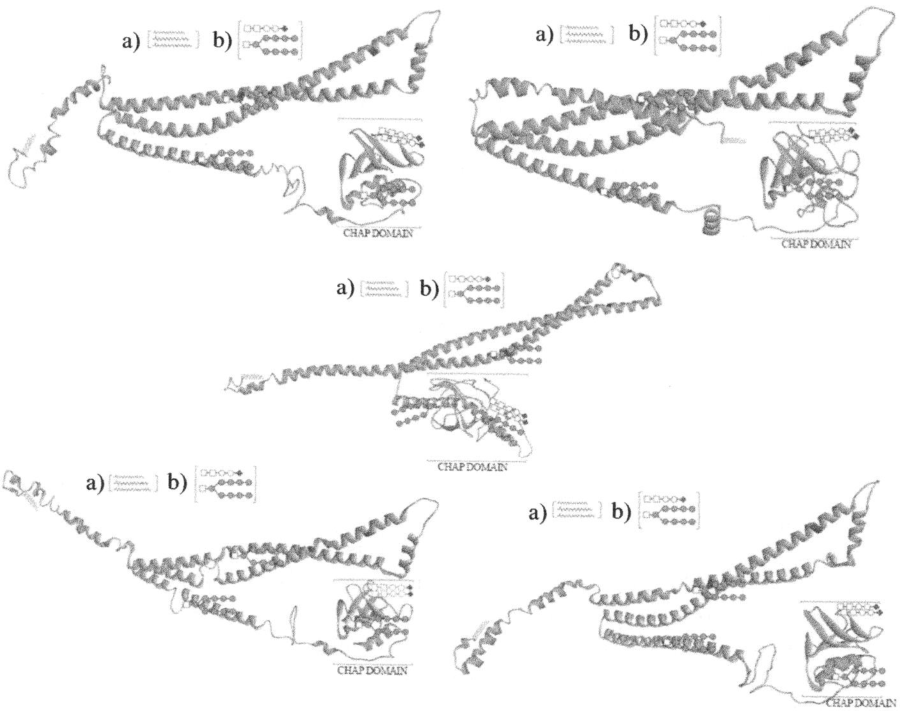

Figure 8.11. *Lipoprotein structure models of* B. longum. *(a) Lipids C-18, C20, C22 (from Scuotto et al. (2016)) and (b) O-glycans. The threonine and serine amino acids, in red, represent potential O-glycosylation sites (developed from I-TASSER and Biovia software)*

This capacity to regulate the translocation of bacteria of intestinal origin is used to prevent cartilage degradation in arthritis and osteoarthritis models (Blareau et al. 2013). Part of the prevention may be linked to the maturation of dendritic cells and the activation of cascades initiated by the recognition of TLR-2 receptors (Hoarau et al. 2006, 2008; Scuotto et al. 2014). However, this recognition is not systematic, with the *B. longum* lipoprotein CBi0703 recognizing neither TLR2 nor TLR6 (Scuotto et al. 2016).

Ultimately, bifidobacterial lipoproteins are probably responsible for stimulating the anti-poliovirus vaccine response observed in infants receiving infant milk fermented with *B. breve* C50 strain (Mullié et al. 2004). The stimulation of fecal anti-poliovirus IgA in infants is linked to an increase in intestinal the *B. breve* and

B. longum species. Although *B. bifidum* is proliferating, it has no effect on the vaccine response.

One potential mechanism is based on the similarity of bifidobacterial lipoproteins to cellular virus receptors. This resemblance either neutralizes infection or activates the natural ligands of these receptors that are present on various immune cells. The hypothesis is corroborated by the neutralizing efficacy of lipoprotein against coxsackievirus B through CAR receptor mimetic action (El Kfoury et al. 2017).

In conclusion, the first industrial applications targeting the prevention of intestinal bacterial infections and immunomodulation of the digestive system appeared rapidly in the form of infant milks (Gallia Lactofidus range 1994; Gallia Calisma 1999), while the support for the activity is still unknown. New applications targeting bacterial translocation have been on the market since 2019 (Arthélio, Bifiboost 1.0). However, applications cannot be limited to this scope. Indeed, the first antiviral results and demonstration of the mimetic mechanism with cell receptors call for in-depth exploration. The risk is to open up the field of drugs, excluding food applications. To maintain the latter, it is crucial to determine the presence of these free lipoproteins in the intestinal biotope, for example, with a view to preventive supplementation in populations suffering from a deficiency of bifidobacteria and, therefore, of bifidobacterial lipoproteins (prematurity, obesity, type 2 diabetes, etc.). The main limitation is the quantity that is potentially present in stools. However, preliminary results from work using viruses as sensors enable lipoproteins to be detected at the intestinal level. An application of the process to breast milk highlights their presence independently of bifidobacterial translocation (Sane et al. 2022). This makes it possible to map free lipoproteins secreted at the intestinal level, and to consider processes for their appropriate supplementation.

8.4.1. *Bifidobacterial bacteriocins and food preservation*

Several compounds inhibiting potentially pathogenic bacteria are isolated from bifidobacterial culture supernatants. Most are heat resistant and protease sensitive, indicating a proteic nature (Touré et al. 2003; Collado et al. 2005a); however, others are lipid based (Liévin et al. 2000). There are two categories of molecular weight, either <10 kDa or between 10 and 30 kDa (Liévin et al. 2000; Collado et al. 2005a).

Table 8.9 lists the various bacteriocins of bifidobacterial-origin described to-date, divided into two classes. Class I includes peptides synthesized through the ribosomal pathway and modified after translation (RiPPs). A 10.2 kb cluster comprising genes encoding a lantibiotic (antibiotic containing lanthionine) is detected in region 12 of the *B. longum* DJO10A chromosome (Lee et al. 2008). This region is deleted in other *B. longum* strains. In fact, out of 31 bifidobacterial genomes analyzed by Bagel3, only four clusters of genes encoding bacteriocin-synthesizing enzymes are detected: two encoding a lanthipeptide II and two encoding a cyclic peptide (Alvarez-Sieiro et al. 2016).

In addition, gene transcription can be sensitive to lantibiotic concentration. Below a minimum concentration, genes are not transcribed (Lee et al. 2011). As production in *B. longum* DJO 10A is random, the transfer of genes involved in the synthesis of this lanthipeptide was carried out in *Lactococcus lactis* (Yu et al. 2018). This being said, the GMO status of the producing bacterium does not facilitate the marketing of this bacteriocin. This class also includes bificin C6165, whose production yield is significantly higher (Pei et al. 2013, 2014). However, the spectrum is restricted to Gram-positive bacteria (Table 8.9).

Of the two class II bacteriocins, bifidocin B, isolated from *B. bifidum* culture supernatant, is active against *Listeria monocytogenes* (Yildirim et al. 1999). Bifidin I, produced by *B. longum subsp. infantis*, has a very broad spectrum, including potentially pathogenic Gram-positive and Gram-negative bacteria (Cheikhyoussef et al. 2009, 2010). Like bificin C6165, the genetic code of class II bifidobacterial bacteriocins is carried by a plasmid (Table 8.9).

Higher molecular mass (25–127 kDa) bifidobacterial molecules inhibiting potentially pathogenic bacteria are also found in culture supernatants (Bifilact Bb-46, Bifilact Bb-12, Bifilong) (Martinez et al. 2013). However, their heat resistance is variable and their structure is unknown.

Ultimately, while the spectrum of activity of the few known bifidobacterial bacteriocins is broad, we can note that some of their targets are lactic acid bacteria, which may be used as ferments in industrial processes. Moreover, although extraction of the bifidobacterial bacteriocin is a prerequisite for its use in the food industry, it is not certain that production costs match the requirements for profitability. However, if the production conditions for bacteriocins are compatible with those for lipoprotein (complex matrix, instead of SRM-type medium), a single bifidobacterial strain would have huge potential for controlling food bacteria affecting product quality. Combining the two metabiotics offers interesting prospects for improving food quality, provided that their sensitivity to oxidation can be controlled.

Class	Bacteriocin	Strain	Location	Spectrum	Peptide sequence	References
Ia (lantibiotics)*	Bisin (~0,1 kDa)	*B. longum* DJO 10A	Chromosome	*S. thermophilus* ST40, *C. perfringens*, *S. epidermidis*, *B. subtilis*, *S. marcescens*, *E. coli* DH5a	MSINEKSIVGESFED LSAADMAMLTGRN DDGVAPASLSFAVS VLSVSFSACSVTVV TRLASCGNCK (pre-peptide)	(Lee et al. 2008, 2011)
	Bificin C6165** (3,395 kDa)	*B. animalis subsp. animalis* CICC 6165	Plasmid	*Alicyclobacillus Spp.*, *S.aureus*, *E. faecium*	KKISGXTLTSDXISL SIXTVSKDXKLATAT XSI	(Pei et al. 2013)
IIa	Bifidocin B (2,880 kDa)	*B. bifidum* NCFB 1454	Plasmid	*L. plantarum*, *L. monocytogenes*	KYYGNGVTCGLHDCR VDRGKATCGIINNGG MWGDIG	(Yildirim et al. 1999)
	Bifidin I (2,880 kDa)	*B. longum subsp. infantis* BCRC 14602	Plasmid	LAB***, *Listeria*, *Staphylococcus*, *Bacillus*, *Streptococcus*, *Salmonella*, *Shigella*, *Escherichia coli*	KYGDVPLY (partial sequence)	(Cheikhyoussef et al. 2010)

*Other class I bacteriocins are potentially produced by bifidobacterial strains (lanthipeptide II and cyclic peptides) (Alvarez-Sieiro et al. 2016).

**Mechanism similar to mersacidin; CTLTSEC consensus pattern with mersacidin and actagardin.

***LAB: lactic acid bacteria.

Table 8.9. *The different classes of* Bifidobacterium *bacteriocins*

9

Lactobacilli: Strain Selection and Probiotic Effects

Jennifer BURGAIN[1] and Marie-Bénédicte ROMOND[2]

[1] *LIBio, Université de Lorraine, Nancy, France*

[2] *Laboratoire de Virologie, Université de Lille, France*

9.1. Species, strain identity and origin

Most probiotic bacteria were first identified in food products or isolated from feces. Lactic acid bacteria have historically been defined by their ability to ferment hexoses into lactate. Lactic acid bacteria are subdivided into several families comprising genera used in food technology, including *Enterococcus*, *Leuconostoc*, *Streptococcus*, *Lactococcus* and *Lactobacillus*. Lactobacilli are classified as phylum Firmicutes, class Bacilli, order Lactobacillales and family *Lactobacillaceae*.

The *Lactobacillaceae* family is mainly found in environmental niches that provide an abundant source of energy, including fruit and dairy ecosystems, wastewater and decomposing plant material, as well as the gastrointestinal tract. Lactobacilli can take the form of rods of variable length, or be organized in chains, as shown in Figure 9.1 for *Lacticaseibacillus rhamnosus* GG (LGG).

Since the first *Lactobacillus delbrueckii* species was named in 1901, taxonomists have attributed over 250 species to the genus *Lactobacillus*. In the age of genomics, the genus appeared to be too heterogeneous. It was important to redefine functionally relevant genera that shared certain physiological, metabolic and lifestyle properties in order to facilitate functional and ecological studies.

Implementation of Enzymatic Processes and Lactic Bacteria in the Food Industries, coordinated by Mohamed GHOUL.

Figure 9.1. *Topographical images of* Lcb. rhamnosus *GG obtained by atomic force microscopy (AFM) after adhesion of the strains to a functionalized substrate. (a) Three-dimensional representation and (b) height image obtained in liquid medium in phosphate buffer at pH 6.8 (from Guérin (2017)). For a color version of this figure, see www.iste.co.uk/ghoul/implementation.zip*

The genus *Lactobacillus* has thus been reclassified into 25 genera, including the modified genus *Lactobacillus*, which includes host-adapted organisms that have been called the *Lactobacillus delbrueckii* group, *Paralactobacillus*, as well as 23 new genera: *Acetilactobacillus*, *Agrilactobacillus*, *Amylolactobacillus*, *Apilactobacillus*, *Bombilactobacillus*, *Companilactobacillus*, *Dellaglioa*, *Fructilactobacillus*, *Furfurilactobacillus*, *Holzapfelia*, *Lacticaseibacillus*, *Lactiplantibacillus*, *Lapidilactobacillus*, *Latilactobacillus*, *Lentilac-tobacillus*, *Levilactobacillus*, *Ligilactobacillus*, *Limosilactobacillus*, *Liquorilactobacillus*, *Loigolactobacilus*, *Paucilactobacillus*, *Schleiferilactobacillus* and *Secundilactobacillus*. Some examples of probiotic species are shown in Table 9.1 (Zheng et al. 2020).

Lactobacilli play a key role in the microbiota of invertebrates, in which they are involved in epithelial growth and differentiation. These observations suggest a co-evolution of lactobacilli with metazoan hosts, resulting in mutually beneficial interactions. Not surprisingly, this bacterial taxon is a significant source of probiotic strains. The description of interactions and potential mechanisms by which lactobacilli can mediate beneficial effects with host cells is the subject of an extensive bibliography, as this information is essential for health claims.

Old nomenclature	New nomenclature
Lactobacillus casei	*Lacticaseibacillus casei*
Lactobacillus paracasei	*Lacticaseibacillus paracasei*
Lactobacillus rhamnosus	*Lacticaseibacillus rhamnosus*
Lactobacillus plantarum	*Lactiplantibacillus plantarum*
Lactobacillus brevis	*Levilactobacillus brevis*
Lactobacillus salivarius	*Ligilactobacillus salivarius*
Lactobacillus fermentum	*Limosilactobacillus fermentum*
Lactobacillus reuteri	*Limosilactobacillus reuteri*
Lactobacillus acidophilus	Unchanged
Lactobacillus delbrueckii subsp. *bulgaricus*	Unchanged
Lactobacillus crispatus	Unchanged
Lactobacillus gasseri	Unchanged
Lactobacillus johnsonii	Unchanged
Lactobacillus helveticus	Unchanged

Table 9.1. *Names of important Lactobacillus probiotic species: old and new names*

9.2. Physiology of probiotic strains

Lactic acid bacteria have complex nutritional requirements. These include vitamins, minerals, carbohydrates, nitrogenous bases and nucleic acid derivatives (Desmazeaud 1983; Bernardeau et al. 2008). Bacteria are thus found in rich ecosystems where these different elements are available (milk and beer, rodent intestines, etc.).

Depending on their glucose metabolism, lactic acid bacteria are classified as homofermentative or heterofermentative (Holzapfel and Wood 2014). Fermentation is a process producing energy by oxidation of organic compounds (mainly carbohydrates), where an electron donor (NADH) yields its electrons to an exogenous acceptor: pyruvate. Lactic acid bacteria use the Embden–Meyerhof and/or phosphoketolase pathways to convert hexoses. Despite their low energy efficiency, these pathways enable the rapid deprivation of hexoses from the surrounding environment, resulting in the production of acids that inhibit the growth of potential competitors.

The homofermentative pathway found in the genera *Pediococcus*, *Streptococcus*, *Lactococcus* and some *Lactobacillus* relies on glycolysis via the Embden–Meyerhof pathway, the end products of which are two molecules of lactate and two molecules of ATP per molecule of glucose consumed. Two specific enzymes are found in homofermentative lactobacilli: fructokinase and aldolase (Zheng et al. 2015).

The heterofermentative pathway, found in certain lactobacilli, is characterized by the presence in the genome of genes encoding acetaldehyde/alcohol dehydrogenase and mannitol dehydrogenase. *Leuconostoc*, another heterofermentative lactic acid bacteria, ferments glucose using the hexose monophosphate pathway. They cleave xylulose-5-phosphate into acetyl-phosphate and glyceraldehyde-3-phosphate through the action of phosphoketolase (Devoyod and Poullain 1988). Under anaerobic conditions, these bacteria degrade hexoses with the quasi-stoichiometric formation of one molecule of lactic acid, one molecule of carbon dioxide and one molecule of ethanol (von Wright and Axelsson 2011). In contrast, through the oxidative pathway, acetate is the main end-product of hexose metabolism. Like *Leuconostoc*, *Weissella* growth is favored under aerobic conditions, which is probably linked to aerobic pyruvate metabolism (Nuraida et al. 1992; Zotta et al. 2018).

The carbohydrate metabolic balance is completed by the pentose phosphate degradation pathway. The pentose concentration will determine the metabolic balance, either with lactate predominating, or acetate, formate and ethanol, to the detriment of lactate (Tanaka et al. 2002).

Depending on fermentation conditions, the metabolic balance of lactic acid bacteria will vary in terms of fatty acids (volatile or otherwise). The initial objective is to protect bacterial vitality, which is compromised by acid production. However, the efficiency of the encapsulation process will be affected differently depending on the composition of the culture medium. Microencapsulation involves drying, resulting in a significant drop in water activity, consequently stopping cellular metabolism. However, drying methods can be aggressive toward bacterial cells, as they may involve significant temperature variations or shearing. The advantage of this process is the elimination of volatile fatty acids. On the other hand, if lactate predominates, not only will the lactate not be eliminated, but the neutralization of the medium will lead to a rise in osmotic pressure upon rehydration, an alternative solution being the use of a protein-type coating. The choice of the encapsulation method and matrix according to culture conditions is therefore a crucial step in the stabilization of lactic acid bacteria (Meng et al. 2008).

9.3. Impact of lactobacilli on digestive health

9.3.1. *General health effects*

Probiotic bacteria are selected to regulate the activity of microorganisms in their host, for example, in mammals, because of their ability to produce pathogen-inhibiting compounds or digestive vitamins and enzymes. Numerous studies have investigated the effect of probiotic bacteria on a variety of conditions, including inflammatory bowel disease, vaginal infections and cirrhosis of the liver. It is these therapeutic approaches that have led to nutritional claims being refused.

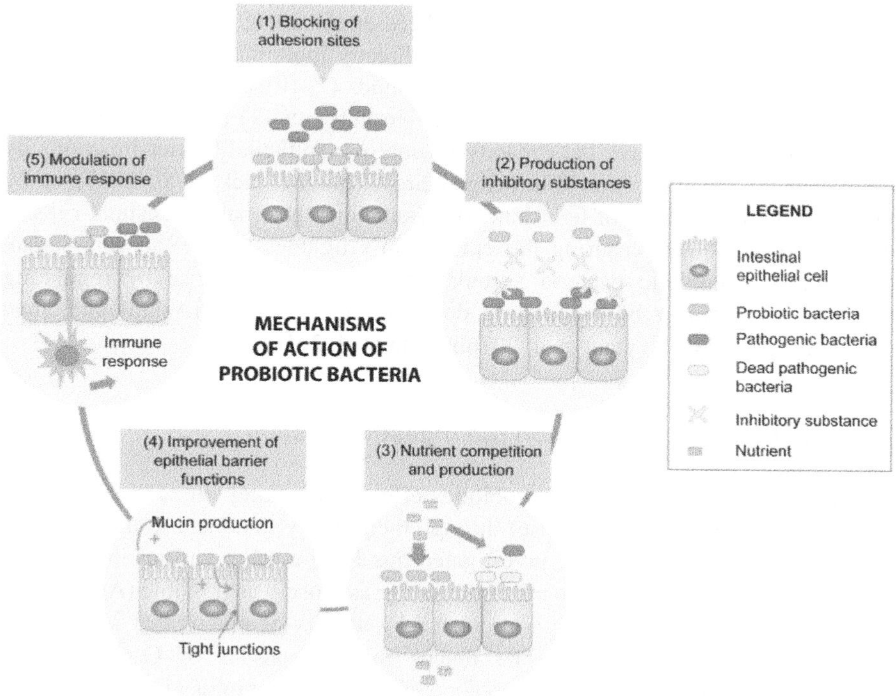

Figure 9.2. *Main mechanisms of action of probiotic bacteria (from Guérin (2017)). For a color version of this figure, see www.iste.co.uk/ghoul/implementation.zip*

Generally speaking, the major effect of probiotics is the maintenance of the physiological functions of the digestive tract and immune system (Hill et al. 2014).

Probiotic bacteria are used in human nutrition to improve digestive health. They are incorporated as ingredients in so-called “functional foods”: dairy desserts, cereals, beverages, chocolate and sauces.

To exert their beneficial effects, three main mechanisms of action are proposed in the literature: exclusion or inhibition of pathogenic bacteria, improvement of epithelial barrier functions or stimulation of the immune system (Corr et al. 2009; Segers and Lebeer 2014). The different mechanisms of action of probiotic bacteria are illustrated in Figure 9.2 and detailed below.

First, probiotic bacteria can exclude pathogenic bacteria by blocking their sites of adhesion to gut epithelial cells (1) (Lebeer et al. 2010; Tytgat et al. 2016a; Campana et al. 2017). They can also act by directly inhibiting pathogenic bacteria through the production of antimicrobial compounds (2) (Rushdy and Gomaa 2013; Arqués et al. 2015) or by competing for certain nutrients (3) (Lebeer et al. 2008). Probiotic bacteria also have the capacity to improve epithelial barrier functions by modulating mucin production following their adhesion (Mack et al. 2003) or by improving the functioning of tight junctions (junctions that enable intestinal cells to adhere to one another, ensuring the epithelium’s tightness) (4) (Rao and Samak 2013). Lastly, probiotic bacteria stimulate the immune response by regulating cytokine expression or by modulating dendritic cell functions (Ashraf and Shah 2014; Fong et al. 2016; Tytgat et al. 2016b; Wan et al. 2016).

9.3.2. *Mechanisms of action on the digestive system*

Bacteria can exert certain probiotic effects (blocking of adhesion sites, modulation of the immune response) through their ability to adhere to the intestinal mucosa. Numerous studies have investigated this feature using various in vitro and in vivo models that resemble the human intestine as closely as possible (Arena et al. 2017). The intestinal epithelium is protected by a layer of mucus that serves as a niche for commensal bacteria. The mucus mainly consists of O-glycosylated proteins called mucins, but also of proteins, lipids and glycolipids (Juge 2012). The ability of probiotic bacteria to adhere to intestinal mucin is made possible by the presence of adhesion factors on their cell surfaces (Nishiyama et al. 2016) (Table 9.2).

The biochemical mechanisms underlying these interactions are complex and insufficiently described, and this is partly due to the heterogeneity of intestinal mucus constituents and the multitude of factors involved in the adhesion process (Juge 2012). However, it is well established that bacterial cell envelope architecture is involved in host adhesion mechanisms.

Adhesion factor	Species	References
Mub family protein	*Lb. Acidophilus*; *Lm. reuteri*	(Roos and Jonsson 2002; Buck et al. 2005)
Mannose-specific adhesin, Msa	*Lp. plantarum*	(Pretzer et al. 2005)
MBF adhesion factor	*Lcb. rhamnosus GG*	(von Ossowski et al. 2011)
CmbA protein	*Lm. reuteri*	(Etzold et al. 2014; Jensen et al. 2014)
***Pili* SpaCBA**	*Lcb. rhamnosus GG*	(Kankainen et al. 2009; von Ossowski et al. 2010; Reunanen et al. 2012)
Elongation factor Tu	*Lb. johnsonii*	(Granato et al. 2004)
GAPDH	*Lp. plantarum*	(Kinoshita et al. 2008)
GroEL chaperone protein	*Lb. johnsonii*	(Bergonzelli et al. 2006)
MapA protein	*Lm. reuteri*	(Miyoshi et al. 2006)
Lam29 protein	*Lm. mucosae*	(Watanabe et al. 2012)

Table 9.2. *Lactobacillus mucin adhesion factor (adapted from Nishiyama et al. (2016))*

The structure and composition of bacterial cell walls influence their ability to adhere. The cell membrane of bacteria (Figure 9.3), particularly Gram-positive bacteria, is enveloped by a thick layer of peptidoglycan, in which teichoic acids, *pili*, proteins and exopolysaccharides (EPS) are embedded.

In some species, the peptidoglycan layer may still be surrounded by a protein envelope forming the S-layer (Chapot-Chartier and Kulakauskas 2014). Each of these compounds thus has the potential to play an important role in the adhesion properties of bacteria to the intestinal mucosa.

Lactobacilli are present in rich environments containing carbohydrate substrates, for example, in intestinal, oral and vaginal mucosa. Bacteria are found in varying proportions along the digestive tract, with increasing concentrations from the oral cavity to the anus (Figure 9.4). The colon is the richest in microbes.

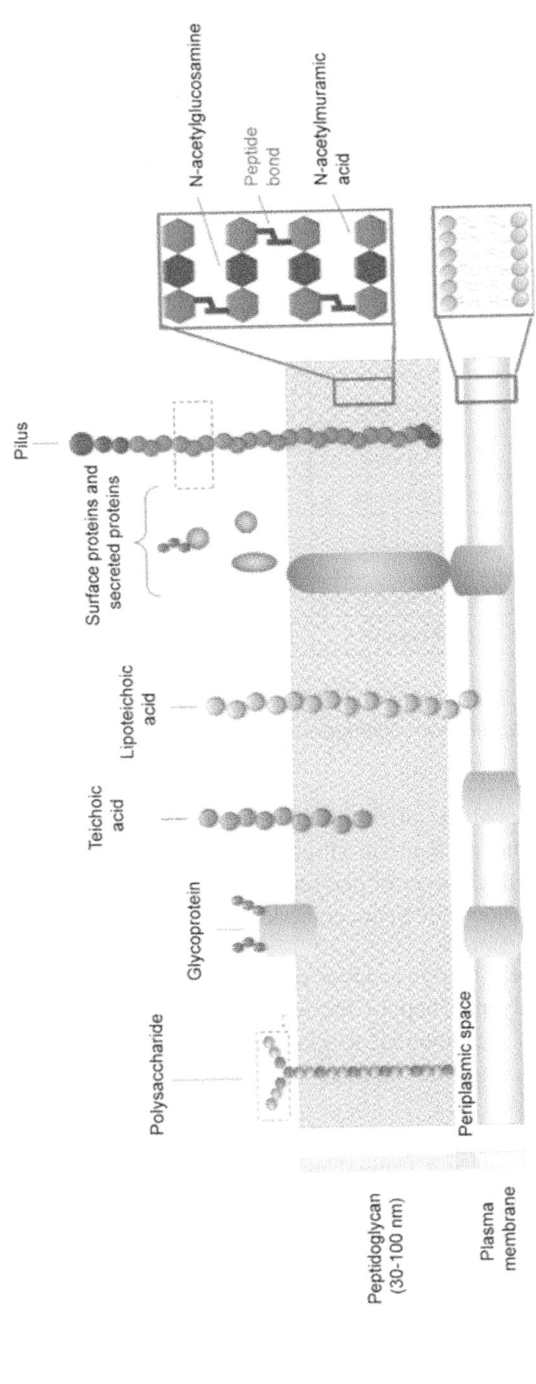

Figure 9.3. *Cell wall composition of Gram-positive bacteria. For a color version of this figure, see www.iste.co.uk/ghoul/implementation.zip*

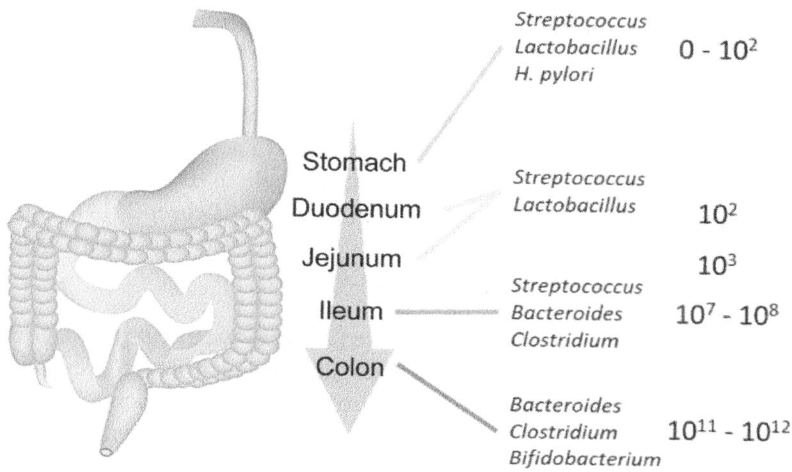

Figure 9.4. *Concentration composition of the main microbial species present in the intestinal microbiota. For a color version of this figure, see www.iste.co.uk/ghoul/implementation.zip*

Very few bacteria are found in the stomach, as this is a highly acidic environment (pH $\approx$ 2) where oxygen is present, which limits lasting colonization of the gastric epithelium. In the small intestine, the microbiota consists of facultative anaerobic bacteria, such as streptococci, lactobacilli and enterobacteria, as well as strict anaerobic bacteria, such as bifidobacteria, *bacteroides* and clostridia. Bacterial concentration is, once again, low, due to peristalsis and the abundance of secretions. In the colon, microbial concentration is higher, but also more varied. This population is dominated by strict anaerobic bacteria (*Bacteroides, Bifidobacterium, Clostridium*), while facultative anaerobic bacteria (such as lactobacilli) are lacking at this level.

Analysis of the microbiome over the last decade has made it possible to compile information about the bacteria colonizing the intestinal microbiota, whether it be healthy or diseased. Dysbiosis of the intestinal microbiota has been correlated with a variety of pathologies, including diabetes, obesity and even autism. It therefore seems essential to better understand how bacteria can mediate and influence human physiology. The aim is to modulate the composition of the microbiota, by encouraging the implantation of species that benefit health, and to ensure a balance within this ecosystem. One possibility is the use of probiotics, such as lactobacilli or bifidobacteria, which are detailed in this work.

9.3.3. Lacticaseibacillus rhamnosus *GG (new nomenclature)*

LGG was first isolated from healthy intestinal microbiota by Gorbach and Goldin (Gorbach 1996), hence the name "GG", and was identified as a potential probiotic bacterium as a result of its high resistance to acidity and bile, its growth capacity and its ability to adhere to the intestinal epithelium (Doron et al. 2005). It is now considered a model probiotic bacterium (Segers and Lebeer 2014). This strain has recently been renamed *Lcb. rhamnosus* GG and was previously known as *Lactobacillus rhamnosus* GG (Zheng et al. 2020).

Numerous clinical trials have demonstrated the positive effects of LGG on host health. For example, LGG efficacy has been demonstrated in the prevention and treatment of numerous gastrointestinal disorders, such as acute diarrhea or antibiotic-associated diarrhoea, as well as in the treatment and prevention of allergy-type diseases or in dental care (Doron et al. 2005; Segers and Lebeer 2014). Initial promising results tend to prove that LGG could help prevent certain diseases such as colon cancer (Goldin et al. 1996; Lim et al. 2002), rheumatoid arthritis (Hatakka et al. 2003; Baharav et al. 2004) or irritable bowel syndrome (Gupta et al. 2000; Gosselink et al. 2004). However, further studies are needed to validate these properties.

Among LGG surface proteins, *fimbriae*, defined as non-flagellar, proteinaceous surface appendages with multiple sub-units, play a major role in the ability to adhere to other bacteria, host cells or abiotic surfaces (Kline et al. 2009). LGG *fimbriae* were first observed by Lebeer et al. (2009) when studying a surface mutant lacking EPS. Whole genome sequencing also identified their genes (Kankainen et al. 2009). More specifically, two clusters of genes encoding two distinct *fimbriae* were identified: the first encoding SpaFED and the second encoding SpaCBA (Kankainen et al. 2009). Currently, under culture conditions, only the gene cluster encoding SpaCBA is expressed (Reunanen et al. 2012; Douillard et al. 2013). Immunostaining techniques have confirmed the presence of SpaCBA. LGG is covered by 10–50 *fimbriae* that are 1 μm in length (Kankainen et al. 2009; Reunanen et al. 2012).

The SpaCBA protein of LGG is composed of three sub-units named SpaA, SpaB and SpaC. The first sub-unt is present all along the *fimbriae* and seems to form the understructure. The SpaB sub-units are found along the length of the appendage in very small numbers, mostly at its base, where it is anchored to the cell wall. The SpaC sub-unit is found at the tip of the *fimbriae* (Kant et al. 2020). It has now been well established that the latter plays a major role in the adhesion of LGG to mucus, and that this adhesion is responsible for the persistence of LGG in the human intestine, compared with lactobacilli lacking fimbrial appendages (Tuomola et al. 2000; von Ossowski et al. 2011).

9.3.4. Lacticaseibacillus casei *and* Lacticaseibacillus paracasei *(new nomenclature)*

Lacticaseibacillus casei and *Lacticaseibacillus paracasei* (*Lcb. paracasei*) have recently been renamed and were previously known as *Lactobacillus casei* and *Lactobacillus paracasei*.

Lacticaseibacillus casei and *Lcb. paracasei* are taxonomically classified in a small clade of facultative heterofermentative lactobacilli.

These strains were isolated from fermented products such as wine and dairy products. Many isolated strains of *Lcb. casei* and *Lcb. paracasei* have been used as single probiotics or as part of a synbiotic formulation. Their role has mainly been reported in digestive diseases, infectious diseases, immunity and depressive disorders.

As early as the 1950s, the *Lcb. casei* Shirota strain was considered a probiotic by the Japanese dairy product manufacturer Yakult. For a long time, however, the scientific community remained sceptical about the beneficial effects attributed to *Lcb. casei*, due to a lack of scientific evidence, a lack of description of the mechanisms involved and incomplete health claim records. Validation of the safety and efficacy of probiotics necessarily requires clinical trials in humans. In this respect, the administration of the *Lcb. casei* Shirota strain demonstrated a reduction in episodes of antibiotic-associated diarrhea, in particular *Clostridioides difficile* diarrhea, in hospitalized patients that were on antibiotics, with no adverse effects being reported (Gao et al. 2010). A number of clinical trials have reported a positive impact of *Lcb. casei* and *Lcb. paracasei* in various intestinal diseases. For example, drinks enriched with probiotic bacteria (*Lcb. casei* DN-114001) have been shown to alleviate antibiotic-associated diarrhea. In an observational study, the frequency of stools in 258 patients was assessed, and it was found that *Lcb. casei* DN-114001 significantly reduced the frequency of diarrhea during antibiotic treatment (Dietrich et al. 2014). Furthermore, to assess the efficacy of a fermented milk drink containing the *Lcb. casei* Shirota strain, a double-blind comparative study was carried out in patients. It was shown that continuous consumption of this fermented milk provided relief for gastrectomized patients, notably reducing symptoms of irregular bowel movements (Aoki et al. 2014).

Lastly, a meta-analysis looking at the effect of probiotics on functional constipation in adults revealed that probiotics had the ability to improve intestinal transit time, stool frequency and consistency. In particular, these beneficial effects were attributed to *Lcb. casei* Shirota (Dimidi et al. 2014). However, caution is required when interpreting these data, due to their high heterogeneity and a risk of bias. Clinical trials are needed to better determine the species, strains, doses, vector (food versus dietary supplement) and duration of probiotic consumption. Nevertheless, the rich bibliography generated from clinical trials assessing the efficacy of *Lcb. casei* or *Lcb. paracasei* is irrefutable, and above all, it is essential to emphasize that none of the studies described adverse effects linked to probiotic supplementation (Jones 2017).

9.3.5. Lactobacillus acidophilus

The *Lactobacillus acidophilus* strain was first isolated in 1900 by Ernst Moro, from the feces and oral cavity of infants (Moro 1900). The whole genome was sequenced in 2005, providing insights into the proteins capable of binding to intestinal mucus and fibronectin, which are thought to be involved in *Lb. acidophilus*' ability to adhere to the mucosa of mammals (Altermann et al. 2005). *Lactobacillus. acidophilus* has long been used in fermented foods, whether dairy or soy products, but also in some functional foods (Anjum et al. 2014). *Lactobacillus. acidophilus* has a post-gastric survival capacity and has been found in feces up to 10 days after ingestion, demonstrating its persistence in the intestinal tract, and therefore its potential as a probiotic (Hütt et al. 2011).

One of the difficulties encountered when assessing the efficacy of *Lb.. acidophilus* is its use in combination with other bacteria. *Lactobacillus acidophilus* is typically used in fermented milks, which require the use of other strains such as *Lb. delbrueckii subsp. bulgaricus*. In addition, there are many different strains of *Lb. acidophilus*. One example is the probiotic preparation VSL#3 (VSL Pharmaceuticals Inc., Fort Lauderdale, USA), which includes four strains of lactobacilli (*Lb. acidophilus*, *Lb. delbrueckii subsp. bulgaricus*, *Lcb. casei*, *Lp. plantarum*), three strains of bifidobacteria (*B. breve*, *B. longum*, *B. infantis*) and one strain of *Streptococcus salivarius subsp. thermophilus*. These highly complex formulations must thus be considered in any assessment of probiotic efficacy (Neish 2017).

As previously mentioned, lactobacilli are part of the intestinal microbiota. As such, they play an important role in the digestive process and can impact intestinal physiology. For example, probiotic preparations containing *Lb. acidophilus* have shown promise in the treatment of functional intestinal disorders. A symbiotic combination of polydextrose, *Lb. acidophilus* NCFM and *Bifidobacterium animalis subsp. lactis* HN019 (in a yogurt-like matrix) demonstrated a shorter transit time and an improvement in the clinical signs of constipation during a two-week treatment (Magro et al. 2014). Similarly, a combination of *Lb. acidophilus* NCFM and *B. animalis subsp. lactis* HN019 (2×10^{11} CFU/day), administered over eight weeks, showed an improvement in bloating in a cohort of 60 patients with functional bowel disorders (with no constipation) (Ringel-Kulka et al. 2011).

Studies including *Lb. acidophilus* thus suggest that it may have beneficial effects on various intestinal disorders. Although, as mentioned above, the use of *Lb. acidophilus* in combination with other microorganisms or with prebiotics complicates the assessment of the role of *Lb. acidophilus*.

10

Encapsulation Processes

Jennifer BURGAIN, Joël SCHER and Claire GAIANI
LIBio, Université de Lorraine, Nancy, France

10.1. Definition and aims of encapsulation

Encapsulation is a process by which substances are entrapped within a dispersed material for immobilization, protection, controlled release and functionalization (Vandamme et al. 2007). There are many reasons for encapsulating ingredients prior to their implementation into food products. First, it allows ingredients to be incorporated into the food matrix without compromising the ingredient's qualities. Encapsulation also makes it possible to protect the ingredient from chemical, physical or biological degradation to mask an odor or to deliver the ingredient to its site of action. It also improves ingredient storage and the shelf life of the food product (McClements 2012). Encapsulation can also avoid affecting the properties of the food matrix by, for example, retaining acids produced by probiotic strains.

Bacterial encapsulation is a widely used technique for trapping and preserving bacteria viability from their incorporation into the food product, through their consumption and passage through the gastrointestinal tract (De Prisco and Mauriello 2016). During the period from production to consumption of a food product, the bacteria contained in the latter need to be protected against various harmful mechanisms encountered during:

– production, such as high temperatures and shearing;

– drying, such as dehydration in the case of dry products;

For a color version of all the figures in this chapter, see www.iste.co.uk/ghoul/implementation.zip.

Implementation of Enzymatic Processes and Lactic Bacteria in the Food Industries,
coordinated by Mohamed GHOUL.

– storage of food products;

– gastrointestinal transit, such as the acid pH encountered in the stomach, proteolytic enzymes and the presence of bile salts in the small intestine (Manojlović et al. 2010).

One of the major difficulties is maintaining bacterial viability, which is determined by the choice of encapsulation method and matrix. The various encapsulation techniques commonly used to encapsulate probiotic bacteria include emulsification, extrusion, spray drying and fluidized bed coating. In the literature, numerous reviews have shown microencapsulation's potential in improving bacterial survival during product manufacture and storage, as well as during digestion (Anal and Singh 2007; Burgain et al. 2011; Cook et al. 2012; Tripathi and Giri 2014).

Depending on the encapsulation techniques used, different types of microparticles can be produced: matrix, reservoir or coated matrix microparticles (Figure 10.1).

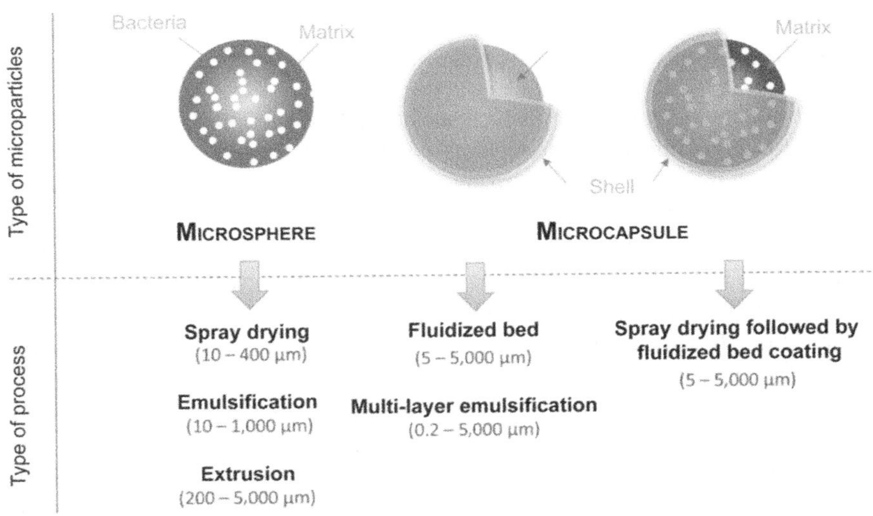

Figure 10.1. *Architecture of microparticles produced according to the type of process (from Guérin (2017))*

Microcapsules are surrounded by a shell that separates the inside of the particle from the surrounding environment. In the case of matrix-type microparticles, the bioactive substance is dispersed throughout the encapsulation matrix. Microparticles can also combine the two types mentioned above, in which case they are referred to as a coated matrix. The matrix type specifically concerns encapsulation processes such as spray drying, emulsification or extrusion. The reservoir type concerns encapsulation processes such as the fluidized bed or multi-layer emulsification. The coated matrix type concerns a matrix-type microparticle, which has then been coated by a fluidized-bed coating process with another material to form a layer surrounding the microparticle.

There are numerous materials used to encapsulate probiotic bacteria. To be used in food products, they must be recognized as "GRAS" (Generally Recognized as Safe). Commonly used polymers include polysaccharides such as alginate, chitosan, starch, κ-carrageenan, gellan gum, xanthan gum, gum arabic, pectins and proteins such as gelatin and milk proteins (Burgain et al. 2011; Sarao and Arora 2017).

10.2. Atomization

Spray drying is a widely used process in the food industry, particularly for the production of milk powder. Spray drying can be divided into four main stages (Broeckx et al. 2016) (Figure 10.2):

– the bacteria are mixed with a liquid matrix, usually a mixture of biopolymers;

– the liquid solution containing the bacteria and encapsulating matrix is sprayed into fine droplets in the drying chamber;

– the fine droplets come into contact with hot, dry air, and are then dried and form particles;

– the solid particles are separated from the cooled, humid air using a cyclone for example. The powder is collected in a container at the cyclone outlet.

Spray drying offers a number of advantages, including low cost, high reproducibility and rapidity. It is also an excellent way of incorporating bacteria into dairy products. The process can easily be scaled up to an industrial scale, enabling large quantities of powder to be produced in a very short time. The particles produced are spherical, with a uniform shape and size distribution, which is favorable from a sensory point of view, if incorporated into a food (Zuidam and Shimoni 2010). However, the protective capacity of small particles may be lower than that of large particles. The main disadvantage of spray drying for bacterial encapsulation is the use of very hot air (around 200°C) for the water evaporation stage, which can be detrimental to bacterial survival. However, even if this drying temperature is very high, the temperature of the droplet does not exceed 60°C under standard conditions, as the temperature of the humid air near the droplet predominates.

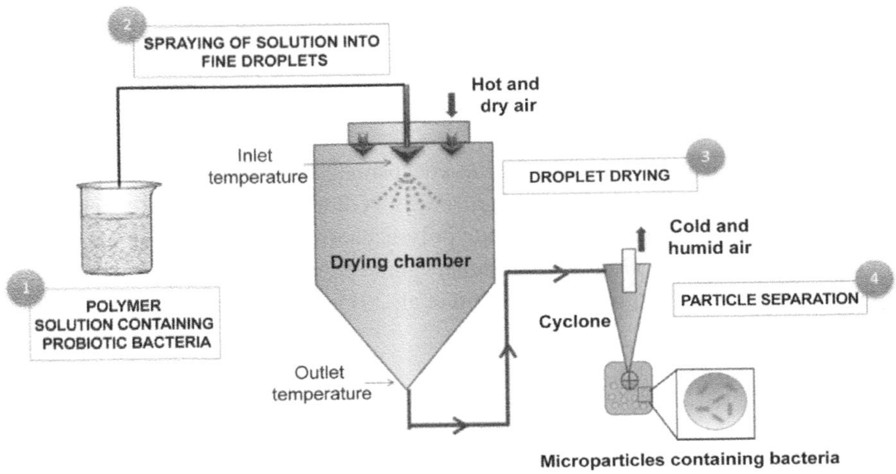

Figure 10.2. *Encapsulation of bacteria by spray drying in a dairy protein matrix (from Guérin (2017))*

Encapsulation of bacteria in milk protein-based matrices by spray drying appears to result in low survival rates (Corcoran et al. 2004; Picot and Lacroix 2004; Jantzen et al. 2013; Maciel et al. 2014; Liu et al. 2015; Khem et al. 2016; Huang et al. 2017a). Bacterial survival after drying depends on the bacterial strain, the drying parameters and the encapsulation matrix used. However, by adapting the drying parameters, in particular by lowering the drying air temperature, a compromise can be found to ensure good bacterial survival after drying.

During spray drying, bacteria undergo numerous unfavorable constraints that can affect their survival. These include heat stress, oxidative stress, osmotic stress, mechanical stress and dehydration stress (Broeckx et al. 2016). In order to minimize these unfavorable constraints as much as possible, drying parameters must be carefully chosen and controlled. It is particularly important to control spraying, the drying temperature, the tower configuration, the residence time of powders in the drying tower and the final water content of the powder produced (Broeckx et al. 2016; Huang et al. 2017a).

– *Spraying*: various systems can be used to spray the solution into the drying tower: centrifugal turbine; liquid pressure nozzle; or two-fluid nozzle. Little is known about the influence of spraying on bacterial survival, which appears to be negligible. However, it is advisable to use a two-fluid nozzle when spray drying bacteria, as this nozzle is suitable for products that are sensitive to high pressures (Schuck et al. 2012). Shear induced during the spraying stage can also have a significant impact on strain functionality: the breaking of LGG chains has been observed (Guerin et al. 2017) and modeled (Gomand et al. 2020).

– *Drying air temperature*: due to the high temperature of the drying air at the tower inlet, spray drying is often considered to be a destructive method for heat-sensitive molecules such as bacteria. Although the inlet drying air temperature can reach an excess of 200°C, bacteria are not subject to such high temperatures. In fact, during spray drying, two drying phases follow one another to form the dry particles: a constant drying phase and a slower drying phase. During the first drying phase, the droplet temperature is limited to the wet temperature, with constant water evaporation (Huang et al. 2017b). At this point, the particle temperature remains low (around 50–60°C) and does not reach the drying air temperature (Zuidam and Shimoni 2010). During the second drying phase, once the water content of the droplets reaches a critical value, a dry crust forms on the surface of the droplets and the evaporation rate rapidly decreases. The particle temperature rises to the temperature of the drying air (between 50 and 80°C) (Zuidam and Shimoni 2010). The temperature reached by the particles is generally close to the outlet air temperature, depending on the residence time of the powder in the drying tower (Gharsallaoui et al. 2007; Huang et al. 2017b). It is in this last stage that bacteria can be inactivated by heat (Broeckx et al. 2016). The lower the outlet temperature, the greater the survival of bacteria.

– *Powder water content*: the drying temperature also influences the quality of the powder obtained, since it affects both water content and water activity (a_w). An a_w and water content that exceeds recommended levels has an impact on powder stability during storage. An a_w of around 0.2 and a water content of 4% are recommended to ensure good powder stability and bacterial survival during storage (Ananta et al. 2005; Abe et al. 2009; Vesterlund et al. 2012). An outlet temperature that is too low can lead to the production of powders with too high an a_w and poor stability during storage. It is therefore important to find an ideal outlet air temperature that allows both low a_w and low water content, thus avoiding loss of bacterial viability during drying (Broeckx et al. 2016).

– *Air flow*: different models are available, depending on the direction the air and liquid take when they enter the drying chamber: co-current, counter-current and mixed flow (Peighambardoust et al. 2011). Depending on the flows used, the powders do not reach the same temperatures. With a co-current flow, product and air enter the drying chamber in the same direction. As a result, the powders are exposed to moderate temperatures ranging from 50 to 80°C below the outlet air temperature, which limits thermal degradation. On the other hand, with a countercurrent flow, the liquid to be sprayed is in the opposite direction to the hot air flow, and the powders are exposed to higher temperatures (Gharsallaoui et al. 2007; Anandharamakrishnan 2017). As bacteria are heat-sensitive organisms, it is important to favor the co-current configuration, so as to limit excessively high temperatures at the particle level.

– *Residence time in the drying tower*: the factor that potentially influences bacterial viability the most is probably the residence time of the powder in the drying chamber. A short drying time is favorable in order to avoid a loss of bacterial viability (Huang et al. 2017b). Indeed, when the particle is dry, its temperature reaches that of the surrounding air. This is why it is important for the powder produced to be evacuated from the drying tower once the water has evaporated. Generally, the drying time is between 15 and 30 s, which limits the negative impact of temperature on bacteria (Gharsallaoui et al. 2007).

To summarize, the following conditions should be favored to ensure the best bacterial viability after spray drying:

– use of a two-fluid nozzle to avoid excessive pressure;

– choosing a drying temperature that is low enough to avoid loss of bacterial viability, but guaranteeing a low water content in the powder, so as to ensure good powder stability during storage;

– use of a co-current air flow to avoid contact with excessively high temperatures;

– choosing the shortest possible residence time in the tower.

10.3. Extrusion

The extrusion method is widely used to encapsulate bacteria, due to its ease of use, low cost and gentle conditions for preserving bacterial viability (Krasaekoopt et al. 2003). Microencapsulation of bacteria by extrusion involves spraying the solution containing the bacterial cells and encapsulation matrix through a high-pressure nozzle or needle. Once formed, the droplets fall directly into a solution to set the matrix (Figure 10.3). The droplet size and shape depend on needle diameter, free-fall distance and the physicochemical properties of the encapsulation matrix (Bugarski et al. 1994; Burgain et al. 2011). The main drawback of this method is that it is difficult to extrapolate at the industrial level, due to the slow formation of particles. Also, the microparticles produced have a relatively large diameter (between 200 and 5,000 μm), which can pose sensory issues when the microparticles are introduced into a food product.

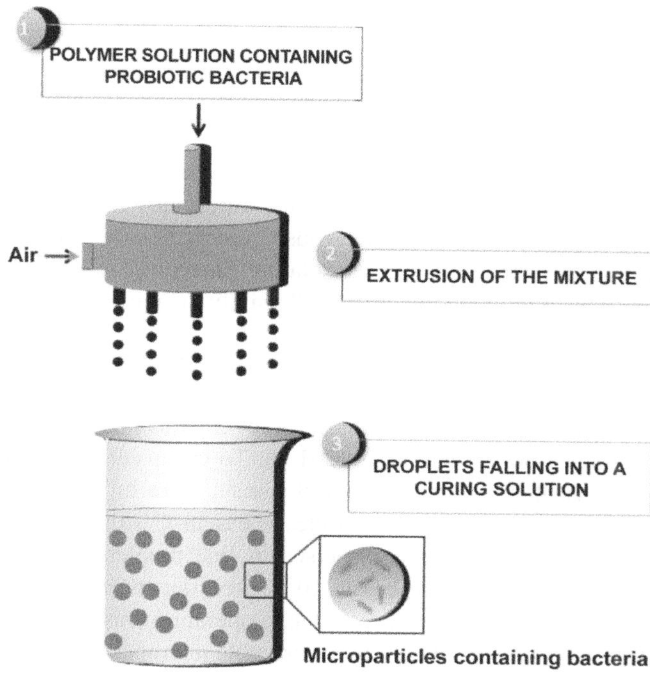

Figure 10.3. *Encapsulation of probiotic bacteria by extrusion (from Guérin (2017))*

Most of the time, the polymer used to encapsulate bacteria by extrusion is alginate (Sultana et al. 2000; Truelstrup Hansen et al. 2002). Only a few studies have used soluble protein isolates to encapsulate probiotic bacteria with this method (Reid et al. 2005; Doherty et al. 2011, 2012). For example, Doherty et al. encapsulated LGG in microparticles produced from a solution of soluble protein isolate using the extrusion technique. To achieve this, the protein and bacterial solution was passed through a 150 μm diameter needle, and the droplets were recovered in 250 mL of curing solution (acetate buffer). Extrusion had no negative impact on bacterial viability, and viability was also maintained during gastrointestinal digestion. The microparticles remained stable at the acidic pH of the stomach and resisted pepsin's action, thus maintaining viable bacteria inside the microparticles. In contrast, the microparticles were disintegrated in the intestinal solution, controlling the release of LGG in the gut (Doherty et al. 2011, 2012).

Another study used soluble protein isolates to encapsulate LGG. This time, the hardening solution was calcium chloride (Reid et al. 2005). Here, extrusion protected the bacteria from the acidic environment of the stomach and the bile salts in the intestine.

10.4. Emulsification

Emulsification involves dispersing the liquid containing the compound to be encapsulated in another immiscible liquid. Most often, this is an aqueous solution and an oily solution. Dispersion of the aqueous liquid in the oily solution leads to the formation of fine droplets. When encapsulating bacteria, low-shear systems are recommended, so as to limit the impact on cell viability (Figure 10.4).

Various bacteria have already been encapsulated by emulsification using milk protein gelation (Heidebach et al. 2009a, 2009b; Burgain et al. 2013a). The encapsulation of LGG by emulsification using chymosin-induced milk protein gelation was achieved by Burgain et al. (2013a). To encapsulate bacteria using this technique, solutions of milk proteins and chymosin were mixed at a temperature below 10°C, so as to allow chymosin to cleave casein-κ without inducing micelle aggregation. The solution was then dispersed in fine droplets in the oil solution, maintaining an initial temperature below 10°C. In order to enable gelation of the droplets containing the probiotic bacteria, the temperature was ramped up to 40°C, bringing the caseins closer together to form insoluble microparticles (Figure 10.5).

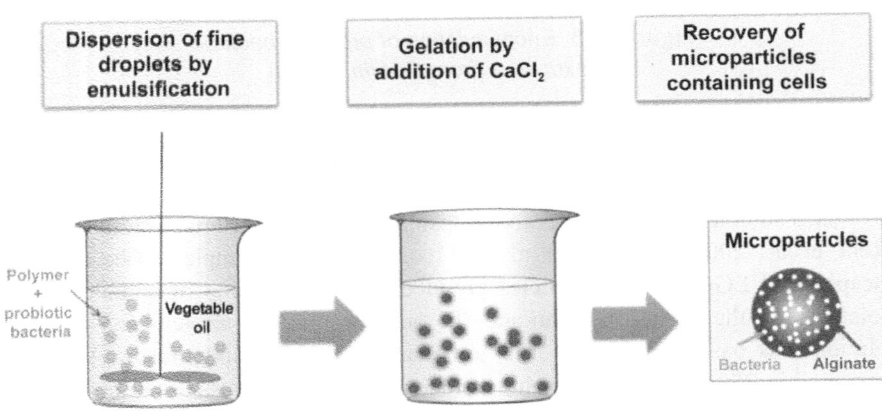

Figure 10.4. *Encapsulation of probiotic bacteria in an alginate matrix by emulsification (from Guérin (2017))*

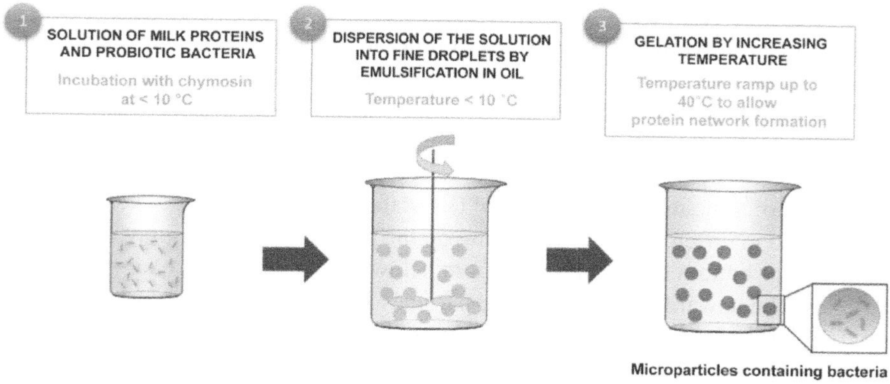

Figure 10.5. *Encapsulation of probiotic bacteria by emulsification using milk protein gelation through the action of chymosin (from Guérin (2017))*

Another enzyme that can be used to produce water-insoluble microparticles by milk protein gelation is transglutaminase. This enzyme was used to encapsulate *Lcb. paracasei* and *Bacillus lactis* in sodium caseinate (Heidebach et al. 2009b, 2010). In order to achieve this, the bacteria were placed in the protein solution in which transglutaminase was active, and the solution was then dispersed in oil. Increasing the temperature produced insoluble microparticles encapsulating the bacteria.

Bacterial encapsulation by emulsification using milk protein gelation (with chymosin or transglutaminase) produces spherical, water-insoluble microparticles with a high encapsulation rate and very good resistance to gastric conditions. However, emulsification leads to the production of a broad size distribution. While large particles can be a negative criterion from a sensory point of view, the accelerated release of bacteria from overly small microparticles is undesirable.

In addition, when recovering microparticles from the oil phase, oil drainage can lead to a loss of bacteria on the particle surface. Another disadvantage is the presence of a thin layer of residual oil on the surface of the microparticles, even after several washes (Burgain et al. 2013a).

10.5. Coating

Coating encapsulation methods apply a superficial layer to particles obtained through spray drying, emulsification or extrusion. This process can also be applied after a bacterial freeze-drying stage (Burgain et al. 2016). In all cases, the sample to be coated must be in particle form. Fluidized bed coating involves passing air through a bed of particles to be coated in order to suspend the powder. The coating medium is also sprayed to form a membrane around the particles. Most of the time, the coating layer does not form immediately, and a large number of successive passes through the spray zone are required. Simultaneously, the coating layer solidifies either by drying, in the case of an aqueous coating solution, or by cooling, in the case of a molten coating material (El Mafadi and Poncelet 2007). Coating can be achieved in different ways, depending on the positioning of the spray nozzle: top, tangential or bottom spray. Spraying from below is best suited to coating individual particles. The upward movement of the particles and the direction of spraying ensure rapid drying and reduce the risk of agglomeration. This spraying also avoids drying the drops before they are deposited onto the particles (Burgain et al. 2016). The example used to illustrate fluidized bed coating (Figure 10.6) shows spraying from above.

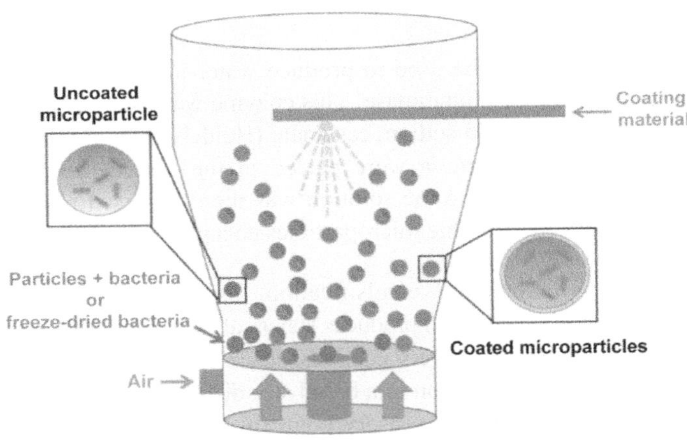

Figure 10.6. *Encapsulation of bacteria by fluidized bed coating (from Guérin (2017))*

A recent study examined the microencapsulation of *Limosilactobacillus reuteri* DSM 20016 in a fluidized bed with sweet whey and shellac as a coating agent (Schell and Beermann 2014). First, the bacterial culture was sprayed from above through a two-fluid nozzle into the fluidized bed, in the presence of sweet whey and various sugars, at a drying temperature of 50°C. The microparticles formed were then coated by spraying a shellac solution from the top of the fluidized bed. The microparticles produced improved *Lm. reuteri* DSM 20016's tolerance to acidity and increase its survival during gastrointestinal transit (Schell and Beermann 2014).

Microencapsulation of bacteria by coating in a fluidized bed has the advantage of using low temperatures, which ensures good survival of the bacteria during the process. The process can also easily be extrapolated to an industrial scale.

11

Encapsulation Matrices

Jennifer BURGAIN, Joël SCHER and Claire GAIANI
LIBio, Université de Lorraine, Nancy, France

The selection of the encapsulation matrix is a preliminary stage that requires a rigorous methodological approach. The selection criteria must be perfectly defined, with the aim of producing microparticles with target properties. It is, for example, important to consider what becomes of the microparticles: stored as a powder or introduced into food, which will itself be preserved. In the latter case, the food properties must be considered when choosing the encapsulation matrix: pH, water content, ionic strength, presence of enzymes, presence of antimicrobial compounds, etc. In the case of probiotic bacteria targeting the intestinal epithelium, the microparticles must be able to withstand acidic pH levels (encountered in the stomach) and also be capable of releasing their contents into an environment with a pH above 6. This pH corresponds to the minimum pH encountered in the intestinal lumen at the entrance to the duodenum. The polymers used for encapsulation must be "GRAS". The most commonly used polymers for encapsulating probiotic bacteria are described below.

11.1. Polysaccharides

– *Starch*: Starch is a natural polymer made up of glucose units. Amylose and amylopectin are the two constituent homopolymers of starch. D-Glucose units are linked together by α (1-4) bonds, while α (1-6) bonds are responsible for branching. The ratio of amylose to amylopectin influences the physicochemical, thermal and rheological properties of starch. Some bacteria, notably bifidobacteria, have the ability to adhere to starch, and this property can therefore be exploited to improve

Implementation of Enzymatic Processes and Lactic Bacteria in the Food Industries, coordinated by Mohamed GHOUL.

cell retention inside microparticles. It is also interesting to note that starch is not completely degraded during digestion, making it a substrate of choice for bacterial colonization of the intestine. The latter will produce short-chain fatty acids that can have positive effects on the host's health.

– *Alginate*: Alginate is a naturally occurring polysaccharide extracted from brown seaweed. This polymer is composed of alternating β-D-mannuronic (M) and α-L-guluronic (G) acid residues. Sodium alginate is used in the food industry as a thickener, gelling agent and colloidal stabilizer. It is an inexpensive, non-toxic and biocompatible additive. One of the special features of alginate, used for micro-encapsulating probiotic bacteria, is its ability to form a gel in the presence of divalent cations (Ca^{2+}). The gel's solubility is highly dependent on the pH of the surrounding medium: it is soluble at neutral or higher pH and insoluble at acid pH. The advantage of this pH-dependent solubility is that the microparticles will preserve their structure during gastric digestion (acid pH), and the particle will be solubilized when the pH rises in the intestine, allowing the contents to be released. The main disadvantage is the porosity of this gel, which will allow the diffusion of compounds that could impair the viability or functionality of the encapsulated bacteria. However, chemical modification can be considered to improve the physicochemical properties of alginate. This is the case, for example, with N-palmitoylated alginate, which has been successfully used for the encapsulation of probiotic bacteria, improving cell viability under simulated gastric conditions.

– *Chitosan*: Chitosan is a cationic polysaccharide obtained by partial N-deacetylation of chitin. Chitin is extracted from the shells of crustaceans (such as crabs and shrimps). This polymer is mainly composed of β-(1-4)-2-amino-2-deoxy-D-glucopyranose units. The overall structure of chitosan is reminiscent of cellulose, but the presence of an amine group on carbon 2 greatly modifies its physicochemical properties, in particular, its solubility and reactivity. Chitosan is biocompatible, biodegradable and non-toxic, making it ideal for a wide range of applications.

Chitosan and alginate are often combined to modify the diffusion rate of compounds through the encapsulation matrix. The electrostatic interactions established between the carboxyl groups of alginate and the amine groups of chitosan enable the formation of spheres capable of isolating encapsulated bacteria. It is important to note, however, that a bacteriostatic effect has been reported for chitosan, which may impact the viability of encapsulated probiotic bacteria.

– *κ-Carrageenan*: There are various types of carrageenan, but κ-carrageenans are the most widely used, notably for their gelling properties. They are polymers of

D-galactose-4-sulfate and 3,6-anhydro-D-galactose from the red algae *Kappaphycus alvarezii*. κ-Carrageenans can form thermo-reversible gels in the presence of monovalent ions. A major constraint, however, is the need for heat to dissolve the polymer, making this step incompatible with the survival of probiotic bacteria.

– *Gellan gum and xanthan gum*: Gellan gum is a highly heat-resistant polysaccharide used in the food industry. It is a linear anionic heteropolyoside. The carbohydrate units found in this polymer are sequences of D-glucose, L-rhamnose and D-glucuronic acid in a 2:1:1 ratio. The sequences are linked together by α (1-3) bonds. Xanthan gum is also used for its gelling properties in the food industry. It is an anionic polymer made up of carbohydrate units: glucose, mannose and glucuronic acid.

The combination of these two gums produces gels that are resistant to gastric conditions, making them a major resource for encapsulating probiotic bacteria.

– *Gum arabic*: Gum arabic is a sap exudate from the stems and branches of acacia trees. It is composed of branched arrangements of carbohydrates and proteins. Encapsulation matrices made from gum arabic have limited capacity against oxidation. However, in combination with other polymers, gum arabic could be a great encapsulation matrix candidate.

– *Pectins*: Pectins are extracted from the walls of fruits and vegetables. These polysaccharides are structurally diverse, with a primary chain composed of galacturonic acid and α (1-4) branches. Pectins can exhibit different degrees of methylation and esterification. Galacturonic acid has the ability to bind to calcium, enabling the formation of a network, but one that remains less stable than that formed by alginate.

11.2. Proteins

– *Milk proteins*: Milk contains two types of protein: lactoserum proteins or soluble proteins (mainly β-lactoglobulin and α-lactalbumin) and caseins. Milk proteins are widely available, inexpensive, natural and recognized as "GRAS" materials, with excellent nutritional potential. Their many functional and structural properties make them highly suitable materials for carrying bioactive compounds (Livney 2010; Tavares et al. 2014). Their ability to bind and form emulsions or gels is regularly used to entrap bioactive compounds. When it comes to encapsulating probiotic bacteria, their ability to gel has been thoroughly studied, as it enables the formation of microparticles large enough to contain bacteria. Protein gels can either be formed by coagulation (enzymatic or acid) of caseins or by gelation of soluble proteins or total milk proteins induced by heat treatment (Livney 2010). Enzymatic coagulation (by chymosin or transglutaminase) of caseins forms gels that retain bacteria. The advantage of this type of system is that the microparticles produced are insoluble in aqueous media, ensuring their stability during storage in food. Second, the network formed is sufficiently resistant to protect the bacteria from stomach conditions, and the increase in pH observed in the intestine enables viable cells to be released.

– *Gelatin*: This is an animal protein derived from the partial hydrolysis of collagen. It is a polymer that is frequently used in the food industry and can be used as an encapsulation matrix if combined with, for example, gellan gum. At a pH below the isoelectric point of gelatin, the positive charges on the polymer interact with the negative charges on the anionic polysaccharide.

11.3. Bacterial adhesion to encapsulation materials

Knowledge of the adhesion phenomena between bacteria and the materials used for encapsulation is fundamental to understanding the structuring of matter. Interactions between bacteria and encapsulation materials mainly depend on their surface properties and have been little studied (Burgain et al. 2014). Two types of adhesion can be distinguished, depending on the distance separating the bacteria from the surface with which they interact.

The first phase of bacterial adhesion to a surface involves reversible, long-distance interactions (between 3 and 50 nm). These interactions include van der Waals and electrostatic forces. This first adhesion phase can be described by a physicochemical approach, known as the DLVO theory (for Derjaguin–Landau–Verwey–Overbeek) (Hermansson 1999; Burgain et al. 2014). According to the DLVO theory, van der Waals and electrostatic interactions govern the initial adhesion of bacteria. These non-specific interactions depend on the physicochemical properties of the bacterial and material surfaces, the distance between the bacteria and the surface, and the ionic strength of the medium (Hermansson 1999).

The second phase of bacterial adhesion is irreversible adhesion. This is created by short-distance (1–5 nm), higher-energy interactions. These include hydrophobic interactions, electrostatic interactions, hydrogen bonds and ligand–receptor interactions (Ofek et al. 2003). Hydrogen bonds are strong interactions between a hydrogen atom and a negatively charged atom. Electrostatic interactions, on the other hand, involve the formation of bridges between a negative charge on the bacterial surface and a negative charge on the material surface by means of divalent cations. These forces are highly dependent on the ionic strength of the medium.

Hydrophobic interactions are formed between two non-polar molecules. Lastly, ligand–receptor interactions correspond to molecular recognition between receptors on the bacterial surface and the material surface. Recognition systems are specific and enable bacteria to adhere to each other or other entities. The molecules involved may either be adhesins, protein or saccharide-type membrane receptors. These irreversible interactions are considered specific. Specific adhesion can therefore be defined as a selective interaction between bacterial surface constituents and a receptor located on the surface of a material (An and Friedman 1998).

The properties of biomolecules present on the bacterial surface have been explored with a view to understanding interactions between bacteria and interfaces (biofilm formation), or between bacteria and intestinal cells (probiotic effect assessment) (Lebeer et al. 2007, 2012; Perea Vélez et al. 2010).

In the case of LGG, atomic force microscopy (Guerin et al. 2018) and high-throughput screening (Gomand et al. 2018) have highlighted LGG's ability to specifically interact with milk proteins. The bacterium interacts specifically with soluble proteins, but not with micellar caseins (Burgain et al. 2013b). Hydrophobic, electrostatic or ligand–receptor interactions must be involved in the interaction. Thanks to the use of LGG surface mutants (*spaCBA* mutant lacking *fimbriae*, EPS-deficient *welE* mutant), it has been possible to demonstrate the indispensable role of SpaCBA in the interaction with soluble proteins (Burgain et al. 2014b; Guerin et al. 2016, 2018). The presence of *pili* improves encapsulation efficiency by emulsification and promotes localization of bacteria inside the microparticle. Indeed, without *fimbriae*, bacteria are observed on the surface of microparticles (Burgain et al. 2014b).

Knowledge of the interactions between bacteria and encapsulation materials is an important factor to control for the improvement of the encapsulation efficiency of bacteria, as well as their localization in the matrix and matrix stability. Numerous studies on the location of bacteria in the cheese matrix, using microscopy, show that bacteria are mainly found on the surface of milk fat globules (Tunick et al. 1993; Laloy et al. 1996; Lopez et al. 2006). These results suggest that adhesion may be possible between bacteria and the milk fat globule membrane (MFGM). A material that could be a good candidate for improving bacterial encapsulation could therefore be the MFGM, or more specifically, the glycoproteins making up this membrane (Guerin et al. 2019).

The adhesion properties of lactic acid bacteria, and lactobacilli strains in particular, on intestinal mucin have already been described (Nishiyama et al. 2016). Numerous proteins anchored in the wall of lactobacilli act as mucin adhesion factors. There are many similarities between the MUC glycoproteins found in the MFGM and the human mucin lining intestinal cells. Both human intestinal mucin and MFGM-derived mucin are high molecular weight, highly glycosylated and predominantly O-glycosylated proteins. They are both composed of N-acetylgalactosamine, N-acetylglucosamine, fucose, galactose, sialic acid and mannose in different proportions (Patton et al. 1995; Bansil and Turner 2006). The similarities found between these two glycosylated proteins suggest that the MUC1 glycoprotein, found in the bovine MFGM, may also interact with lactic acid bacteria and thus regulate their localization in the food matrix.

The study of adhesion between LGG and milk proteins has enabled us to highlight various phenomena (Burgain et al. 2015; Guerin et al. 2016, 2018):

– The wild-type strain LGG interacts specifically with β-lactoglobulin and not with caseins, α-lactalbumin and bovine serum albumin. The use of LGG surface mutants demonstrated the key role of SpaCBA in the interaction of this strain with β-lactoglobulin.

– Other surface molecules may interact with β-lactoglobulin, α-lactalbumin and bovine serum albumin when SpaCBA is absent. These molecules could be long and/or short EPS, and/or surface proteins (MabA or MBF).

– LGG interacts specifically with the MFGM. The use of surface mutants also demonstrates the key role of the SpaCBA pilus in the interaction with the MFGM.

Preferential localization of bacteria in the encapsulation matrix prior to droplet/particle formation can impact process efficiency, leading to the production of heterogeneous particles (in terms of bacteria/particle quantity). Bacterial adhesion to matrix components can also modulate the functional properties of strains: milk proteins can, in some cases, hide bacterial adhesion sites to intestinal cells (in vitro observations on Caco-2 cells).

Indeed, the ability of milk constituents, and the MFGM in particular, to inhibit the attachment of LGG to intestinal epithelial cells has been demonstrated. Using two complementary techniques (biophysical and cellular), a mechanism was proposed to explain the impact of the presence of milk constituents on LGG adhesion to intestinal cells (Figure 11.1). Surface molecules such as *pili* are involved in adhesion phenomena with both milk constituents (such as β-lactoglobulin and the MFGM) and intestinal epithelial cells. When the bacteria reach the intestine in the presence of a dairy matrix (in native form, as is the case for the MFGM glycoproteins), the sites of *pili* adhesion to intestinal cells are already occupied by the dairy matrix. As a result, bacterial adhesion to the intestine can be inhibited or limited.

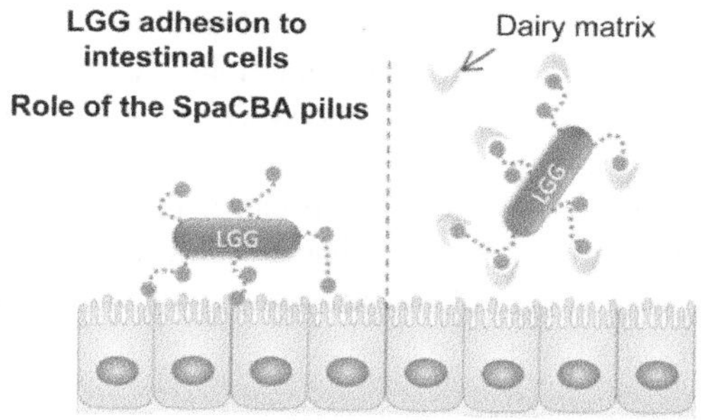

Figure 11.1. *Competition between LGG adhesion to the MFGM and LGG adhesion to Caco-2 intestinal epithelial cells. For a color version of this figure, see www.iste.co.uk/ghoul/implementation.zip*

However, in vivo studies are essential to quantify the impact of the dairy matrix, as an encapsulation matrix, on the ability of bacteria to adhere to intestinal cells, that is, their site of action. The formulation of the encapsulation matrix therefore plays a major role in determining:

– the resistance of microparticles to environmental conditions, and therefore the ability to isolate bacteria from a medium that could affect their vitality;

– release of bacteria at their target site under defined physicochemical conditions;

– the ability to preserve strain functionality by preserving the role of surface adhesions.

11.4. Conclusion

Microencapsulation is the technique of choice for preserving the vitality of probiotic bacteria. Nowadays, the use of prebiotics, starch, gelatin and milk proteins as encapsulation matrices offers greater functionality. These components not only protect bacteria during food processing and storage, as well as gastrointestinal conditions, but also have their own health benefits. Further research is needed to improve the formulation of encapsulation matrices, and clinical trials involving a diversity of populations must be conducted, in order to prove the system's efficacy and host health benefits.

By combining the various processes developed to-date and the numerous matrices available, it is possible to develop multiple vectors. There is an extensive literature on this subject. However, tests are systematically carried out on a single strain or a defined mixture of strains. A better understanding of encapsulation mechanisms (process and formulation) and bacteria–matrix interactions will enable us to optimize the protection of probiotic bacteria in order to preserve their vitality and vectorize them to their site of action, where they will be able to exert their beneficial effect.

12

Fermented Foods

Frank PIVA and Marie-Bénédicte ROMOND
Laboratoire de Virologie, Université de Lille, France

As presented in the introduction, food fermentation is a natural process that can improve food preservation and have health benefits; however, conversely, it can induce harmful effects, depending on the proliferating bacteria.

From the very beginning, human populations have tried to control this phenomenon, first empirically, then, with advances in bacteriology, by developing strategies based on a growing knowledge of fermentation biotopes. This rational approach has been revived by the development of the concept of functional food, which aims to protect human and animal health, as well as improve food safety.

Two areas of research are currently being developed:

– implementation of traditional fermented foods and associated microbiota;

– monitored fermentation of foods or ingredients, or even waste products, for the development of functional food.

As we will see, the two approaches do not currently coincide, as the choice of bacteria used in monitored fermentation is focused on lactic acid bacteria and/or bifidobacteria, for which knowledge of safety aspects has progressed.

Implementation of Enzymatic Processes and Lactic Bacteria in the Food Industries,
coordinated by Mohamed GHOUL.

12.1. Traditional fermented foods

In addition to the fermented milks already described, traditional fermented foods not only include cereals, but also many fruits and vegetables (Figures 12.1 and 12.2). With regards to cereals, the distribution of traditional fermented products depends on the type of cereal grown. Thus, Africa consumes fermented products based on sorghum and millet, while Western Europe and the Maghreb mainly use wheat (for example, in the form of sourdough bread), in addition to the consumption of fermented foods based on spelt (Italy), corn (Portugal), teff (Ireland) or barley (Maghreb) (Figure 12.1(a)). Rye-based fermented foods (bread-making) are found in the Nordic countries and Russia. Poland and Turkey are consumers of traditional fermented products, based on buckwheat and einkorn. China, Japan and South-East Asia are characterized by the widespread use of rice as a base for traditional fermented foods. Traditional products in India and Thailand are divided between fermented rice and wheat.

Lastly, the Americas use corn as a base for fermentation, with Central America also using sorghum. For some countries, information on traditional foods is patchy, particularly in Australia, where British colonization in the late 18th century led to the disappearance of aboriginal practices.

Traditional foods fermented from cereals also take the form of beverages, such as rye-based kvass in Russia, barley-based beer in Western Europe, North America and Australia, and rice-based spirits, including sake in Asia (Figure 12.1(b)).

These beverages are widely produced industrially, but a renewed interest in fermentation processes is encouraging artisanal and even family production. Boza is obtained by fermenting breads of different cereals (millet, wheat, corn, rye, barley and rice). This type of drink is well known around the Mediterranean (for example, in Turkey and Egypt). Bulgaria also consumes a similar drink. The Andean countries have drinks based on corn pre-cooked in boiling water, with the dough being fermented after it is wrapped in banana leaves. In Africa, sorghum, millet, teff and corn are used as fermentation bases for drinks such as tchoukoutou, tchapalo or dolo, tella (more or less alcoholic) and cheka (a mixture of cereals and plants from the *Brassicaceae* family). All of these beverages are not necessarily highly concentrated in alcohol, since yeast fermentation also leads to the production of CO_2.

Vegetables and fruits are also traditionally eaten in fermented form (Figure 12.2). A wide variety of foods and condiments are used. In Europe, cabbage (sauerkraut), turnips, radishes, olives, cucumbers and gherkins are traditionally fermented (Figure 12.2). Pickled gherkins and cucumbers are also commonly consumed in Russia. Fermented sea fennel can be found in countries around the Mediterranean.

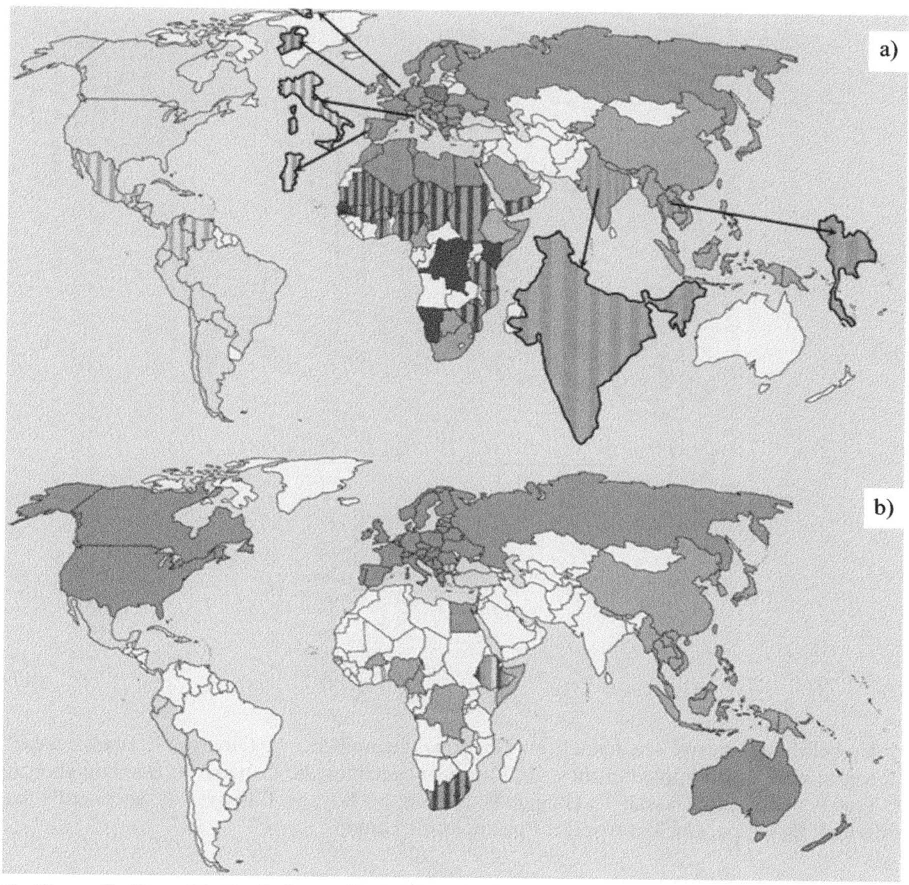

■: Corn; ■: Rye; ■: Buckwheat; ■: Teff; ■: Wheat; ■: Spelt; ■: Einkorn ; ■: Rice; ■: Sorghum; ■: Millet; ■: Barley

Figure 12.1. *Global distribution of fermented cereal-based foods/condiments and beverages. Traditional fermented cereals: (a) breads and (b) beverages (Franz et al. 2014; Ashaolu and Reale 2020; Neela and Fanta 2020). For a color version of this figure, see www.iste.co.uk/ghoul/implementation.zip*

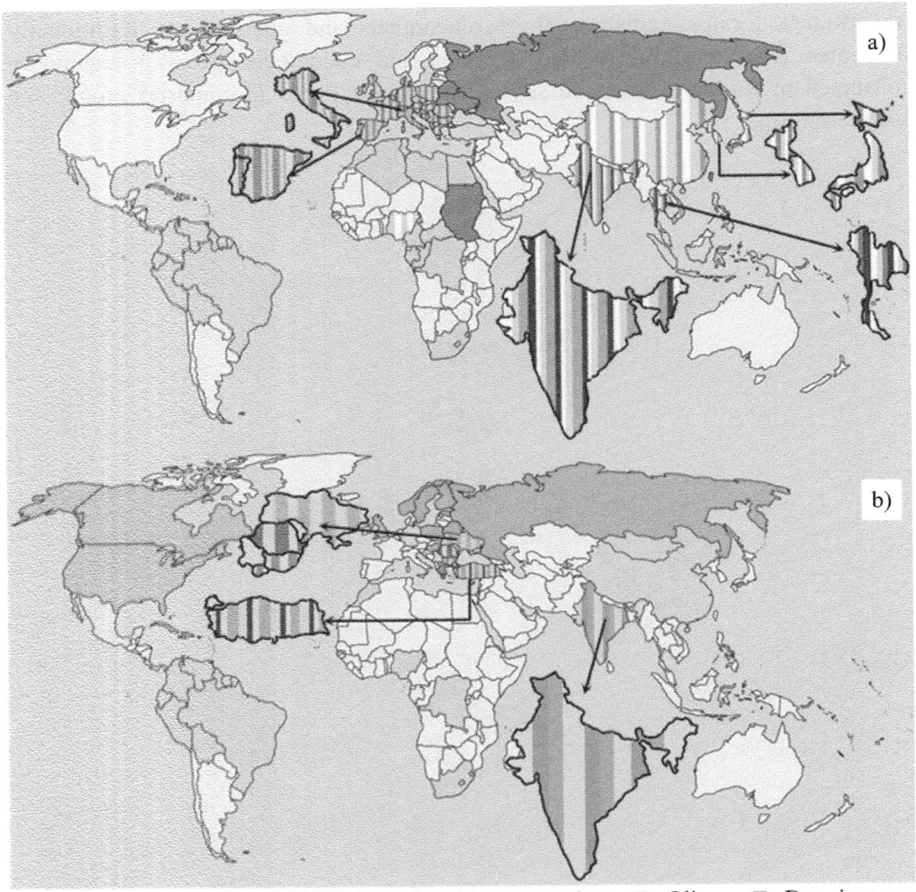

■: Umbelliferae: carrot, sea fennel; ■: Gherkins, cucumbers; ■: Olives ; ■: Brassicaceae: cabbages, mustard, turnips, radishes; ■: Various vegetables; ■: Capers; ■: Bamboo shoots; ■: Tea; ■ : Birch; ■: Cassia; ■: Ginger; ■: Sureau; ■: Soy; ■: Cassava; ■: Shepherd's tea (sideritis); ■: Grape, Guelder rose; ■: Pineapple; ■: Lemon

Figure 12.2. *Worldwide distribution of traditional fermented non-cereal plant-based foods/condiments and beverages. For a color version of this figure, see www.iste.co.uk/ghoul/implementation.zip*

COMMENT ON FIGURE 12.2. – *(a) Food/condiments; (b) beverages (Ramasamy et al. 2012; Satish Kumar et al. 2013; Franz et al. 2014; de la Fuente-Salcido et al. 2015; Tamang et al. 2016; Baschali et al. 2017; Abakar Idriss et al. 2019; Ibrahim et al. 2019; Ashaolu and Reale 2020; Pérez-Armendáriz and Cardoso-Ugarte 2020).*

The range expands in Asia, with soybeans, bamboo shoots, peppers, eggplants, mustard greens, etc. Preparations based on fermented vegetables can also be found in Turkey, including rutabagas and beetroots. The Maghreb and parts of the Mashreq are characterized by their use of pickled lemons. In Africa and Latin America, cassava is the basis of many fermented foods. In Sudan, cassava leaves are fermented (kawal) and used as a meat substitute. Thailand, where the cassia flower is a national symbol, also uses fermented cassia.

Lastly, several fermented beverages are also widely consumed worldwide: non-alcoholic fermented grape juice (sira or hardaliye) and Guelder rose drink (gilaburu) in Turkey; birch and elderberry in Russia, Ukraine and neighboring countries; teas in Asia; pineapple (tepache) in Mexico; cassava in Africa and Latin America; and ginger in North America. In India, carrot juice is consumed in fermented form.

These foods and beverages are either produced by traditional methods, or industrially, especially in countries where sanitary requirements demand strict contamination control. The human populations of Eurasia, Africa and Latin America have long practiced food fermentation, initially in a random way, with inoculation involving bacteria from the environment and manipulators; then, from the 20th century onward, by rationalizing the fermentative process for better reproducibility of taste, color and odor.

One approach is to isolate bacteria by culture for use as an inoculant (alone or combined) (Mokoena et al. 2016). However, the absence of a universal medium makes it impossible to draw up an exhaustive catalog of strains that are present in a fermented food (Table 12.1). Inoculation with defined strains (starter) does not always ensure reproduction of the flavors developed in traditionally fermented foods. On the other hand, the advantage of controlled inoculation is the reduced contamination by potentially pathogenic strains found in traditional foods (Wilfrid Padonou et al. 2009; Keisam et al. 2019). The role of enterococci, both lactic acid bacteria and potential pathogens, is an issue. Table 12.1 shows the presence of enterococci in tepache, kinema, tungrymbai, fermented capers, jiang-gua, bamboo shoots (soiboum, soidon) and sauerkraut. Cooking can prevent the risk of infection, but new tendencies to consume raw products are leading to fears of increased transmission of mobile genetic elements that carry antibiotic resistance genes (plasmids and/or phages).

The advent of molecular methods is broadening our knowledge of the spectrum of bacteria fermenting a food (Du et al. 2019; Østlie et al. 2021). These methods confirm that the bacteria involved in traditional processes fluctuate significantly, even within a single production (Saubade et al. 2017), and that the place of production of raw materials is a prime factor in influencing the bacterial composition of finished foods (Østlie et al. 2021).

In finished products, cereal fermentations (sourdough breads, boza-type drinks, etc.) combine acid-producing bacteria (including lactic acid), such as lactobacilli (*Companilactobacillus*, *Fructilactobacillus*, *Lactobacillus*, *Lacticaseibacillus*, *Lactiplantibacillus*, *Latilactobacillus*, *Lentilactobacillus*, *Levilactobacillus*, *Limosilactobacillus*, *Ligilactobacillus*, *Loigolactobacillus*), lactococci, *Pediococcus*, *Leuconostoc*, *Weisella*, enterococci, streptococci and yeasts, in particular *Saccharomyces*, but also *Candida*, *Torulaspora*, *Kluyveromyces*, *Debaryomyces* and *Schizosaccharomyces* (Tamang et al. 2016).

Product (origin)	Fermented substrate	Microorganisms described as participating in fermentation
Fermented olives (Spain, Portugal, Italy)	Green olives	*Lcb. casei*, *Lo. coryniformis*, *Lp. pentosus*, *Lp. plantarum*, *Le. mesenteroides*, *Pc. pentosaceus*, *E. casseliflavus*
Fermented capers (Greece, Italy, Spain)	Capers	*Lp. plantarum*, *Lp. pentosus*, *Lm. fermentum*, *Lv. brevis*, *Lp. paraplantarum*, *E. faecium*, *Pc. pentosaceus*
Gherkins, fermented cucumbers (Europe, Poland)	Gherkins, cucumbers	*Lactobacillus spp.*, *Leuconostoc spp.*, *Pediococcus spp.*
Chrourk Tror Sok, Chrourk Sapei (Cambodia)	Cucumber, cabbage	*Pc. pentosaceus*
Oiji (Korea)	Cucumber, salt, water	*Le. mesenteroides*, *Lv. brevis*, *Lp. plantarum*, *Pc.cerevisiae*
Jiang-Gua (Taiwan)	Cucumber	*E. casseliflavus*, *Le. lactis*, *Le. mesenteroides*, *Lp. pentosus*, *Lp. plantarum*, *Lp. paraplantarum*, *Lc. lactis subsp. lactis*, *Ws. cibaria*, *Ws. hellenica*
Soibum (India)	Bamboo shoots	*Lp. plantarum*, *Lv. brevis*, *Lo. coryniformis*, *Lb. delbrueckii subsp. Bulgaricus*, *Le. fallax*, *Lc. lactis*, *Le. mesenteroides*, *E. durans*, *K. lactis*, *B. subtilis*, *B. licheniformis*, *Wz. coagulans*
Soidon (India)	Bamboo tips	*Lv. brevis*, *Le. fallax*, *Lc. lactis*, *Lp. plantarum*, *Carnobacterium spp.*, *E. faecium*
Naw-Mai-Dong (Thailand)	Bamboo shoots	*Le. mesenteroides*, *Pc. cerevisiae*, *Lp. plantarum*, *Lm. fermentum*, *Lv. brevis*

Table 12.1. *Fermented non-cereal plant-based products*

COMMENT ON TABLE 12.1.– *Abbreviations: Ca.:* Carnobacterium*; Fr.:* Fructilactobacillus*; K.:* Kluyveromyces*; Lc.:* Lactococcus*; Le.:* Leuconostoc*; La.:* Latilactobacillus*; Ln.:* Lentilactobacillus*; Lo.:* Loigolactobacillus*; Lq.:* Liquorilactobacillus*; Lv.:* Levilactobacillus*; Mc.:* Mammaliicoccus*; Pc.:* Pediococcus*; Pl.:* Paucilactobacillus*; Se.:* Secundilactobacillus*; Sc.:* Schleiferilactobacillus*; T.:* Tetragenococcus*; Ws.:* Weissella*; Wz.:* Weizmannia*; LAB: lactic acid bacteria (Ramasamy et al. 2012; Satish Kumar et al. 2013; Franz et al. 2014; Sagdic et al. 2014; de la Fuente-Salcido et al. 2015; Tamang et al. 2016; Baschali et al. 2017; Abakar Idriss et al. 2019; Ibrahim et al. 2019; Ashaolu and Reale 2020; Pérez-Armendáriz and Cardoso-Ugarte 2020).*

One constant is the variety of bacterial and fungal strains involved in lactic fermentation (Table 12.1). The lactic biotope of traditionally prepared kimchi comprises 20–30 lactic acid bacteria, most of which participate in the fermentation process by generating fermentation compounds that include biogenic amines (Nam et al. 2009; Won et al. 2020; Kim et al. 2021).

Product (origin)	Fermented substrate	Microorganisms described as participating in fermentation
Brovada, Sauerrüben (Italy, Germany)	Salted root vegetable	*Lp. plantarum, Ln. Hilgardii, Lo. coryniformis, Ca. maltaromaticum corrig., Ws. viridescens, Pc. parvulus*
Sauerkraut (Europe)	Cabbage	*Lv. brevis, Lp. plantarum, Le. mesenteroides, Pc. cerevisiae, E. faecalis*
Sinki (India)	Radishes	*Lv. brevis, Lp. plantarum, Lm. fermentum*
Sayur Asin (Indonesia)	Mustard greens, cabbage, salt, coconut	*Le. mesenteroides, Lp. plantarum, Lv. brevis, Ws. confusa corrig., Pc. pentosaceus*
Burong Mustala (Philippines)	Mustard	*Lv. brevis, Pc.cerevisiae*
Phak-Dong (Philippines)	Cabbage	*B. siamensis B44v*
Pao Cai (China)	Cabbage	*Le. mesenteroides, Lp. plantarum, Lcb. casei, Lc. lactis*
Kimchi (Korea)	Cabbage, green onion, bell pepper, ginger	*Ws. cibaria, Ws. confusa, Ws. koreensis*
Pak gua Dong (Thailand)	Mustard greens	*Lp. plantarum*
Dha Muoi (Vietnam)	Cabbage, mixed vegetables	*Le. mesenteroides, Lp. Plantatum*

Table 12.2. *Fermented non-cereal plant-based products*

COMMENT ON TABLE 12.2.– *Abbreviations: Ca.:* Carnobacterium*; Fr.:* Fructilactobacillus*; K.:* Kluyveromyces*; Lc.:* Lactococcus*; Le.:* Leuconostoc*; La.:* Latilactobacillus*; Ln.:* Lentilactobacillus*; Lo.:* Loigolactobacillus*; Lq.:* Liquorilactobacillus*; Lv.:* Levilactobacillus*; Mc.:* Mammaliicoccus*; Pc.:* Pediococcus*; Pl.:* Paucilactobacillus*; Se.:* Secundilactobacillus*; Sc.:* Schleiferilactobacillus*; T.:* Tetragenococcus*; Ws.:* Weissella*; Wz.:* Weizmannia*; LAB: lactic acid bacteria (Ramasamy et al. 2012; Satish Kumar et al. 2013; Franz et al. 2014; Sagdic et al. 2014; de la Fuente-Salcido et al. 2015; Tamang et al. 2016; Baschali et al. 2017; Abakar Idriss et al. 2019; Ibrahim et al. 2019; Ashaolu and Reale 2020 ; Pérez-Armendáriz and Cardoso-Ugarte 2020).*

Product (origin)	Fermented substrate	Microorganisms described as participating in fermentation
Dha Muoi (Vietnam)	Mustard, beetroot	*Lm. fermentum, Lp. pentosus, Lp. plantarum, Pc. pentosaceus*
Fermented mustard (Vietnam)	Mustard	*Lm. fermentum, Lp. plantarum, Lv. brevis*
Pickled turnip, Sunki, Tsuda (Japan)	Otaki turnips, Tsuda leaves, root	*Lp. plantarum, Lv. brevis, Wz. coagulans, Pc. pentosaceus*
Takuanzuke (Japan)	Japanese radish, salt, sugar, Shochu	*Lb. acetotolerans, Lv. brevis, Lcb. casei, Se. collinoides, La. curvatus, Fr. fructivorans, Lv. namurensis, La. sakei, Lb. delbrueckii subsp. Bulgaricus, Lm. fermentum, Lp. plantarum, Le. citreum, Pc. pentosaceus, Pc. damnosus*
Suan-Cai (China)	Cabbage, mustard	*Pc. pentosaceus, T. halophilus*
Dha Muoi (Vietnam)	Eggplant	*Lm. fermentum, Lp. pentosus, Lv. brevis*
Dua Chua (Dua Muoi) (Vietnam)	Vegetables	*Lm. fermentum, Lp. pentosus, Lp. plantarum, Pc. pentosaceus, Lv. brevis*
Hom-Dong (Thailand)	Red onion	*Le. mesenteroides, Pc. cerevisiae, Lp. plantarum, Lm. fermentum, Ln. buchneri*
Tsukemono (Japan)	Daikon radish, carrot, turnip, cabbage	*Se. collinoides, Fr. fructivorans, Lcb. casei, Lcb. paracasei, Lb. delbrueckii subsp. Bulgaricus, Lm. fermentum, Lp. plantarum, Le. citreum, Pc. pentosaceus, Pc. damnosus*

Table 12.3. *Fermented non-cereal plant-based products*

COMMENT ON TABLE 12.3.– *Abbreviations: Ca.:* Carnobacterium*; Fr.:* Fructilactobacillus*; K.:* Kluyveromyces*; Lc.:* Lactococcus*; Le.:* Leuconostoc*; La.:* Latilactobacillus*; Ln.:* Lentilactobacillus*; Lo.:* Loigolactobacillus*; Lq.:* Liquorilactobacillus*; Lv.:* Levilactobacillus*; Mc.:* Mammaliicoccus*; Pc.:* Pediococcus*; Pl.:* Paucilactobacillus*; Se.:* Secundilactobacillus*; Sc.:* Schleiferilactobacillus*; T.:* Tetragenococcus*; Ws.:* Weissella*; Wz.:* Weizmannia*; LAB: lactic acid bacteria (Ramasamy et al. 2012; Satish Kumar et al. 2013; Franz et al. 2014; Sagdic et al. 2014; de la Fuente-Salcido et al. 2015; Tamang et al. 2016; Baschali et al. 2017; Abakar Idriss et al. 2019; Ibrahim et al. 2019; Ashaolu and Reale 2020; Pérez-Armendáriz and Cardoso-Ugarte 2020).*

Product (origin)	Fermented substrate	Microorganisms described as participating in fermentation
Kocho (Ethiopia)	Abyssinian banana (*Ensette ventricosum*)	*LAB, yeast*
Kawal (Sudan)	Cassia leaves (*Cassia obtusifolia*)	*B. subtilis, Lp. plantarum, Mc. scuiri*
Kinema, Tungrymbai (India)	Fermented soy	*E. faecium*
Tempeh (Malaysia)	Fermented soy	*Lp. plantarum*
Fermented sea fennel (Italy, Greece, Turkey, Spain)	Sea fennel	*LAB*
Kanji (India)	Carrots, beetroot, rice, mustard	*Lp. pentosus, Lp. paraplantarum, Lp. plantarum*
Salgam Suyu (Turkey)	Rutabaga, beetroot, turnip, chickpea, flour	*LAB, yeast*
Hardaliye (Turkey)	Grapes, mustard seeds, cherry leaves	*Lb. acetotolerans, Lv. brevis, Lcb. casei, Lcb. paracasei, Lm. pontis, Fr. sanfranciscensis, Pl. vaccinostercus*
Gilaburu juice (Turkey)	Guelder rose fruit	*Lp. plantarum, Lcb. casei, Lv. brevis, Le. mesenteroides, Lq. hordei, Lp. paraplantarum, Lo. coryniformis, Ln. buchneri, Ln. parabuchneri, Lcb. pantheris, Le. pseudomesenteroides, Sc. harbinensis*
Tepache (Mexico)	Pineapple	*Lc. lactis, E. faecium*

Table 12.4. *Fermented non-cereal plant-based products*

COMMENT ON TABLE 12.4.– *Abbreviations: Ca.:* Carnobacterium*; Fr.:* Fructilactobacillus*; K.:* Kluyveromyces*; Lc.:* Lactococcus*; Le.:* Leuconostoc*; La.:* Latilactobacillus*; Ln.:* Lentilactobacillus*; Lo.:* Loigolactobacillus*; Lq.:* Liquorilactobacillus*; Lv.:* Levilactobacillus*; Mc.:* Mammaliicoccus*; Pc.:* Pediococcus*; Pl.:* Paucilactobacillus*; Se.:* Secundilactobacillus*; Sc.:* Schleiferilactobacillus*; T.:* Tetragenococcus*; Ws.:* Weissella*; Wz.:* Weizmannia*; LAB: lactic acid bacteria (Ramasamy et al. 2012; Satish Kumar et al. 2013; Franz et al. 2014; Sagdic et al. 2014; de la Fuente-Salcido et al. 2015; Tamang et al. 2016; Baschali et al. 2017; Abakar Idriss et al. 2019; Ibrahim et al. 2019; Ashaolu and Reale 2020; Pérez-Armendáriz and Cardoso-Ugarte 2020).*

12.2. New approaches to fermentation

The profusion of bacterial species found in traditional fermented foods is difficult to reproduce on an industrial scale. First and foremost, it is imperative to control processes, so as to avoid accidents related to the proliferation of potentially pathogenic bacteria, such as *Listeria monocytogenes*. This cold-resistant bacterium can survive several winters in soils contaminated by flocks of infected sheep. It also tolerates the extreme pH and salinity conditions that are characteristic of sauerkraut. It can therefore develop on cabbages intended for sauerkraut and cause infections in susceptible individuals (abortion in pregnant women, sepsis in immunocompromised people). Trying to control this with a bacteriocin such as nisin does not solve the problem, as the lactobacilli of the acidifying flora are sensitive. To ensure fermentation, the biotope needs to be reconditioned with nisin-resistant strains (Breidt et al. 1995).

Traditional fruit fermentations, such as tempoyak, also fail to prevent the survival of salmonella, *S. aureus* and *L.monocytogenes*, by more than 10 days (Chuah et al. 2016). Currently available bacteriocins are also ineffective against Gram-negative bacteria.

It is therefore necessary to find countermeasures by controlling the bacterial seeding of plants and fruit. The challenge for the 21st century is to develop fermented foods that provide a diversity of bacterial and fungal strains similar to those found in traditional foods. Indeed, the higher the level of hygiene, the more inflammatory and/or autoimmune diseases develop (inflammatory bowel disease, rheumatism, type 1 diabetes, multiple sclerosis, etc.) (Wasko et al. 2020; Murdaca et al. 2021). The microbiome plays a pivotal role. However, recurrent antibiotic treatments, as well as the presence of biocides and/or antibiotics in food, deplete the richness and diversity of intestinal bacteria. Traditionally fermented foods partially compensate for this deficit by adding a multitude of bacteria, yeasts and molds, forcing the immune system to be flexible in adapting to the vast range of antigens supplied in varying concentrations over time. Industrial food products are formulated to preserve sanitary quality, which implies less microbial diversity. In this context, reproducing the microbial diversity of traditional products is difficult to achieve by strain seeding, and secondary to sanitary treatments (such as sterilization). Our experience with milk fermentation shows that it is extremely difficult to ensure reproducibility of processes beyond three fermentations, both in terms of bacterial numbers and the genesis of fermentation-derived compounds. However, the fermentation of plants and fruits should be explored to increase bacterial inputs, particularly in populations with low dairy consumption. A secondary objective is to optimize the nutritional value of fruit and vegetable juices (Ricci et al. 2020). This being said, their characteristics, such as low pH, high sugar concentration and the presence of polyphenols, define an unfavorable environment for the growth of lactic acid bacteria.

With these limits in mind, it should be noted that some laboratory trials are encouraging. Apricot juice, for example, has been shown to support the growth of lactobacilli and bifidobacteria without supplementation (Bujna et al. 2017). Pineapple juice, after neutralization of its highly acidic pH, is also favorable for the growth of lactobacilli and *B. animalis subsp lactis*. Bacterial populations are maintained for over a month at +4°C (Nguyen et al. 2019). Cherimoya juice, sterilized before inoculation with *Lv. brevis*, *Lcb. rhamnosus* and *Lp. plantarum* strains isolated from fruit, is also an excellent culture medium, without adjusting the acid pH. The strains are resistant to adverse conditions after 21 days of storage at +4°C (Isas et al. 2020). In addition, blueberry juice fermented with *Lp. plantarum* isolated from dairy products shows enhanced antioxidant protection (Wu et al. 2021). Lastly, co-encapsulation tests are effective in preserving the survival of lactic acid bacteria in blackberry juice (Colín-Cruz et al. 2019). Co-encapsulation is also of interest in the case of passion fruit juice (Monteiro et al. 2020). However, the improvement in nutritional qualities may come at the expense of taste, as shown by the results for orange juice (Miranda et al. 2019).

Part 2 References

Abakar Idriss, L., Guira, F., Tapsoba, F., Zongo, C., Abdoullahi Hissein, O., Tidjani, A., Savadogo, A. (2019). Le kawal, un condiment à base de feuilles fermentées de *Senna obtusifolia* : technologies et valeurs nutritionnelles. *Afr. J. Food Agric. Nutr. Dev.*, 19(2), 14244–14260.

Abe, F., Miyauchi, H., Uchijima, A., Yaeshima, T., Iwatsuki, K. (2009). Effects of storage temperature and water activity on the survival of bifidobacteria in powder form. *Int. J. Dairy Technol.*, 62(2), 234–239.

Afzaal, M., Saeed, F., Ateeq, H., Ahmad, A., Tabussam Tufail, T., Ismail, Z., Anjum, F.M. (2020). Encapsulation of *Bifidobacterium bifidum* by internal gelation method to access the viability in cheddar cheese and under simulated gastrointestinal conditions. *Food Sci. Nutr.*, 8(6), 2739–2747.

Ahmed, S., Macfarlane, G.T., Fite, A., McBain, A.J., Gilbert, P., Macfarlane, S. (2007). Mucosa-associated bacterial diversity in relation to human terminal ileum and colonic biopsy samples. *Appl. Environ. Microbiol.*, 73(22), 7435–7442.

Alberoni, D., Gaggìa, F., Baffoni, L., Modesto, M.M., Biavati, B., Di Gioia, D. (2019). *Bifidobacterium xylocopae Sp. nov.* and *Bifidobacterium aemilianum sp. nov.*, from the carpenter bee (*Xylocopa violacea*) digestive tract. *Syst. Appl. Microbiol.*, 42(2), 205–216.

Albrecht, S., Lane, J.A., Karina Mariño, K., Al Busadah, K.A., Carrington, S.D., Hickey, R.M., Rudd, P.M. (2014). A comparative study of free oligosaccharides in the milk of domestic animals. *Br. J. Nutr.*, 111(7), 1313–1328.

Altermann, E., Russell, W.M., Azcarate-Peril, M.A., Barrangou, R., Buck, B.L., McAuliffe, O., Souther, N., Dobson, A., Duong, T., Callanan, M. (2005). Complete genome sequence of the probiotic lactic acid bacterium *Lactobacillus acidophilus* NCFM. *Proc. Natl. Acad. Sci. USA*, 102(11), 3906–3912.

Alvarez-Sieiro, P., Montalbán-López, M., Mu, D., Kuipers, O.P. (2016). Bacteriocins of lactic acid bacteria: Extending the family. *Appl. Microbiol. Biotechnol.*, 100(7), 2939–2951.

Implementation of Enzymatic Processes and Lactic Bacteria in the Food Industries, coordinated by Mohamed GHOUL.

An, Y.H., Friedman, R.J. (1998). Concise review of mechanisms of bacterial adhesion to biomaterial surfaces. *J. Biomed. Mater. Res. (Appl. Biomater.)*, 43, 338–348.

Anal, A.K. and Singh, H. (2007). Recent advances in microencapsulation of probiotics for industrial applications and targeted delivery. *Trends Food Sci. Technol.*, 18(5), 240–251.

Anandharamakrishnan, C. (ed.). (2017). *Handbook of Drying for Dairy Products*. Wiley Blackwell, Chichester/Hoboken.

Ananta, E., Volkert, M., Knorr, D. (2005). Cellular injuries and storage stability of spray-dried *Lactobacillus rhamnosus* GG. *Int. Dairy J.*, 15(4), 399–409.

Andresen, V., Gschossmann, J., Layer, P. (2020). Heat-inactivated *Bifidobacterium bifidum* MIMBb75 (SYN-HI-001) in the treatment of irritable bowel syndrome: A multicentre, randomised, double-blind, placebo-controlled clinical trial. *Lancet Gastroenterol. Hepatol.*, 5(7), 658–666.

Anjum, N., Maqsood, S., Masud, T., Ahmad, A., Sohail, A., Momin, A. (2014). *Lactobacillus acidophilus*: Characterization of the species and application in food production. *Crit. Rev. Food Sci. Nutr.*, 54(9), 1241–1251.

Aoki, T., Asahara, T., Matsumoto, K., Takada, T., Chonan, O., Nakamori, K., Nonaka, C., Yamaji, I., Hisamoto, T., Sato, M. (2014). Effects of the continuous intake of a milk drink containing *Lactobacillus casei* strain Shirota on abdominal symptoms, fecal microbiota, and metabolites in gastrectomized subjects. *Scand. J. Gastroenterol.*, 49(5), 552–563.

Arboleya, S., Watkins, C., Stanton, C., Ross, R.P. (2016). Gut bifidobacteria populations in human health and aging. *Front. Microbiol.*, 7, 1204.

Arena, M.P., Capozzi, V., Spano, G., Fiocco, D. (2017). The potential of lactic acid bacteria to colonize biotic and abiotic surfaces and the investigation of their interactions and mechanisms. *Appl. Microbiol. Biotechnol.*, 101(7), 2641–2657.

Arqués, J.L., Rodrigues, E., Langa, S., Landete, J., Medina, M. (2015). Antimicrobial activity of lactic acid bacteria in dairy products and gut: Effect on pathogens. *BioMed Res. Int.*, e584183.

Asemi, Z., Zare, Z., Shakeri, H., Sabihi, S., Esmaillzadeh, A. (2013). Effect of multispecies probiotic supplements on metabolic profiles, hs-CRP, and oxidative stress in patients with type 2 diabetes. *Ann. Nutr. Metab.*, 63, 1–9.

Ashaolu, T.J. and Reale, A. (2020). Review: A holistic review on euro-asian lactic acid bacteria fermented cereals and vegetables. *Microorganisms*, 8(8), 1176.

Ashraf, R. and Shah, N.P. (2014). Immune system stimulation by probiotic microorganisms. *Crit. Rev. Food Sci. Nutr.*, 54, 938–956.

Baffoni, L., Stenico, V., Strahsburger, E., Gaggìa, F., Di Gioia, D., Modesto, M., Mattarelli, P., Biavati, B. (2013). Identification of species belonging to the *Bifidobacterium* genus by PCR-RFLP analysis of a hsp60 gene fragment. *BMC Microbiol.*, 13, 149.

Baharav, E., Mor, F., Halpern, M., Weinberger, A. (2004). *Lactobacillus* GG bacteria ameliorate arthritis in Lewis rats. *J. Nutr.*, 134(8), 1964–1969.

Bansil, R. and Turner, B.S. (2006). Mucin structure, aggregation, physiological functions and biomedical applications. *Curr. Opin. Colloid Interface Sci.*, 11, 164–170.

Barberis, C.M., Cittadini, R.M., Almuzara, M.N., Feinsilberg, A., Famiglietti, A.M., Ramírez, M.S., Vay, C.A. (2012). Recurrent urinary infection with *Bifidobacterium scardovii. J. Clin. Microbiol.*, 50(3), 1086–1088.

Barroso, E., Martín, V., Martínez-Cuesta, M.C., Peláez, C., Requena, T. (2015). Stability of saliva microbiota during moderate consumption of red wine. *Arch. Oral Biol.*, 60(12), 1763–1768.

Baschali, A., Tsakalidou, E., Kyriacou, A., Karavasiloglou, N., Matalas, A.-L. (2017). Traditional low-alcoholic and non-alcoholic fermented beverages consumed in European countries: A neglected food group. *Nutr. Res. Rev.*, 30(1), 1–24.

Beighton, D., Al-Haboubi, M., Mantzourani, M., Gilbert, S.C., Clark, D., Zoitopoulos, L., Gallagher, J.E. (2010). Oral Bifidobacteria: Caries-associated bacteria in older adults. *J. Dent. Res.*, 89(9), 970–974.

Berche, P. (2012). Louis Pasteur, from crystals of life to vaccination. *Clin. Microbiol. Infect.*, 18, 1–6.

Bergonzelli, G.E., Granato, D., Pridmore, R.D., Marvin-Guy, L.F., Donnicola, D., Corthésy-Theulaz, I.E. (2006). GroEL of *Lactobacillus johnsonii* La1 (NCC 533) is cell surface associated: Potential role in interactions with the host and the gastric pathogen. *Helicobacter pylori. Infect. Immun.*, 74(1), 425–434.

Bernardeau, M., Vernoux, J.P., Henri-Dubernet, S., Gueguen, M. (2008). Safety assessment of dairy microorganisms: The *Lactobacillus* genus. *Int. J. Food Microbiol.*, 126, 278–285.

Bertelli, C., Pillonel, T., Torregrossa, A., Prod'hom, G., Fischer, C.J., Greub, G., Giannoni, E. (2014). *Bifidobacterium longum* bacteremia in preterm infants recei-ving probiotics. *Clin. Infect. Dis.*, 60(6), 924–927.

Biavati, B. and Mattarelli, P. (1991). *Bifidobacterium ruminantium sp. nov.* and *Bifidobac-terium merycicum sp. nov.* from the rumens of cattle. *Int. J. Syst. Bacteriol.*, 41(1), 163–168.

Biavati, B., Scardovi, V., Moore, W.E.C. (1982). Electrophoretic patterns of proteins in the genus *Bifidobacterium* and proposal of four new species. *Int. J. Syst. Bacteriol.*, 32, 358–373.

Binnendijk, K.H. and Rijkers, G.T. (2013). What is a health benefit? An evaluation of EFSA opinions on health benefits with reference to probiotics. *Benef. Microbes*, 4(3), 223–230.

Blareau, J.-P., Romond, M.-B., Romond, C., Lecroix, F., Gontier, C. (2008). Method for the production of an immunostimulant milk product and uses thereof. Patent, US7410653B1.

Blareau, J.-P., Colavizza, M., Huguet, F., Romond, C. (2013). Macromolecular complex of bacterial origin and use of said macromolecular complex for preventing and treating inflammatory rheumatism. Patent, US20130130988A1.

Breidt, F., Crowley, K.A., Fleming, H.P. (1995). Controlling cabbage fermentations with nisin and nisin-resistant *Leuconostoc mesenteroides*. *Food Microbiol.*, 12, 109–116.

Broeckx, G., Vandenheuvel, D., Claes, I.J.J., Lebeer, S., Kiekens, F. (2016). Drying techniques of probiotic bacteria as an important step towards the development of novel pharmabiotics. *Int. J. Pharm.*, 505(1/2), 303–318.

Buck, B.L., Altermann, E., Svingerud, T., Klaenhammer, T.R. (2005). Functional analysis of putative adhesion factors in *Lactobacillus acidophilus* NCFM. *Appl. Environ. Microbiol.*, 71(12), 8344–8351.

Bugarski, B., Li, Q., Goosen, M.F.A., Poncelet, D., Neufeld, R.J., Vunjak, G. (1994). Electrostatic droplet generation: Mechanism of polymer droplet formation. *AIChE J.*, 40(6), 1026–1031.

Bujna, E., Farkas, N.A., Tran, A.M., Dam, M.S., Nguyen, Q.D. (2017). Lactic acid fermentation of apricot juice by mono- and mixed cultures of probiotic *Lactobacillus* and *Bifidobacterium* strains. *Food Sci. Biotechnol.*, 27(2), 547–554.

Bunesova, V., Vlkova, E., Rada, V., Killer, J., Kmet, V. (2013). Identification of bifidobacteria isolated from Asian elephant (*Elephas maximus*). *J. Biosci.*, 38(2), 239–43.

Burgain, J., Gaiani, C., Linder, M., Scher, J. (2011). Encapsulation of probiotic living cells: From laboratory scale to industrial applications. *J. Food Eng.*, 104(4), 467–483.

Burgain, J., Gaiani, C., Cailliez-Grimal, C., Jeandel, C., Scher, J. (2013a). Encapsulation of *Lactobacillus rhamnosus* GG in microparticles: Influence of casein to whey protein ratio on bacterial survival during digestion. *Innov. Food Sci. Emerg. Technol.*, 19, 233–242.

Burgain, J., Gaiani, C., Francius, G., Revol-Junelles, A.M., Cailliez-Grimal, C., Lebeer, S., Tytgat, H.L.P., Vanderleyden, J., Scher, J. (2013b). In vitro interactions between probiotic bacteria and milk proteins probed by atomic force microscopy. *Colloids Surf. B*, 104, 153–162.

Burgain, J., Scher, J., Francius, G., Borges, F., Corgneau, M., Revol-Junelles, A.M., Cailliez-Grimal, C., Gaiani, C. (2014a). Lactic acid bacteria in dairy food: Surface characterization and interactions with food matrix components. *Adv. Colloid Interface Sci.*, 213, 21–35.

Burgain, J., Scher, J., Lebeer, S., Vanderleyden, J., Cailliez-Grimal, C., Corgneau, M., Francius, G., Gaiani, C. (2014b). Significance of bacterial surface molecules interactions with milk proteins to enhance microencapsulation of *Lactobacillus rhamnosus* GG. *Food Hydrocoll.*, 41, 60–70.

Burgain, J., Scher, J., Lebeer, S., Vanderleyden, J., Corgneau, M., Guerin, J., Caillet, C., Duval, J.F., Francius, G., Gaiani, C. (2015). Impacts of pH-mediated EPS structure on probiotic bacterial *pili*–whey proteins interactions. *Colloids Surf. B*, 134, 332–338.

Burgain, J., Scher, J., Gaiani, C. (2016). Microencapsulation de bactéries probiotiques. *Techniques de l'ingénieur* [Online]. Available at: https://www.techniques-ingenieur.fr/base-documentaire/procedes-chimie-bio-agro-th2/procedes-biochimiques-et-chimiques-en-agroalimentaire-42431210/ microencapsulation-de-bacteries-probiotiques-f3800/ [Accessed 30 November 2020].

Campana, R., van Hemert, S., Baffone, W. (2017). Strain-specific probiotic properties of lactic acid bacteria and their interference with human intestinal pathogens invasion. *Gut Pathog.*, 6(9), 12.

Carlstedt-Duke, B., Midtvedt, T., Nord, C.E., Gustafsson, B.E. (1986). Isolation and characterization of a mucin-degrading strain of *Peptostreptococcus* from rat intestinal tract. *Acta Pathol. Microbiol. Immunol. Scand. B.*, 94(5), 293–300.

Celik, O.F. and O'Sullivan, D.J. (2013). Factors influencing the stability of freeze-dried stress-resilient and stress-sensitive strains of bifidobacteria. *J. Dairy Sci.*, 96(6), 3506–3516.

Chapot-Chartier, M.-P. and Kulakauskas, S. (2014). Cell wall structure and function in lactic acid bacteria. *Microb. Cell Factories*, 13(S1), S9.

Cheikhyoussef, A., Pogori, N., Chen, H., Tian, F., Chen, H., Tang, J., Zhang, H. (2009). Antimicrobial activity and partial characterization of bacteriocin-like inhibitory substances (BLIS) produced by *Bifidobacterium infantis* BCRC 14602. *Food Control*, 20, 553–559.

Cheikhyoussef, A., Cheikhyoussef, N., Chen, H., Zhao, J., Tang, J., Zhang, H., Chen, W. (2010). Bifidin I – A new bacteriocin produced by *Bifidobacterium infantis* BCRC 14602: Purification and partial amino acid sequence. *Food Control*, 21, 746–753.

Chen, J., Wang, J., Zheng, H. (2021). Characterization of *Bifidobacterium apousia sp. nov.*, *Bifidobacterium choladohabitans sp. nov.*, and *Bifidobacterium polysaccharolyticum sp. nov.*, three novel species of the genus *Bifidobacterium* from honey bee gut. *Syst. Appl. Microbiol.*, 44, 126247.

Choi, J.-H., Lee, K.M., Lee, M.-K., Cha, C.-J., Kim, G.-B. (2014). *Bifidobacterium fae-cale sp. nov.*, isolated from human faeces. *Int. J. Syst. Evol. Microbiol.*, 64(9), 3134–3139.

Chuah, L.-O., Shamila-Syuhada, A.-K., Liong, M.-T., Rosma, A., Thong, K.L., Rusul, G. (2016). Physio-chemical, microbiological properties of tempoyak and molecular characterisation of lactic acid bacteria isolated from tempoyak. *Food Microbiol.*, 58, 95–104.

Cleusix, V., Lacroix, C., Dasen, G., Leo, M., Le Blay, G. (2010). Comparative study of a new quantitative real-time PCR targeting the xylulose-5-phosphate/fructose-6-phosphate phosphoketolase bifidobacterial gene (xfp) in faecal samples with two fluorescence in situ hybridization methods. *J. Appl. Microbiol.*, 108(1), 181–193.

Colín-Cruz, M.A., Pimentel-González, D.J., Carrillo-Navas, H., Alvarez-Ramírez, J., Guadarrama-Lezama, A.Y. (2019). Co-encapsulation of bioactive compounds from blackberry juice and probiotic bacteria in biopolymeric matrices. *LWT. Food Sci. Technol.*, 110, 94–101.

Collado, M.C., Gonzalez, A., Gonzalez, R., Hernandez, M., Ferrus, M.A., Sanza, Y. (2005a). Antimicrobial peptides are among the antagonistic metabolites produced by *Bifidobacterium* against *Helicobacter pylori*. *Int. J. Antimicrob. Agents*, 25(5), 385–391.

Collado, M.C., Hernandez, M., Sanza, Y. (2005b). Production of bacteriocin-like inhibitory compounds by human fecal *Bifidobacterium* strains. *J. Food Prot.*, 68(5), 1034–1040.

Colombel, J.-F., Cortot, A., Neut, C., Romond, C. (1987). Yoghurt with *Bifidobacterium longum* reduces erythromycin-induced gastrointestinal effects. *Lancet*, 2(8549), 43.

Cook, M.T., Tzortzis, G., Charalampopoulos, D., Khutoryanskiy, V.V. (2012). Microencapsulation of probiotics for gastrointestinal delivery. *J. Control. Release*, 162(1), 56–67.

Corcoran, B.M., Ross, R.P., Fitzgerald, G.F., Stanton, C. (2004). Comparative survival of probiotic lactobacilli spray-dried in the presence of prebiotic substances. *J. Appl. Microbiol.*, 96(5), 1024–1039.

Corr, S.C., Hill, C., Gahan, C.G. (2009). Understanding the mechanisms by which probiotics inhibit gastrointestinal pathogens. *Adv. Food Nutr. Res.*, 56, 1–15.

Crociani, F., Alessandrini, A., Mucci, M.M., Biavati, B. (1994). Degradation of complex carbohydrates by *Bifidobacterium spp*. *Int. J. Food Microbiol.*, 24(1/2), 199–210.

Cui, J.-H., Cao, Q.-R., Lee, B.-J. (2007). Enhanced delivery of bifidobacteria and fecal changes after multiple oral administrations of bifidobacteria-loaded alginate poly-l-lysine microparticles in human volunteers. *Drug Deliv.*, 14(5), 265–271.

De Bruyn, F., Beauprez, J., Maertens, J., Soetaert, W., De Mey, M. (2013). Unraveling the Leloir pathway of *Bifidobacterium bifidum*: Significance of the uridylyltransferases. *Appl. Environ. Microbiol.*, 79(22), 7028–7035.

De Prisco, A. and Mauriello, G. (2016). Probiotication of foods: A focus on microencapsulation tool. *Trends Food Sci. Technol.*, 48, 27–39.

Delcenserie, V., Gavini, F., Beerens, H., Tresse, O., Franssen, C., Daube, G. (2007). Description of a new species, *Bifidobacterium crudilactis sp. nov.*, isolated from raw milk and raw milk cheeses. *Syst. Appl. Microbiol.*, 30(5), 381–389.

Desmazeaud, M. (1983). L'état des connaissances en matière de nutrition des bactéries lactiques. *Lait*, 63, 267–316.

Devoyod, J.J. and Poullain, F. (1988). Les Leuconostocs. Propriétés : leur rôle en technologie laitière. *Lait*, 68(3), 249–279.

Dietrich, C.G., Kottmann, T., Alavi, M. (2014). Commercially available probiotic drinks containing *Lactobacillus casei* DN-114001 reduce antibiotic-associated diarrhea. *World J. Gastroenterol.*, 20(42), 15837.

Dimidi, E., Christodoulides, S., Fragkos, K.C., Scott, S.M., Whelan, K. (2014). The effect of probiotics on functional constipation in adults: A systematic review and meta-analysis of randomized controlled trials. *Am. J. Clin. Nutr.*, 100(4), 1075–1084.

Distrutti, E., Monaldi, L., Ricci, P., Fiorucci, S. (2016). Gut microbiota role in irritable bowel syndrome: New therapeutic strategies. *World J. Gastroenterol.*, 22(7), 2219–2241.

Dlugosz, A., Björn Winckler, B., Lundin, E., Zakikhany, K., Sandström, G., Ye, W., Engstrand, L., Lindberg, G. (2015). No difference in small bowel microbiota between patients with irritable bowel syndrome and healthy controls. *Sci. Rep.*, 5, 8508.

Doherty, S.B., Gee, V.L., Ross, R.P., Stanton, C., Fitzgerald, G.F., Brodkorb, A. (2011). Development and characterisation of whey protein micro-beads as potential matrices for probiotic protection. *Food Hydrocoll.*, 25(6), 1604–1617.

Doherty, S.B., Auty, M.A., Stanton, C., Ross, R.P., Fitzgerald, G.F., Brodkorb, A. (2012). Survival of entrapped *Lactobacillus rhamnosus* GG in whey protein micro-beads during simulated ex vivo gastro-intestinal transit. *Int. Dairy J.*, 22(1), 31–43.

Dong, X., Xin, Y., Jian, W., Liu, X., Ling, D. (2000). *Bifidobacterium thermacidophilum sp. nov.*, isolated from an anaerobic digester. *Int. J. Syst. Evol. Microbiol.*, 50, 119–125.

Donovan, S.M. and Comstock, S.S. (2016). Human milk oligosaccharides influence neonatal mucosal and systemic immunity. *Ann. Nutr. Metab.*, 69, 41–51.

Doron, S., Snydman, D.R., Gorbach, S.L. (2005). *Lactobacillus* GG: Bacteriology and clinical applications. *Gastroenterol. Clin. North Am.*, 34(3), 483–498.

Douillard, F.P., Ribbera, A., Järvinen, H.M., Kant, R., Pietilä, T.E., Randazzo, C., Paulin, L., Laine P.K., Caggia, C., Ossowski, I.V. et al. (2013). Comparative genomic and functional analysis of Lactobacillus casei and Lactobacillus rhamnosus strains marketed as probiotics. *Appl. Environ. Microbiol.*, 79(6), 1923–1933.

Du, H., Wang, X., Zhang, Y., Xu, Y. (2019). Exploring the impacts of raw materials and environments on the microbiota in Chinese Daqu starter. *Int. J. Food Microbiol.*, 297, 32–40.

Duranti, S., Gaiani, F., Mancabelli, L., Milani, C., Grandi, A., Bolchi, A., Santoni, A., Lugli, G.A., Ferrario, C., Mangifesta, M. et al. (2016). Elucidating the gut microbiome of ulcerative colitis: Bifidobacteria as novel microbial biomarkers. *FEMS Microbiol. Ecol.*, 92(12).

Duranti, S., Mangifesta, M., Lugli, G.A., Turroni, F., Anzalone, R., Milani, C., Mancabelli, L., Ossiprandi, M.C., Ventura, M. (2017). *Bifidobacterium vansinderenii sp. nov.*, isolated from faeces of emperor tamarin (*Saguinus imperator*). *Int. J. Syst. Evol. Microbiol.*, 67(10), 3987–3995.

Duranti, S., Lugli, G.A., Napoli, S., Anzalone, R., Milani, C., Mancabelli, L., Alessandri, G., Turroni, F., Ossiprandi, M.C., van Sinderen, D. et al. (2019). Characterization of the phylogenetic diversity of five novel species belonging to the genus *Bifidobacterium*: *Bifidobacterium castoris sp. nov.*, *Bifidobacterium callimi-conis sp. nov.*, *Bifidobacterium goeldii sp. nov.*, *Bifidobacterium samirii sp. nov.* and *Bifidobacterium dolichotidis sp. nov. Int. J. Syst. Evol. Microbiol.*, 69(5), 1288–1298.

Duranti, S., Lugli, G.A., Viappiani, A., Mancabelli, L., Alessandri, G., Anzalone, R., Longhi, G., Milani, C., Ossiprandi, M.C., Turroni, F. et al. (2020). Characterization of the phylogenetic diversity of two novel species belonging to the genus *Bifidobacterium*: *Bifidobacterium cebidarum sp. nov.* and *Bifidobacterium leonto-pitheci sp. nov. Int. J. Syst. Evol. Microbiol.*, 70(4), 2288–2297.

Eckel, V.P.L., Ziegler, L.-M., Vogel, R.F., Ehrmann, M. (2020). *Bifidobacterium tibiigranuli sp. nov.* isolated from homemade water kefir. *Int. J. Syst. Evol. Microbiol.*, 70(3), 1562–1570.

EFSA Panel on Dietetic Products, Nutrition and Allergies (NDA) (2010). Scientific Opinion on the substantiation of health claims related to live yoghurt cultures and improved lactose digestion (ID 1143, 2976) pursuant to Article 13(1) of Regulation (EC) No. 1924/2006. *EFSA Journal*, 8(10), 1763.

EFSA Panel on Dietetic Products, Nutrition and Allergies (NDA) (2014). Scientific Opinion on the substantiation of a health claim related to AlphaGOS® and a reduction of post-prandial glycaemic responses pursuant to Article 13(5) of Regulation (EC) No. 1924/2006. *EFSA Journal*, 12(10), 3838.

EFSA Panel on Dietetic Products, Nutrition and Allergies (NDA) (2015). Scientific Opinion on the substantiation of a health claim related to "native chicory inulin" and maintenance of normal defecation by increasing stool frequency pursuant to Article 13.5 of Regulation (EC) No. 1924/20061. *EFSA Journal*, 13(1), 3951.

Egan, M., Motherway, M.O., Ventura, M., van Sinderen, D. (2014). Metabolism of sialic acid by *Bifidobacterium breve* UCC2003. *Appl. Environ. Microbiol.*, 80(14), 4414–4426.

El Kfoury, K.A., Romond, M.-B., Scuotto, A., Alidjinou, E.K., Dabboussi, F., Hamze, M., Engelmann, I., Sane, F., Hober, D. (2017). Bifidobacteria-derived lipoproteins inhibit infection with coxsackievirus B4 in vitro. *Int. J. Antimicrob. Agents*, 50(2), 177–185.

El Mafadi, S. and Poncelet, D. (2007). L'enrobage en lit fluidisé pour la production de microcapsules. In *Microencapsulation: des sciences aux technologies*, Vandamme, T., Poncelet, D., Subra-Paternault, P. (eds). Lavoisier, Paris.

El-Hawiet, A., Kitova, E.N., Klassen, J.S. (2015). Recognition of human milk oligosaccharides by bacterial exotoxins. *Glycobiology*, 25, 845–854.

Endo, A., Futagawa-Endo, Y., Schumann, P., Pukall, R., Dicks, L.M.T. (2012). *Bifidobacterium reuteri sp. nov.*, *Bifidobacterium callitrichos sp. nov.*, *Bifidobacterium saguini sp. nov.*, *Bifidobacterium stellenboschense sp. nov.* and *Bifidobacterium biavatii sp. nov.* isolated from faeces of common marmoset (*Callithrix jacchus*) and red-handed tamarin (*Saguinus midas*). *Syst. Appl. Microbiol.*, 35(2), 92–97.

Eriksson, L., Holgerson, P., Johansson, I. (2017). Saliva and tooth biofilm bacterial microbiota in adolescents in a low caries community. *Sci. Rep.*, 7(1), 5861.

Esaiassen, E., Cavanagh, P., Hjerde, E., Simonsen, G.S., Støen, R., Klingenberg, C. (2016). *Bifidobacterium longum subspecies infantis* bacteremia in 3 extremely preterm infants receiving probiotics. *Emerg. Infect. Dis.*, 22(9), 1664–1666.

Eshaghi, M., Bibalan, M.H., Rohani, M., Esghaei, M., Douraghi, M., Talebi, M., Reza Pourshafie, M. (2017). Bifidobacterium obtained from mother's milk and their infant stool; a comparative genotyping and antibacterial analysis. *Microb. Pathog.*, 111, 94–98.

Etzold, S., MacKenzie, D.A., Jeffers, F., Walshaw, J., Roos, S., Hemmings, A.M., Juge, N. (2014). Structural and molecular insights into novel surface-exposed mucus adhesins from *Lactobacillus reuteri* human strains. *Mol. Microbiol.*, 92(3), 543–556.

Fakhry, S., Manzo, N., D'Apuzzo, E., Pietrini, L., Sorrentini, I., Ricca, E., De Felice, M., Baccigalupi, L. (2009). Characterization of intestinal bacteria tightly bound to the human ileal epithelium. *Res. Microbiol.*, 160(10), 817–823.

Fong, F.L.Y., Shah, N.P., Kirjavainen, P., El-Nezami, H. (2016). Mechanism of action of probiotic bacteria on intestinal and systemic immunities and antigen-presenting cells. *Int. Rev. Immunol.*, 35, 179–188.

Forbes, J.D., Chen, C.-Y., Knox, N.C., Marrie, R.-A., El-Gabalawy, H., Kievit, T., Alfa, M., Bernstein, C.N., Van Domselaar, G. (2018). A comparative study of the gut microbiota in immune-mediated inflammatory diseases. Does a common dysbiosis exist? *Microbiome*, 6(1), 221.

Franz, C.M., Huch, M., Mathara, J.M., Abriouel, H., Benomar, N., Reid, G., Galvez, A., Holzapfel, W.H. (2014). African fermented foods and probiotics. *Int. J. Food Microbiol.*, 190, 84–96.

Freitas, A.C. and Hill, J.E. (2017). Quantification, isolation and characterization of *Bifidobacterium* from the vaginal microbiomes of reproductive aged women. *Anaerobe*, 47, 145–156.

Freitas, A.C. and Hill, J.E. (2018). Bifidobacteria isolated from vaginal and gut microbiomes are indistinguishable by comparative genomics. *PLoS One*, 13(4), e0196290.

de la Fuente-Salcido, N.M., Castañeda-Ramírez, J.C., García-Almendárez, B.E., Bideshi, D.K., Salcedo-Hernández, R., Barboza-Corona, J.E. (2015). Isolation and characterization of bacteriocinogenic lactic bacteria from M-Tuba and Tepache, two traditional fermented beverages in Mexico. *Food Sci. Nutr.*, 3(5), 434–442.

Fuller, R. (1989). Probiotics in man and animals. *Afrc. J. Appl. Microbiol.*, 66, 365–378.

Fuller, R. (1992). History and development of probiotics. In *Probiotics*, Fuller, R. (ed.). Springer, Dordrecht.

Fushinobu, S. (2010). Unique sugar metabolic pathways of bifidobacteria. *Biosci. Biotechnol. Biochem.*, 74(12), 2374–2384.

Gao, X.W., Mubasher, M., Fang, C.Y., Reifer, C., Miller, L.E. (2010). Dose-response efficacy of a proprietary probiotic formula of *Lactobacillus acidophilus* CL1285 and *Lactobacillus casei* LBC80R for antibiotic-associated diarrhea and *Clostridium difficile*-associated diarrhea prophylaxis in adult patients. *Am. J. Gastroenterol.*, 105(7), 1636–1641.

Garrido, D., Ruiz-Moyano, S., Lemay, D.G., Sela, D.A., German, J.B., Mills, D.A. (2015). Comparative transcriptomics reveals key differences in the response to milk oligosaccharides of infant gut-associated bifidobacteria. *Sci. Rep.*, 5(13517).

Gavini, F., Pourcher, A.-M., Neut, C., Monget, D., Romond, C., Oger C., Izar, D. (1991). Phenotypic differentiation of bifidobacteria of human and animal origins. *Int. J. Syst. Bacteriol.*, 41(4), 548–557.

Gavini, F., Delcenserie, V., Kopeinig, K., Pollinger, S., Beerens, H., Bonaparte, C., Upmann, M. (2006). *Bifidobacterium* species isolated from animal feces and from beef and pork meat. *J. Food Prot.*, 69(4), 871–877.

Gharsallaoui, A., Roudaut, G., Chambin, O., Voilley, A., Saurel, R. (2007). Applications of spray-drying in microencapsulation of food ingredients: An overview. *Food Res. Int.*, 40(9), 1107–1121.

Gibson, G.R. and Roberfroid, M.B. (1995). Dietary modulation of the human colonic microbiota: Introducing the concept of prebiotics. *J. Nutr.*, 125(6), 1401–1412.

Gibson, G.R., Hutkins, R., Sanders, M.E., Prescott, S.L., Reimer, R.A., Salminen, S.J., Scott, K., Stanton, C., Swanson, K.S., Cani, P.D. (2017). Expert consensus document: The International Scientific Association for Probiotics and Prebiotics (ISAPP) consensus statement on the definition and scope of prebiotics. *Nat. Rev. Gastroenterol. Hepatol.*, 14(8), 491.

Gilliland, S.E., Morelli, L., Reid, G. (2001). Joint FAO/WHO Expert Consultation on Evaluation of Health and Nutritional Properties of Probiotics in Food including Powder Milk with Live Lactic Acid Bacteria. Report, World Health Organization, Geneva/Food and Agriculture Organization of the United Nations, Rome.

Gleinser, M., Grimm, V., Zhurina, D., Yuan, J., Riedel, C.U. (2012). Improved adhesive properties of recombinant bifidobacteria expressing the *Bifidobacterium bifidum*-specific lipoprotein BopA. *Microb. Cell Fact.*, 11, 80.

Goldin, B.R., Gualtieri, L.J., Moore, R.P. (1996). The effect of *Lactobacillus* GG on the initiation and promotion of DMH-induced intestinal tumors in the rat. *Nutr. Cancer.*, 25(2), 197–204.

Gomand, F., Borges, F., Salim, D., Burgain, J., Guerin, J., Gaiani, C. (2018). High-throughput screening approach to evaluate the adhesive properties of bacteria to milk biomolecules. *Food Hydrocoll.*, 84, 537–544.

Gomand, F., Borges, F., Burgain, J., Guerin, J., Revol-Junelles, A.-M., Gaiani, C. (2019). Food matrix design for effective lactic acid bacteria delivery. *Annu. Rev. Food Sci. Technol.*, 10(1), 285–310.

Gomand, F., Mitchell, W.H., Burgain, J., Petit, J., Borges, F., Spagnolie, S.E., Gaiani, C. (2020). Shaving and breaking bacterial chains with a viscous flow. *Soft Matter*, 16, 9273–9291.

Gorbach, S.L. (1996). The discovery of *Lactobacillus* GG. *Nutr. Today*, 31, 2S–4S.

Gosselink, M.P., Schouten, W.R., van Lieshout, L.M.C., Hop, W.C.J., Laman, J.D., Ruseler-van Embden, J.G.H. (2004). Delay of the first onset of pouchitis by oral intake of the probiotic strain *Lactobacillus rhamnosus* GG. *Dis. Colon Rectum*, 47(6), 876–884.

Gotoh, A., Katoh, T., Sakanaka, M., Ling, Y., Yamada, C., Asakuma, S., Urashima, T., Tomabechi, Y., Katayama-Ikegami, A., Kurihara, S. et al. (2018). Sharing of human milk oligosaccharides degradants within bifidobacterial communities in faecal cultures supplemented with *Bifidobacterium bifidum*. *Sci. Rep.*, 8, 13958.

Granato, D., Bergonzelli, G.E., Pridmore, R.D., Marvin, L., Rouvet, M., Corthésy-Theulaz, I.E. (2004). Cell surface-associated elongation factor Tu mediates the attachment of *Lactobacillus johnsonii* NCC533 (La1) to human intestinal cells and mucins. *Infect. Immun.*, 72(4), 2160–2169.

Grill, J.-P., Crociani, J., Ballongue J. (1995). Characterization of fructose 6 phosphate phosphoketolases purified from *Bifidobacterium* species. *Curr. Microbiol.*, 31(1), 9–54.

Guerin, J. (2017). Influence de l'ajout d'ingrédients fonctionnels laitiers sur l'encapsulation de *Lcb. rhamnosus* GG. Alimentation et Nutrition. PhD Thesis, Université de Lorraine, Nancy.

Guerin, J., Bacharouche, J., Burgain, J., Lebeer, S., Francius, G., Borges, F., Scher, J., Gaiani, C. (2016). *Pili* of *Lactobacillus rhamnosus* GG mediate interaction with β-lactoglobulin. *Food Hydrocoll.*, 58, 35–41.

Guerin, J., Petit, J., Burgain, J., Borges, F., Bhandari, B., Perroud, C., Desobry, S., Scher, J., Gaiani, C. (2017). *Lactobacillus rhamnosus* GG encapsulation by spray-drying: Milk proteins clotting control to produce innovative matrices. *J. Food Eng.*, 193, 10–19.

Guerin, J., Burgain, J., Francius, G., El-Kirat-Chatel, S., Beaussart, A., Scher, J., Gaiani, C. (2018). Adhesion of *Lactobacillus rhamnosus* GG surface biomolecules to milk proteins. *Food Hydrocoll.*, 82, 296–303.

Guerin, J., Burgain, J., Gomand, F., Scher, J., Gaiani, C. (2019). Milk fat globule membrane glycoproteins: Valuable ingredients for lactic acid bacteria encapsulation? *Crit. Rev. Food Sci. Nutr.*, 59, 639–651.

Guevara Nuñez, D., De Paulis, A.N., Bertona, E., Gutiérrez, M.A., Vay, C.A., Suárez, J.P., Predari, S.C. (2017). *Bifidobacterium scardovii* isolated from 2 consecutive urine samples. *Rev. Argent. Microbiol.*, 49(4), 328–331.

Guglielmetti, S., Ciranna, A., Mora, D., Parini, P., Karp, M. (2008). Construction, characterization and exemplificative application of bioluminescent *Bifidobacterium longum* biovar *longum*. *Int. J. Food Microbiol.*, 124, 285–290.

Guglielmetti, S., Mora, D., Gschwender, M., Popp, K. (2011). Randomised clinical trial: *Bifidobacterium bifidum* MIMBb75 significantly alleviates irritable bowel syndrome and improves quality of life–a double-blind, placebo-controlled study. *Aliment Pharmacol. Ther.*, 33(10), 1123–1132.

Guglielmetti, S., Balzaretti, S., Taverniti, V., Miriani, M., Milani, C., Scarafoni, A., Corona, S., Ciranna, A., Arioli, S., Santala, V. et al. (2014). TgaA, a VirB1-like component belonging to a putative type IV secretion system of *Bifidobacterium bifidum* MIMBb75. *Appl. Environ. Microbiol.*, 80(17), 5161–5169.

Guo, W., Mao, B., Cui, S., Tang, X., Zhang, Q., Zhao, J., Zhang, H. (2022). Protective effects of a novel probiotic *Bifidobacterium pseudolongum* on the intestinal barrier of colitis mice via modulating the Pparγ/STAT3 pathway and intestinal microbiota. *Foods*, 11(11), 1551.

Gupta, P., Andrew, H., Kirschner, B.S., Guandalini, S. (2000). Is *Lactobacillus* GG helpful in children with Crohn's disease? Results of a preliminary, open-label study. *J. Pediatr. Gastroenterol. Nutr.*, 31(4), 453–457.

Gyllenberg, H. and Niemelä, S. (1959). A selective method for the demonstration of bifid bacteria (*Lactobacillus bifidus*) in materials tested for faecal contamination. *Agr. Food Sci.*, 31(1), 94–97.

György, P., Kuhn, R., Rose, C.S., Zilliken, F. (1954). *Bifidus* factor. II. Its occurrence in milk from different species and in other natural products. *Arch. Biochem. Biophys.*, 48(1), 202–208.

Haarman, M. and Knol, J. (2005). Quantitative real-time PCR assays to identify and quantify fecal *Bifidobacterium* species in infants receiving a prebiotic infant formula. *Appl. Environ. Microbiol.*, 71(5), 2318–2324.

Hamilton-Miller, J.M.T., Gibson, G.R., Bruck, W. (2003). Some insights into the derivation and early uses of the word "probiotic". *Br. J. Nutr.*, 90, 845–845.

Hassinen, J.B., Durbin, G.T., Tomarelli, R.M., Bernhart, F.W. (1951). The minimal nutritional requirements of *Lactobacillus bifidus*. *J. Bacteriol.*, 62(6), 771–777.

Hatakka, K., Martio, J., Korpela, M., Herranen, M., Poussa, T., Laasanen, T., Saxelin, M., Vapaatalo, H., Moilanen, E., Korpela, R. (2003). Effects of probiotic therapy on the activity and activation of mild rheumatoid arthritis. A pilot study. *Scand. J. Rheumatol.*, 32(4), 211–215.

Heidebach, T., Först, P., Kulozik, U. (2009a). Transglutaminase-induced caseinate gelation for the microencapsulation of probiotic cells. *Int. Dairy J.*, 19(2), 77–84.

Heidebach, T., Först, P., Kulozik, U. (2009b). Microencapsulation of probiotic cells by means of rennet-gelation of milk proteins. *Food Hydrocoll.*, 23(7), 1670–1677.

Heidebach, T., Först, P., Kulozik, U. (2010). Influence of casein-based microencapsulation on freeze-drying and storage of probiotic cells. *J. Food Eng.*, 98(3), 309–316.

Herías, M.V., Midtvedt, T., Hanson, L.A., Wold, A.E. (1998). Increased antibody production against gut-colonizing *Escherichia coli* in the presence of the anaerobic bacterium *Peptostreptococcus*. *Scand. J. Immunol.*, 48(3), 277–282.

Hermansson, M. (1999). The DLVO theory in microbial adhesion. *Colloids Surf. B*, 14, 105–119.

Heuvelin, E., Lebreton, C., Bichara, M., Cerf-Bensussan, N., Heyman, M. (2010). A *Bifidobacterium* probiotic strain and its soluble factors alleviate chloride secretion by human intestinal epithelial cells. *J. Nutr.*, 140(1), 7–11.

Hill, C., Guarner, F., Reid, G., Gibson, G.R., Merenstein, D.J., Pot, B., Morelli, L., Canani, R.B., Flint, H.J., Salminen, S. (2014). Expert consensus document: The International Scientific Association for Probiotics and Prebiotics consensus statement on the scope and appropriate use of the term probiotic. *Nat. Rev. Gastroenterol. Hepatol.*, 11(8), 506.

Hoarau, C., Lagaraine, C., Martin, L., Velge-Roussel, F., Lebranchu, Y. (2006). Supernatant of *Bifidobacterium breve* induces dendritic cell maturation, activation, and survival through a Toll-like receptor 2 pathway. *J. Allergy Clin. Immunol.*, 117(3), 696–702.

Hoarau, C., Martin, L., Faugaret, D., Baron, C., Dauba, A., Aubert-Jacquin, C., Velge-Roussel, F., Lebranchu, Y. (2008). Supernatant from *Bifidobacterium* differentially modulates transduction signaling pathways for biological functions of human dendritic cells. *PLoS One*, 3(7), e2753.

Hojo, K., Nagaoka, S., Murata, S., Taketomo, N., Ohshima, T., Maeda, N. (2007). Reduction of vitamin K concentration by salivary *Bifidobacterium* strains and their possible nutritional competition with *Porphyromonas gingivalis*. *J. Appl. Microbiol.*, 103(5), 1969–1974.

Holzapfel, W.H. and Wood, B.J.B. (eds) (2014). *Lactic Acid Bacteria: Biodiversity and Taxonomy*. Wiley, Hoboken.

Hoyles, L., Inganäs, E., Falsen, E., Drancourt, M., Weiss, N., McCartney, A.L., Collins, M.D. (2002). *Bifidobacterium scardovii sp. nov.*, from human sources. *Int. J. Syst. Evol. Microbiol.*, 52(3), 995–999.

Huang, S., Méjean, S., Rabah, H., Dolivet, A., Le Loir, Y., Chen, X.D., Jan, G., Jeantet, R., Schuck, P. (2017a). Double use of concentrated sweet whey for growth and spray drying of probiotics: Towards maximal viability in pilot scale spray dryer. *J. Food Eng.*, 196, 11–17.

Huang, S., Vignolles, M.-L., Chen, X.D., Le Loir, Y., Jan, G., Schuck, P., Jeantet, R. (2017b). Spray drying of probiotics and other food-grade bacteria: A review. *Trends Food Sci. Technol.*, 63, 1–17.

Hütt, P., Koll, P., Stsepetova, J., Alvarez, B., Mändar, R., Krogh-Andersen, K., Marcotte, H., Hammarström, L., Mikelsaar, M. (2011). Safety and persistence of orally administered human *Lactobacillus* sp. Strains in healthy adults. *Benef. Microbes*, 2(1), 79–90.

Ibrahim, N., Elgasim, E., Amustafa, S. (2019). Isolation and identification of microorganisms in Kawal (*Cassia obtusifolia*) fermented leaves. *Int. J. Curr. Res.*, 11(10), 7956–7960.

Ipar, N., Aydogdu, S.D., Yildirim, G.K., Inal, M., Gies, I., Vandenplas, Y., Dinleyici, E.C. (2015). Effects of synbiotic on anthropometry, lipid profile and oxidative stress in obese children. *Benef. Microbes*, 6, 775–781.

Isas, A.S., Mariotti Celis, M.S., Pérez Correa, J.R., Fuentes, E., Rodríguez, L., Palomo, I., Mozzi, F., Van Nieuwenhove, C. (2020). Functional fermented cherimoya (*Annona cherimola Mill.*) juice using autochthonous lactic acid bacteria. *Food Res. Int.*, 138, 109729.

James, K., Motherway, M.O., Bottacini, F., van Sinderen, D. (2016). *Bifidobacterium breve* UCC2003 metabolises the human milk oligosaccharides lacto-N-tetraose and lacto-N-neo-tetraose through overlapping, yet distinct pathways. *Sci. Rep.*, 6, 38560.

Jantzen, M., Gopel, A., Beermann, C. (2013). Direct spray drying and microencapsulation of probiotic *Lactobacillus reuteri* from slurry fermentation with whey. *J. Appl. Microbiol.*, 115(4), 1029–1036.

Jarocki, P., Podleśny, M., Komoń-Janczara, E., Kucharska, J., Glibowska, A., Targoński, Z. (2016). Comparison of various molecular methods for rapid differentiation of intestinal bifidobacteria at the species, subspecies and strain level. *BMC Microbiol.*, 16(1), 159.

Jensen, H., Roos, S., Jonsson, H., Rud, I., Grimmer, S., van Pijkeren, J.-P., Britton, R.A., Axelsson, L. (2014). Role of *Lactobacillus reuteri* cell and mucus-binding protein A (CmbA) in adhesion to intestinal epithelial cells and mucus in vitro. *Microbiology (Reading, Engl.)*, 160, 671–681.

Jian, W., Zhu, L., Dong, X. (2001). New approach to phylogenetic analysis of the genus *Bifidobacterium* based on partial HSP60 gene sequences. *Int. J. Syst. Evol. Microbiol.*, 51(5), 1633–1638.

Jones, R.M. (2017). The use of *Lactobacillus casei* and *Lactobacillus paracasei* in clinical trials for the improvement of human health. In *The Microbiota in Gastrointestinal Pathophysiology*, Martin Floch, M., Ringel, Y., Walker, W.A. (eds). RELX Group/Elsevier, Amsterdam.

Juge, N. (2012). Microbial adhesins to gastrointestinal mucus. *Trends Microbiol.*, 20(1), 30–39.

Kankainen, M., Paulin, L., Tynkkynen, S., Ossowski, I.V., Reunanen, J., Partanen, P., Satokari, R., Vesterlund, S., Hendrickx, A.P.A., Lebeer, S. et al. (2009). Comparative genomic analysis of Lactobacillus rhamnosus GG reveals pili containing a human mucus binding protein. *Proc. Natl. Acad. Sci. USA*, 106(40), 17193–17198.

Kant, A., Palva, A., von Ossowski, I., Krishnan, V. (2020). Crystal structure of lactobacillar SpaC reveals an atypical five-domain *pilus* tip adhesin: Exposing its substrate-binding and assembly in SpaCBA. *Pili. J. Struct. Biol.*, 211, 107571.

Karav, S., Le Parc, A., Leite Nobrega de Moura Bell, J.M., Frese, S.A., Kirmiz, N., Block, D.E., Barile, D., Mill, D.A. (2016). Oligosaccharides released from milk glycoproteins are selective growth substrates for infant-associated bifidobacteria. *Appl. Environ. Microbiol.*, 82(12), 3622–3630.

Karav, S., Cohen, J.L., Barile, D., Leite Nobrega de Moura Bell, J.M. (2017). Recent advances in immobilization strategies for glycosidases. *Biotechnol. Prog.*, 33(1), 104–112.

Kato, K., Odamaki, T., Mitsuyama, E., Sugahara, H., Xiao, J.-Z., Osawa, R. (2017). Age-related changes in the composition of gut *Bifidobacterium* species. *Curr. Microbiol.*, 74(8), 987–995.

Katoh, T., Maeshibu, T., Kikkawa, K.-I., Gotoh, A., Tomabechi, Y., Nakamura, M., Liao, W.-H., Yamaguchi, M., Ashida, H., Yamamoto, K., et al. (2017). Identification and characterization of a sulfoglycosidase from *Bifidobacterium bifidum* implicated in mucin glycan utilization. *Biosci. Biotechnol. Biochem.*, 81(10), 2018–2027.

Katoh, T., Ojima, M.N., Sakanaka, M., Ashida, H., Gotoh, A., Katayama, T. (2020). Enzymatic adaptation of *Bifidobacterium bifidum* to host glycans, viewed from glycoside hydrolyases and carbohydrate-binding modules. *Microorganisms*, 8(4), 481.

Kaur, R., Gilbert, S.C., Sheehy, C., Beighton, D. (2013). Salivary levels of bifidobacteria in caries-free and caries-active children. *Int. J. Paediatr. Dent.*, 23(1), 32–38.

Keisam, S., Tuikhar, N., Ahmed, G., Jeyaram, K. (2019). Toxigenic and pathogenic potential of enteric bacterial pathogens prevalent in the traditional fermented foods marketed in the Northeast region of India. *Int. J. Food Microbiol.*, 296, 21–30.

Kelly, W.J., Cookson, A.L., Altermann, E., Lambie, S.C., Perry, R., Teh, K.H., Otter, D.E., Shapiro, N., Woyke, T., Leahy, S.C. (2016). Genomic analysis of three *Bifidobacterium* species isolated from the calf gastrointestinal tract. *Sci. Rep.*, 6, 30768.

Kerckhoffs, A.P.M., Samsom, M., van der Rest, M.E., de Vogel, J., Knol, J., Ben-Amor, K., Akkermans, L.M.A. (2009). Lower bifidobacteria counts in both duodenal mucosa-associated and fecal microbiota in irritable bowel syndrome patients. *World J. Gastroenterol.*, 15(23), 2887–2892.

Khem, S., Bansal, V., Small, D.M., May, B.K. (2016). Comparative influence of pH and heat on whey protein isolate in protecting *Lactobacillus plantarum* A17 during spray drying. *Food Hydrocoll.*, 54, 162–169.

Killer, J. and Marounek, M. (2011). Fermentation of mucin by bifidobacteria from rectal samples of humans and rectal and intestinal samples of animals. *Folia Microbiol.*, 56(2), 85–89.

Killer, J., Kopecný, J., Mrázek, J., Rada, V., Benada, O., Koppová, I., Havlík, J., Straka, J. (2009). *Bifidobacterium bombi sp. nov.,* from the bumblebee digestive tract. *Int. J. Syst. Evol. Microbiol.*, 59(8), 2020–2024.

Killer, J., Kopečný, J., Mrázek, J., Koppová, I., Havlík, J., Benada, O., Kott, T. (2011). *Bifidobacterium actinocoloniiforme sp. nov.* and *Bifidobacterium bohemicum sp. nov.*, from the bumblebee digestive tract. *Int. J. Syst. Evol. Microbiol.*, 61(6), 1315–1321.

Kim, B.J., Kim, H.-Y., Yun, Y.-J., Kim, B.-J., Kook, Y.-H. (2010). Differentiation of *Bifidobacterium* species using partial RNA polymerase {beta}-subunit (rpoB) gene sequences. *Int. J. Syst. Evol. Microbiol.*, 60(12), 2697–2704.

Kim, J.H., An, H.J., Garrido, D., German, J.B., Lebrilla, C.B., Mills, D.A. (2013). Proteomic analysis of *Bifidobacterium longum subsp. infantis* reveals the metabolic insight on consumption of prebiotics and host glycans. *PLoS One*, 8(2), e57535.

Kim, E., Cho, E.-J., Yang, S.-M., Kim, M.-J., Kim, H.-Y. (2021). Novel approaches for the identification of microbial communities in kimchi: MALDI-TOF MS analysis and high-throughput sequencing. *Food Microbiol.*, 94, 103641.

Kinoshita, H., Uchida, H., Kawai, Y., Kawasaki, T., Wakahara, N., Matsuo, H., Watanabe, M., Kitazawa, H., Ohnuma, S., Miura, K. et al. (2008). Cell surface *Lactobacillus plantarum* LA 318 glyceraldehyde-3-phosphate dehydrogenase (GAPDH) adheres to human colonic mucin. *J. Appl. Microbiol.*, 104(6), 1667–1674.

Kline, K.A., Falker, S., Dahlberg, S., Normark, S., Henriques-Normark, B. (2009). Bacterial adhesins in host-microbe interactions. *Cell Host Microbe*, 5(6), 580–592.

Kordy, K., Gaufin, T., Mwangi, M., Li, F., Cerini, C., Lee, D.J., Adisetiyo, H., Woodward, C., Pannaraj, P.S., Tobin, N.H. et al. (2020). Contributions to human breast milk microbiome and enteromammary transfer of *Bifidobacterium breve. PLoS One*, 15(1), e0219633.

Korgan, A.C., Foxx, C.L., Hashmi, H., Sago, S.A., Stamper, C.E., Heinze, J.D., O'Leary, E., King, J.L., Perrot, T.S., Lowry, C.A., et al. (2022). Effects of paternal high-fat diet and maternal rearing environment on the gut microbiota and behavior. *Sci. Rep.*, 12(1), 10179.

Krasaekoopt, W., Bhandari, B., Deeth, H. (2003). Evaluation of encapsulation techniques of probiotics for yoghurt. *Int. Dairy J.*, 13(1), 3–13.

Laloy, E., Vuillemard, J.-C., El Soda, M., Simard, R.E. (1996). Influence of the fat content of cheddar cheese on retention and localization of starters. *Int. Dairy J.*, 6, 729–740.

Langkamp-Henken, B., Rowe, C.C., Ford, A.L., Christman, M.C., Nieves Jr, C., Khouri, L., Specht, G.J., Girard, S.-A., Spaiser, S.J., Dahl, W.J. (2015). *Bifidobacterium bifidum* R0071 results in a greater proportion of healthy days and a lower percentage of academically stressed students reporting a day of cold/flu: A randomised, double-blind, placebo-controlled study. *Br. J. Nutr.*, 113(3), 426–434.

Lauer, E. (1990). *Bifidobacterium gallicum sp. nov.* isolated from human feces. *Int. J. Syst. Bacteriol.*, 40(1), 100–102.

Laureys, D., Cnockaert, M., De Vuyst, L., Vandamme, P. (2016). *Bifidobacterium aquikefiri sp. nov.*, isolated from water kefir. *Int. J. Syst. Evol. Microbiol.*, 66(3), 1281–1286.

Lebeer, S., Verhoeven, T.L.A., Velez, M.P., Vanderleyden, J., De Keersmaecker, S.C.J. (2007). Impact of environmental and genetic factors on Biofilm formation by the pro-biotic strain *Lactobacillus rhamnosus* GG. *Appl. Environ. Microbiol.*, 73, 6768–6775.

Lebeer, S., Vanderleyden, J., De Keersmaecker, S.C.J. (2008). Genes and molecules of lactobacilli supporting probiotic action. *Microbiol. Mol. Biol. Rev.*, 72, 728–764.

Lebeer, S., Verhoeven, T.L., Francius, G., Schoofs, G., Lambrichts, I., Dufrene, Y., Vanderleyden, J., De Keersmaecker, S.C. (2009). Identification of a gene cluster for the biosynthesis of a long galactose-rich exopolysaccharide in *Lactobacillus rhamnosus* GG and functional analysis of the priming glycosyltransferase. *Appl. Environ. Microbiol.*, 75, 3554–3563.

Lebeer, S., Vanderleyden, J., De Keersmaecker, S.C.J. (2010). Host interactions of probiotic bacterial surface molecules: Comparison with commensals and pathogens. *Nat. Rev. Micro.*, 8, 171–184.

Lebeer, S., Claes, I., Tytgat, H.L., Verhoeven, T.L., Marien, E., von Ossowski, I., Reunanen, J., Palva, A., Vos, W.M., Keersmaecker, S.C., Vanderleyden, J. et al. (2012). Functional analysis of *Lactobacillus rhamnosus* GG *pili* in relation to adhesion and immunomodulatory interactions with intestinal epithelial cells. *Appl. Environ. Microbiol.*, 78, 185–93.

Lee, J.-H., Karamychev, V.N., Kozyavkin, S., Mills, D., Pavlov, A., Pavlova, N., Polouchine, N., Richardson, P., Shakhova, V., Slesarev, A. et al. (2008). Comparative genomic analysis of the gut bacterium bifidobacterium longum reveals loci susceptible to deletion during pure culture growth. *BMC Genomics*, 9, 247.

Lee, J.-H., Li, X., O'Sullivan, D.J. (2011). Transcription analysis of a lantibiotic gene cluster from *Bifidobacterium longum* DJO10A. *Appl. Environ. Microbiol.*, 77(17), 5879–5887.

Li, T.T., Zhang, H.X., Gu, C.T. (2022). *Bifidobacterium mizhiense sp. nov.*, isolated from the gut of honeybee (*Apis mellifera*). *Int. J. Syst. Evol. Microbiol.*, 72, 005390.

Liévin, V., Peiffer, I., Hudault, S., Rochat, F., Brassart, D., Neeser, J.-R., Servin, A.L. (2000). *Bifidobacterium* strains from resident infant human gastrointestinal microflora exert antimicrobial activity. *Gut*, 47, 646–652.

Lilly, D.M. and Stillwell, R.H. (1965). Probiotics: Growth-promoting factors produced by microorganisms. *Science*, 147(3659), 747–748.

Lim, B.-K., Mahendran, R., Lee, Y.-K., Bay, B.-H. (2002). Chemopreventive effect of *Lactobacillus rhamnosus* on growth of a subcutaneously implanted bladder cancer cell line in the mouse. *Jpn. J. Cancer Res.*, 93(1), 36–41.

Liu, H., Gong, J., Chabot, D., Miller, S.S., Cui, S.W., Ma, J., Zhong, F., Wang, Q. (2015). Protection of heat-sensitive probiotic bacteria during spray-drying by sodium caseinate stabilized fat particles. *Food Hydrocoll.*, 51, 459–467.

Livney, Y.D. (2010). Milk proteins as vehicles for bioactives. *Curr. Opin. Colloid Interface Sci.*, 15(1/2), 73–83.

Lomonaco, S., Furumoto, E.J., Loquasto, J.R., Morra, P., Grassi, A., Roberts, R.F. (2015). Development of a rapid SNP-typing assay to differentiate *Bifidobacterium animalis ssp. lactis* strains used in probiotic-supplemented dairy products. *J. Dairy Sci.*, 98(2), 804–812.

Lopez, C., Maillard, M.-B., Briard-Bion, V., Camier, B., Hannon, J.A. (2006). Lipolysis during ripening of emmental cheese considering organization of fat and preferential localization of bacteria. *J. Agric. Food Chem.*, 54, 5855–5867.

Lugli, G.A., Mangifesta, M., Duranti, S., Anzalone, R., Milani, C., Mancabelli, L., Alessandri, G., Turroni, F., Ossiprandi, M.C., van Sinderen, D., et al. (2018). Phylogenetic classification of six novel species belonging to the genus *Bifidobacte-rium* comprising *Bifidobacterium anseris sp. nov.*, *Bifidobacterium criceti sp. nov.*, *Bifidobacterium imperatoris sp. nov.*, *Bifidobacterium italicum sp. nov.*, *Bifidobacte-rium margollesii sp. nov.* and *Bifidobacterium parmae sp. nov. Syst. Appl. Microbiol.*, 41(3), 173–183.

Lugli, G.A., Duranti, S., Milani, C., Mancabelli, L., Turroni, F., Alessandri, G., Longhi, G., Anzalone, R., Viappinai, A., Tarracchini, C. et al. (2020). Investigating bifidobacteria and human milk oligosaccharide composition of lactating mothers. *FEMS Microbiol. Ecol.*, 96(5).

Lugli, G.A., Calvete-Torre, I., Alessandri, G., Milani, C., Turroni, F., Laiolo, P., Ossiprandi, M.C., Margolles, A., Ruiz, L., Ventura, M. (2021). Phylogenetic classification of ten novel species belonging to the genus *Bifidobacterium* comprising *B. phasiani sp. nov.*, *B. pongonis sp. nov.*, *B. saguinibicoloris sp. nov.*, *B. colobi sp. nov.*, *B. simiiventris sp. nov.*, *B. santillanense sp. nov.*, *B. miconis sp. nov.*, *B. amazo-nense sp. nov.*, *B. pluvialisilvae sp. nov.*, and *B. miconisargentati sp. nov. Syst. Appl. Microbiol.*, 44, 126273.

Maciel, G.M., Chaves, K.S., Grosso, C.R.F., Gigante, M.L. (2014). Microencapsulation of *Lactobacillus acidophilus* La-5 by spray-drying using sweet whey and skim milk as encapsulating materials. *J. Dairy Sci.*, 97(4), 1991–1998.

Mack, D.R., Ahrne, S., Hyde, L., Wei, S., Hollingsworth, M.A. (2003). Extracellular MUC3 mucin secretion follows adherence of *Lactobacillus* strains to intestinal epithelial cells in vitro. *Gut*, 52, 827–833.

Magro, D.O., de Oliveira, L.M.R., Bernasconi, I., de Souza Ruela M., Credidio, L., Barcelos, I.K., Leal, R.F., de Lourdes Stesuko Ayrizono, M., Fagundes, J.J., de B Teixeira, L. et al. (2014). Effect of yogurt containing polydextrose, *Lactobacillus acidophilus* NCFM and *Bifidobacterium lactis* HN019: A randomized, double-blind, controlled study in chronic constipation. *Nutr. J.*, 13(1), 1–5.

Manojlović, V., Nedović, V.A., Kailasapathy, K., Zuidam, N.J. (2010). Encapsulation of probiotics for use in food products. In *Encapsulation Technologies for Active Food Ingredients and Food Processing*, Zuidam, N.J. and Nedovic, V. (eds). Springer, New York.

Manome, A., Abiko, Y., Kawashima, J., Washio, J., Fukumoto, S., Takahashi, N. (2019). Acidogenic potential of oral bifidobacterium and its high fluoride tolerance. *Front. Microbiol.*, 16(10), 1099-1109.

Mantzourani, M., Fenlon, M., Beighton, D. (2009). Association between *Bifidobacteriaceae* and the clinical severity of root caries lesions. *Oral Microbiol. Immunol.*, 24(1), 32–37.

Mantzourani, M., Gilbert, S.C., Fenlon, M., Beighton, D. (2010). Non-oral bifidobacteria and the aciduric microbiota of the denture plaque biofilm. *Mol. Oral Microbiol.*, 25(3), 190–199.

Markowiak, P. and Śliżewska, K. (2017). Effects of probiotics, prebiotics, and synbiotics on human health. *Nutrients*, 9, 1021.

Martens, M., Dawyndt, P., Coopman, R., Gillis, M., De Vos, P., Willems, A. (2008). Advantages of multilocus sequence analysis for taxonomic studies: A case study using 10 housekeeping genes in the genus *Ensifer* (including former *Sinorhizobium*). *Int. J. Syst. Evol. Microbiol.*, 58, 200–214.

Martinez, F.A.C., Balciunas, E.M., Converti, A., Cotter, P.D., de Souza Oliveira, R.P. (2013). Bacteriocin production by *Bifidobacterium spp.* A review. *Biotechnol. Adv.*, 31(4), 482–488.

Masco, L., Ventura, M., Zink, R., Huys, G., Swings, J. (2004). Polyphasic taxonomic analysis of *Bifidobacterium animalis* and *Bifidobacterium lactis* reveals relatedness at the subspecies level: Reclassification of *Bifidobacterium animalis* as *Bifidobacterium animalis subsp. animalis subsp. nov.* and *Bifidobacterium lactis* as *Bifidobacterium animalis subsp. lactis subsp. nov. Int. J. Syst. Evol. Microbiol.*, 54, 1137–1143.

Mathys, S., Lacroix, C., Mini, R., Meile, L. (2008). PCR and real-time PCR primers developed for detection and identification of *Bifidobacterium thermophilum* in faeces. *BMC Microbiol.*, 8, 179.

Mattarelli, P., Brandi, G., Calabrese, C., Fornari, F., Prati, G.M., Biavati, B., Sgorbati, B. (2014). Occurrence of bifidobacteriaceae in human hypochlorhydria stomach. *Microb. Ecol. Health Dis.*, 9(25), 21379–21386.

Mattarelli, P., Felis, G.E., Pot, B., Holzapfel, W.H., Franz, C.M.A.P. (2020). International committee on systematics of prokaryotes. Subcommittee on the taxonomy of *Bifidobacterium*, *Lactobacillus* and related organisms. Minutes of the closed meeting, 20 June 2019, Prague. *Int. J. Syst. Evol. Microbiol.*, 70(4), 2949–2951.

McClements, D.J. (2012). Requirements for food ingredient and nutraceutical delivery systems. In *Encapsulation Technologies and Delivery Systems for Food Ingredients and Nutraceuticals*, Garti, N. and McClements, D.J. (eds). Woodhead Publishing, Sawston.

Meddah, A.T., Yazourh, A., Desmet, I., Risbourg, B., Verstraete, W., Romond, M.-B. (2001). The regulatory effects of whey retentate from bifidobacteria fermented milk on the microbiota of the Simulator of the Human Intestinal Microbial Ecosystem (SHIME). *J. Appl. Microbiol.*, 91(6), 1110–1117.

Mekadim, C., Bunešová, V., Vlková, E., Hroncová, Z., Killer, J. (2019). Genetic marker-based multi-locus sequence analysis for classification, genotyping, and phylogenetics of the family *Bifidobacteriaceae* as an alternative approach to phylogenomics. *Antonie van Leeuwenhoek*, 112, 1785–1800.

Meng, X.C., Stanton, C., Fitzgerald, G.F., Daly, C., Ross, R.P. (2008). Anhydrobiotics: The challenges of drying probiotic cultures. *Food Chem.*, 106(4), 1406–1416.

Metchnikoff, E. (1907). *Essais optimistes*. Maloine, Paris.

Michael, D.R., Jack, A.A., Masetti, G., Davies, T.S., Loxley, K.E., Kerry-Smith, J., Plummer, J.F., Marchesi, J.R., Mullish, B.H., McDonald, J.A.K. et al. (2020). A randomised controlled study shows supplementation of overweight and obese adults with lactobacilli and bifidobacteria reduces bodyweight and improves well-being. *Sci. Rep.*, 10(1), 4183.

Michelini, S., Oki, K., Yanokura, E., Shimakawa, Y., Modesto, M., Mattarelli, P., Biavati, B., Watanabe, K. (2016a). *Bifidobacterium myosotis sp. nov., Bifidobacterium tissieri sp. nov.* and *Bifidobacterium hapali sp. nov.*, isolated from faeces of baby common marmosets (*Callithrix jacchus L.*). *Int. J. Syst. Evol. Microbiol.*, 66(1), 255–265.

Michelini, S., Modesto, M., Pisi, A.M., Filippini, G., Sandri, C., Spiezio, C., Biavati, B., Sgorbati, B., Mattarelli, P. (2016b). *Bifidobacterium eulemuris sp. nov.*, isolated from faeces of black lemurs (*Eulemur macaco*). *Int. J. Syst. Evol. Microbiol.*, 66(3), 1567–1576.

Michelini, S., Modesto, M., Filippini, G., Spiezio, C., Sandri, C., Biavati, B., Pisi, A., Mattarelli, P. (2016c). *Bifidobacterium aerophilum sp. nov., Bifidobacterium avesanii sp. nov.* and *Bifidobacterium ramosum sp. nov.*: Three novel taxa from the faeces of cotton-top tamarin (*Saguinus oedipus L.*). *Syst. Appl. Microbiol.*, 39(4), 229–236.

Milani, C., Duranti, S., Lugli, G.A., Bottacini, F., Strati, F., Arioli, S., Foroni, E., Turroni, F., van Sinderen, D., Ventura, M. (2013). Comparative genomics of *Bifidobacterium animalis* subsp. lactis reveals a strict monophyletic bifidobacterial taxon. *Appl. Environ. Microbiol.*, 79, 4304–4315.

Milani, C., Mangifesta, M., Mancabelli, L., Lugli, G.A., James, K., Duranti, S., Turroni, F., Ferrario, C., Ossiprandi, M.C., van Sinderen, D., et al. (2017). Unveiling bifidobacterial biogeography across the mammalian branch of the tree of life. *ISME J.*, 11(12), 2834–2847.

Milani, C., Duranti, S., Napoli, S., Alessandri, G., Mancabelli, L., Anzalone, R., Longhi, G., Viappiani, A., Mangifesta, M., Lugli, D.V. et al. (2019). Colonization of the human gut by bovine bacteria present in Parmesan cheese. *Nat. Commun.*, 10(1), 1286.

Miranda, R.F., de Paula, M.M., da Costa, G.M., Barão, C.E., da Silva, A.C.R., Raices, R.S.L., Gomes, R.G., Pimentel, T.C. (2019). Orange juice added with *L. casei*: Is there an impact of the probiotic addition methodology on the quality parameters? *LWT. Food Sci. Technol.*, 106, 186–193.

Mitsuoka, T. (1969). Comparative studies on bifidobacteria isolated from the alimentary tract of man and animals (including descriptions of *Bifidobacterium thermophilum nov. spec.* and *Bifidobacterium pseudolongum nov. spec.*). *Zentralbl. Bakteriol. Orig.*, 210(1), 52–64.

Miyoshi, Y., Okada, S., Uchimura, T., Satoh, E. (2006). A mucus adhesion promoting protein, MapA, mediates the adhesion of *Lactobacillus reuteri* to Caco-2 human intestinal epithelial cells. *Biosci. Biotechnol. Biochem.*, 70(7), 1622–1628.

Modesto, M., Michelini, S., Stefanini, I., Ferrara, A., Tacconi, S., Biavati, B., Mattarelli, P. (2014). *Bifidobacterium aesculapii sp. nov.*, from the faeces of the baby common marmoset (*Callithrix jacchus*). *Int. J. Syst. Evol. Microbiol.*, 64(8), 2819–2827.

Modesto, M., Michelini, S., Stefanini, I., Sandri, C., Spiezio, C., Pisi, A., Filippini, G., Biavati, B., Mattarelli, P. (2015). *Bifidobacterium lemurum sp. nov.*, from faeces of the ring-tailed lemur (*Lemur catta*). *Int. J. Syst. Evol. Microbiol.*, 65(6), 1726–1734.

Modesto, M., Michelini, S., Sansosti, M.C., De Filippo, C., Cavalieri, D., Qvirist, L., Andlid, T., Spiezio, C., Sandri, C., Pascarelli, S. et al. (2018a). *Bifidobacterium callitrichidarum sp. nov.* from the faeces of the emperor tamarin (*Saguinus imperator*). *Int. J. Syst. Evol. Microbiol.*, 68(1), 141–148.

Modesto, M., Michelini, S., Oki, K., Biavati, B., Watanabe, K., Mattarelli, P. (2018b). *Bifidobacterium catulorum sp. nov.*, a novel taxon from the faeces of the baby common marmoset (*Callithrix jacchus*). *Int. J. Syst. Evol. Microbiol.*, 68(2), 575–581.

Modesto, M., Puglisi, E., Bonetti, A., Michelini, S., Spiezio, C., Sandri, C., Sgorbati, B., Morelli, L., Mattarelli, P. (2018c). *Bifidobacterium primatium sp. nov.*, *Bifidobacterium scaligerum sp. nov.*, *Bifidobacterium felsineum sp. nov.* and *Bifidobacterium simiarum sp. nov.*: Four novel taxa isolated from the faeces of the cotton top tamarin (*Saguinus oedipus*) and the emperor tamarin (*Saguinus imperator*). *Syst. Appl. Microbiol.*, 41(6), 593–603.

Modesto, M., Watanabe, K., Arita, M., Satti, M., Oki, K., Sciavilla, P., Patavino, C., Cammà, C., Michelini, S., Sgorbati, B. et al. (2019a). *Bifidobacterium jac-chi sp. nov.*, isolated from the faeces of a baby common marmoset (*Callithrix jacchus*). *Int. J. Syst. Evol. Microbiol.*, 69(8), 2477–2485.

Modesto, M., Satti, M., Watanabe, K. Puglisi, E., Morelli, L., Huang, C.H., Liou, J.S., Miyashita, M., Tamura, T., Saito, S. et al. (2019b). Characterization of *Bifidobacterium* species in feaces of the egyptian fruit bat: Description of *B. vespertilionis sp. nov. and B. rousetti sp. nov. Syst. Appl. Microbiol.*, 42(6), 126017.

Modesto, M., Satti, M., Watanabe, K., Huang, C.H., Liou, J.H., Tamura, T., Saito, S., Mori, K., Huang, L., Sandri, C. et al. (2020a). Bifidobacteria in two-toed sloths (*Choloepus didactylus*): Phylogenetic characterization of the novel taxon *Bifidobacterium choloepi sp. nov. Int. J. Syst. Evol. Microbiol.*, 70, 6115–6125.

Modesto, M., Satti, M., Watanabe, K., Scarafile, D., Huang, C.H., Liou, J.S., Tamura, T., Saito, S., Watanabe, M., Mori, K. et al. (2020b). Phylogenetic characterization of two novel species of the genus *Bifidobacterium: Bifidobacterium saimiriisciurei sp. nov.* and *Bifidobacterium platyrrhinorum sp. nov. Syst. Appl. Microbiol.*, 43, 126111.

Mokoena, M.P., Mutanda, T., Olaniran, A.O. (2016). Perspectives on the probiotic potential of lactic acid bacteria from African traditional fermented foods and beverages. *Food Nutr. Res.*, 60, 29630.

Monteiro, S.S., Beserra, A.Y.S., Oliveira, H.M.L., Pasquali, M.A.B. (2020). Production of probiotic passion fruit (*Passiflora edulis Sims f. flavicarpa Deg.*) drink using *Lactobacillus reuteri* and microencapsulation via spray drying. *Foods*, 9(3), 335.

Moro, E. (1900). Ueber die nach Gram farbbaren Bacillen des Säuglingsstuhles. *Wien. Klin. Wochenschr.*, 13, 114–115.

Moroti, C., Magri, L.F.S., de Rezende Costa, M., Cavallini, D.C., Sivieri, K. (2012). Effect of the consumption of a new symbiotic shake on glycemia and cholesterol levels in elderly people with type 2 diabetes mellitus. *Lipids Health Dis.*, 11, 29.

Mullié, C., Yazourh, A., Singer, E., Lecroix, F., Blareau, J.P., Romond, M.B., Romond, C. (2002). Partial characterization of *Bifidobacterium breve* C50 cell-free whey compounds inducing modifications to the intestinal microflora. *J. Dairy Sci.*, 85(6), 1383–1389.

Mullié, C., Odou, M.-F., Singer, E., Romond, M.-B., Izard, D. (2003). Multiplex PCR using 16S rRNA gene-targeted primers for the identification of bifidobacteria from human origin. *FEMS Microbiol. Lett.*, 222(1), 129–136.

Mullié, C., Yazourh, A., Thibault, H., Odou, M.F., Singer, E., Kalach, N., Kremp, O., Romond, M.B. (2004). Increased poliovirus-specific intestinal antibody response coincides with promotion of *Bifidobacterium longum-infantis* and *Bifidobacterium breve* in infants: A randomized, double-blind, placebo-controlled trial. *Pediatr. Res.*, 56(5), 791–795.

Murdaca, G., Greco, M., Borro, M., Gangemi, S. (2021). Hygiene hypothesis and autoimmune diseases: A narrative review of clinical evidences and mechanisms. *Autoimmun. Rev.*, 20(7), 102845.

Nakajo, K., Takahashi, N., Beighton, D. (2010). Resistance to acidic environments of caries-associated bacteria: *Bifidobacterium dentium* and *Bifidobacterium longum*. *Caries Res.*, 44(5), 431–437.

Nakazawa, T., Kaneko, K., Takahashi, H., Inoue, S. (1996). Neonatal meningitis caused by *Bifidobacterium breve*. *Brain Dev.*, 18(2), 160–162.

Nam, Y.-D., Chang, H.-W., Kim, K.-H., Roh, S.W., Bae, J.-W. (2009). Metatranscriptome analysis of lactic acid bacteria during kimchi fermentation with genome-probing microarrays. *Int. J. Food Microbiol.*, 130(2), 140–146.

Neela, S. and Fanta, S.W. (2020). Injera (an ethnic, traditional staple food of Ethiopia): A review on traditional practice to scientific developments. *J. Ethn. Foods*, 7(32), 15.

Neish, A.S. (2017). Probiotics of the *Acidophilus* Group: *Lactobacillus acidophilus, delbrueckii subsp. Bulgaricus* and *Johnsonii*. In *The Microbiota in Gastrointestinal Pathophysiology*, Martin Floch, M., Ringel, Y., Walker, W.A. (eds). RELX Group/Elsevier, Amsterdam.

Neuzil-Bunesova, V., Lugli, G.A., Modrackova, N., Makovska, M., Mrazek, J., Mekadim, C., Musilova, S., Svobodova, I., Spanek, R., Ventura, M., et al. (2020). *Bifidobacterium canis sp. nov.*, a novel member of the *Bifidobacterium pseudolon-gum* phylogenetic group isolated from faeces of a dog (*Canis lupus f. familiaris*). *Int. J. Syst. Evol. Microbiol.*, 70, 5040–5047.

Neuzil-Bunesova, V., Lugli, G.A., Modrackova, N., Vlkova, E., Bolechova, P., Burtscher, J., Longhi, G., Mancabelli, L., Killer, J., Domig, K., et al. (2021). Five novel bifidobacterial species isolated from faeces of primates in two Czech zoos: *Bifidobacterium erythrocebi sp. nov., Bifidobacterium moraviense sp. nov., Bifidobacterium oedipodis sp. nov., Bifidobacterium olomucense sp. nov.* and *Bifidobacterium panos sp. nov. Int. J. Syst. Evol. Microbiol.*, 71, 4573.

Nguyen, B.T., Bujna, E., Fekete, N., Tran, A.T.M., Rezessy-Szabo, J.M., Prasad, R., Nguyen, Q.D. (2019). Probiotic beverage from pineapple juice fermented with *Lactobacillus* and *Bifidobacterium* strains. *Front. Nutr.*, 6, 54.

Nishino, K., Nishida, A., Inoue, R., Kawada, Y., Ohno, M., Sakai, S., Inatomi, O., Bamba, S., Sugimoto, M., Kawahara, M., et al. (2018). Analysis of endoscopic brush samples identified mucosa-associated dysbiosis in inflammatory bowel disease. *J. Gastroenterol.*, 53(1), 95–106.

Nishiyama, K., Sugiyama, M., Mukai, T. (2016). Adhesion properties of lactic acid bacteria on intestinal mucin. *Microorganisms*, 4(3), 34.

Nuraida, L., Grigolava, I., Owens, J.D., Campbell-Platt, G. (1992). Oxygen and pyruvate as external electron acceptors for *Leuconostoc spp. J. Appl. Microbiol.*, 72(6), 517–522.

O'Toole, P.W., Marchesi, J.R., Hill, C. (2017). Next-generation probiotics: The spectrum from probiotics to live biotherapeutics. *Nat. Microbiol.*, 2(5), 1–6.

Oda, H., Wakabayashi, H., Yamauchi, K., Abe, F. (2014). Lactoferrin and bifidobacteria. *Biometals*, 27(5), 915–922.

Ofek, I., Hasty, D.L., Doyle, R.J. (eds) (2003). Basic concepts in bacterial adhesion. In *Bacterial Adhesion to Animal Cells and Tissues*. ASM Press, Washington.

Ohishi, A., Takahashi, S., Ito, Y., Ohishi, Y., Tsukamoto, K., Nanba, Y., Ito, N., Kakiuchi, S., Saitoh, A., Morotomi, M., et al. (2010). *Bifidobacterium* septicemia associated with postoperative probiotic therapy in a neonate with omphalocele. *Pediatr.*, 56(4), 679–681.

Okamoto, M., Benno, Y., Leung, K.-P., Maeda, N. (2008). *Bifidobacterium tsurumiense sp. nov.*, from hamster dental plaque. *Int. J. Syst. Evol. Microbiol.*, 58(1), 144–148.

Oren, A. and Garrity, G.M. (2022). Valid publication of new names and new combinations effectively published outside the IJSEM. *Int. J. Syst. Evol. Microbiol.*, 72(3), 005260.

Orla-Jensen, S. (1924). Classification des bactéries lactiques. *Lait*, 4, 468–474.

von Ossowski, I., Reunanen, J., Satokari, R., Vesterlund, S., Kankainen, M., Huhtinen, H., Tynkkynen, S., Salminen, S., de Vos, W.M., Palva, A. (2010). Mucosal adhesion properties of the probiotic *Lactobacillus rhamnosus* GG SpaCBA and SpaFED pilin subunits. *Appl. Environ. Microbiol.*, 76(7), 2049–2057.

von Ossowski, I., Satokari, R., Reunanen, J., Lebeer, S., De Keersmaecker, S.C.J., Vanderleyden, J., de Vos, W.M., Palva, A. (2011). Functional characterization of a mucus-specific LPXTG surface adhesin from probiotic *Lactobacillus rhamnosus* GG. *Appl. Environ. Microbiol.*, 77(13), 4465–4472.

Østlie, H.M., Porcellato, D., Kvam, G., Wicklund, T. (2021). Investigation of the microbiota associated with ungerminated and germinated Norwegian barley cultivars with focus on lactic acid bacteria. *Int. J. Food Microbiol.*, 341, 109059.

Pandey, K.R., Naik, S.R., Vakil, B.V. (2015). Probiotics, prebiotics and synbiotics – A review. *J. Food Sci. Technol.*, 52, 7577–7587.

Patton, S., Gendler, S.J., Spicer, A.P. (1995). The epithelial mucin, MUC1, of milk, mammary gland and other tissues. *Biochim. Biophys. Acta.*, 1241, 407–423.

Pechar, R., Killer, J., Salmonová, H., Geigerová, M., Švejstil, R., Švec, P., Sedláček, I., Rada, V., Benada, O. (2017). *Bifidobacterium apri sp. nov.*, a thermophilic *Actinobacterium* isolated from the digestive tract of wild pigs (*Sus scrofa*). *Int. J. Syst. Evol. Microbiol.*, 67(7), 2349–2356.

Pei, C.X., Liu, Q., Dong, C.S., Li, H., Jiang, J.B., Gao, W.J. (2010). Diversity and abundance of the bacterial 16S rRNA gene sequences in forestomach of alpacas (*Lama pacos*) and sheep (*Ovis aries*). *Anaerobe*, 16(4), 426–432.

Pei, J., Yuan, Y., Yue, T. (2013). Characterization of bacteriocin bificin C6165: A novel bacteriocin. *J. Applied Microbiol.*, 114, 1273–1284.

Pei, J., Yue, T., Yuan, Y. (2014). Control of *Alicyclobacillus acidoterrestris* in fruit juices by a newly discovered bacteriocin. *World J. Microbiol. Biotechnol.*, 30(3), 855–863.

Peighambardoust, S.H., Golshan Tafti, A., Hesari, J. (2011). Application of spray drying for preservation of lactic acid starter cultures: A review. *Trends Food Sci. Technol.*, 22(5), 215–224.

Perea Vélez, M., Petrova, M.I., Lebeer, S., Verhoeven, T.L.A., Claes, I., Lambrichts, I., Tynkkynen, S., Vanderleyden, J., De Keersmaecker, S.C.J. (2010). Characterization of MabA, a modulator of *Lactobacillus rhamnosus* GG adhesion and biofilm formation. *FEMS Immunol. Med. Microbiol.*, 59, 386–398.

Pérez-Armendáriz, B. and Cardoso-Ugarte, G.A. (2020). Traditional fermented beverages in Mexico: Biotechnological, nutritional, and functional approaches. *Food Res. Int.*, 136, 109307.

Petay, V., Lecroix, F., Perrin, E., Gontier, C., Blareau, J.-P., Romond, M.-B., Singer, E., Odou, M.-F., Demailly-Mullie, C. (2008). Producto inmunomodulador obtenido a partir de un cultivo de *Bifidobacterium* y composiciones que lo contienen. Patent, ES2291896T3.

Picot, A. and Lacroix, C. (2004). Encapsulation of bifidobacteria in whey protein-based microcapsules and survival in simulated gastrointestinal conditions and in yoghurt. *Int. Dairy J.*, 14(6), 505–515.

Pokusaeva, K., Fitzgerald, G.F., van Sinderen, D. (2011). Carbohydrate metabolism in bifidobacteria. *Genes Nutr.*, 6(3), 285–306.

Polonovski, M. and Lespagnol, A. (1930). Un nouveau constituant du lait de femme : le gynolactose. *C. R. Soc. Biol.*, 104, 553–555.

Post, J. (2013). Acceptability and feasibility of probiotic and prebiotic supplementation in alleviating symptoms of lactose maldigestion in lactose intolerant subjects. Master's Thesis, Florida State University, Tallahassee.

Poxton, I.R., Brown, R., Sawyerr, A., Ferguson, A. (1997). Mucosa-associated bacterial flora of the human colon. *J. Med. Microbiol.*, 46(1), 85–91.

Praet, J., Meeus, I., Cnockaert, M., Aerts, M., Smagghe, G., Vandamme, P. (2015). *Bifidobacterium commune sp. nov.* isolated from the bumble bee gut. *Antonie van Leeuwenhoek*, 107(5), 1307–1313.

Pratt, C. and Campbell, M.D. (2020). The effect of *Bifidobacterium* on reducing symptomatic abdominal pain in patients with irritable bowel syndrome: A systematic review. *Probiotics Antimicrob. Proteins*, 12(3), 834–839.

Pretzer, G., Snel, J., Molenaar, D., Wiersma, A., Bron, P.A., Lambert, J., de Vos, W.M., van der Meer, R., Smits, M.A., Kleerebezem, M. (2005). Biodiversity-based identification and functional characterization of the mannose-specific adhesin of *Lactobacillus plantarum*. *J. Bacteriol.*, 187(17), 6128–6136.

Pruccoli, G., Erika Silvestro, E., Pace Napoleone, C., Aidala, E., Garazzino, S., Scolfaro, C. (2019). Are probiotics safe? *Bifidobacterium* bacteremia in a child with severe heart failure. *Infez. Med.*, 27(2), 175–178.

Pulverer, G., Schütt-Gerowitt, H., Schaal, K.P. (2003). Human cervicofacial actinomycoses: Microbiological data for 1997 cases. *Clin. Infect. Dis.*, 37(4), 490–497.

Pyclik, M., Srutkova, D., Schwarzer, M., Górska, S. (2020). Bifidobacteria cell wall-derived exo-polysaccharides, lipoteichoic acids, peptidoglycans, polar lipids and proteins – Their chemical structure and biological attributes. *Int. J. Biol. Macromol.*, 147, 333–349.

Ramasamy, K., Abdul Rahman, N.Z., Sieo, C.C., Alitheen, N.J., Abdullah, N., Ho, Y.W. (2012). Probiotic potential of lactic acid bacteria from fermented Malaysian food or milk products. *Int. J. Food Sci. Technol.*, 47, 2175–2183.

Rao, R.K. and Samak, G. (2013). Protection and restitution of gut barrier by probiotics: Nutritional and clinical implications. *Curr. Nutr. Food Sci.*, 9, 99–107.

Reid, A.A., Vuillemard, J.C., Britten, M., Arcand, Y., Farnworth, E., Champagne, C.P. (2005). Microentrapment of probiotic bacteria in a Ca2+-induced whey protein gel and effects on their viability in a dynamic gastro-intestinal model. *J. Microencapsul.*, 22(6), 603–619.

Reunanen, J., von Ossowski, I., Hendrickx, A.P., Palva, A., de Vos, W.M. (2012). Characterization of the SpaCBA pilus fibers in the probiotic *Lactobacillus rhamnosus* GG. *Appl. Environ. Microbiol.*, 78(7), 2337–2344.

Reuter, G. (1963). Comparative studies on the *Bifidus* flora in the feces of infants and adults. With a contribution to classification and nomenclature of *Bifidus* strains. *Zentralbl. Bakteriol. Orig.*, 191, 486–507.

Ricci, A., Allende, A., Bolton, D., Chemaly, M., Davies, R., Girones, R., Herman, L., Koutsoumanis, K., Lindqvist, R., Nørrung, B. et al. (2018). Update of the list of QPS-recommended biological agents intentionally added to food or feed as notified to EFSA 7: Suitability of taxonomic units notified to EFSA until September 2017. *EFSA Journal*, 16(1), 5131.

Ricci, A., Marrella, M., Hadj Saadoun, J., Bernini, V., Godani, F., Dameno, F., Neviani, E., Lazzi, C. (2020). Development of lactic acid-fermented tomato products. *Microorganisms*, 8, 1192.

Ringel-Kulka, T., Palsson, O.S., Maier, D., Carroll, I., Galanko, J.A., Leyer, G., Ringel, Y. (2011). Probiotic bacteria *Lactobacillus acidophilus* NCFM and *Bifidobacterium lactis* Bi-07 versus placebo for the symptoms of bloating in patients with functional bowel disorders: A double-blind study. *J. Clin. Gastroenterol.*, 45(6), 518–525.

Rodríguez, J.M. (2014). The origin of human milk bacteria: Is there a bacterial enteromammary pathway during late pregnancy and lactation? *Adv. Nutr.*, 5(6), 779–784.

Romond, M.-B., Hamze, M., Romond, C., Bourlioux, P. (1990a). Host-Bifidobacterium interactions in the axenic mouse: Partial characterization of bifidogenic factors in the intestinal contents. *Can. J. Microbiol.*, 36(4), 286–291.

Romond, M.-B., Chadli, N.-E., Mitsuoka, T., Romond, C. (1990b). Host-intestinal microflora interactions: Role of the mucus and gynolactose. *Bifidobacteria and Microflora*, 9(2), 119–130.

Romond, M.-B., Haddou, Z., Mielcareck, C., Romond, C. (1997a). Bifidobacteria and human health: Regulatory effect of indigenous bifidobacteria on *Escherichia coli* intestinal colonization. *Anaerobe*, 3(2/3), 131–136.

Romond, M.-B., Ais, A., Yazourh, A., Romond, C. (1997b). Cell-free wheys from bifidobacteria fermented milks exert a regulatory effect on the intestinal microflora of mice and humans. *Anaerobe*, 3(2/3), 137–143.

Romond, M.-B., Ais, A., Guillemot, F., Bounouader, R., Cortot, A., Romond, C. (1998). Cell-free whey from milk fermented with *Bifidobacterium breve* C50 used to modify the colonic microflora of healthy subjects. *J. Dairy Sci.*, 81(5), 1229–1235.

Romond, M.-B., Colavizza, M., Mullié, C., Kalach, N., Kremp, O., Mielcarek, C., Izard, D. (2008). Does the intestinal bifidobacterial colonisation affect bacterial translocation? *Anaerobe*, 14(1), 43–48.

Romond, M.-B., Colavizza, M., Romond, P.C., Bezirtzoglou, E. (2011). Selective decontamination by antimicrobials during long term treatment: Perspectives for saving host indigenous microbiota. *Anaerobe*, 17(6), 468–473.

Roos, S. and Jonsson, H. (2002). A high-molecular-mass cell-surface protein from *Lactobacillus reuteri* 1063 adheres to mucus components. *Microbiology (Reading, Engl.)*, 148, 433–442.

Rozenfeld, K.L., Nitzan, O., Peretz, A. (2018). Presence of anaerobic bacteria in the urinary tract of catheterized ICU patients. *Eur. J. Clin. Microbiol. Infect. Dis.*, 37(11), 2131–2136.

Rushdy, A.A. and Gomaa, E.Z. (2013). Antimicrobial compounds produced by probiotic *Lactobacillus brevis* isolated from dairy products. *Ann. Microbiol.*, 63, 81–90.

Saez-Lara, M.J., Gomez-Llorente, C., Plaza-Diaz, J., Gil, A. (2015). The role of probiotic lactic acid bacteria and bifidobacteria in the prevention and treatment of inflammatory bowel disease and other related diseases: A systematic review of randomized human clinical trials. *Biomed. Res. Int.*, e505878.

Safavi, M., Farajian, S., Kelishadi, R., Mirlohi, M., Hashemipour, M. (2013). The effects of synbiotic supplementation on some cardio-metabolic risk factors in overweight and obese children: A randomized triple-masked controlled trial. *Int. J. Food Sci. Nutr.*, 64, 687–693.

Sagdic, O., Ozturk, I., Yapar, N., Yetim, H. (2014). Diversity and probiotic potentials of lactic acid bacteria isolated from gilaburu, a traditional Turkish fermented European cranberrybush (*Viburnum opulus L.*) fruit drink. *Food Res. Int.*, 64, 537–545.

Sakanaka, M., Gotoh, A., Yoshida, K., Odamaki, T., Koguchi, H., Xiao, J.-Z., Kitaoka, M., Katayama, T. (2019). Varied pathways of infant gut-associated bifidobacterium to assimilate human milk oligosaccharides: Prevalence of the gene set and its correlation with bifidobacteria-rich microbiota formation. *Nutrients*, 12(1), 71.

Salem, F., Kindt, N., Marchesi, J.R., Netter, P., Lopez, A., Kokten, T., Danese, S., Jouzeau, J.-Y., Peyrin-Biroulet, L., Moulin, D. (2019). Gut microbiome in chronic rheumatic and inflammatory bowel diseases: Similarities and differences. *United European Gastroenterol. J.*, 7(8), 1008–1032.

Sanchez, M., Darimont, C., Drapeau, V., Emady-Azar, S., Lepage, M., Rezzonico, E., Ngom-Bru, C., Berger, B., Philippe, L., Ammon-Zuffrey, C. (2014). Effect of *Lactobacillus rhamnosus* CGMCC1. 3724 supplementation on weight loss and maintenance in obese men and women. *Br. J. Nutr.*, 111, 1507–1519.

Sane, F., Piva, F., Pierrat, V., Bray, F., Devassine, S., Rolando, C., Alidjinou, E.K., Romond, M.B. (2022). Detection of free bifidobacterial lipoproteins independently of *Bifidobacterium* in human milk. *Pediatrics*, 88, 149–159.

Sarao, L.K. and Arora, M. (2017). Probiotics, prebiotics, and microencapsulation: A review. *Crit. Rev. Food Sci. Nutr.*, 57(2), 344–371.

Sarkar, A. and Mandal, S. (2016). Bifidobacteria-insight into clinical outcomes and mechanisms of its probiotic action. *Microbiol. Res.*, 192, 159–171.

Satish Kumar, R., Kanmani, P., Yuvaraj, N., Paari, K.A., Pattukumar, V., Arul V. (2013). Traditional Indian fermented foods: A rich source of lactic acid bacteria. *Int. J. Food Sci. Nutr.*, 64(4), 415–428.

Saubade, F., Humblot, C., Hemery, Y.M., Guyot, J.-P. (2017). PCR screening of an African fermented pearl-millet porridge metagenome to investigate the nutritional potential of its microbiota. *Int. J. Food Microbiol.*, 244, 103–110.

Scardovi, V. (1986). The genus *Bifidobacterium* Orla-Jensen 1924, 472AL. In *Bergey's Manual of Systematic Bacteriology*, Sneath, P.H.A., Mair, N.S., Sharpe, M.E., Holt, J.G. (eds). Williams and Wilkins, Baltimore.

Scardovi, V. and Crociani, F. (1974). *Bifidobacterium catenulatum*, *Bifidobacterium dentium*, and *Bifidobacterium angulatum*: Three new species and their deoxyribonucleic acid homology relationships. *Int. J. Syst. Bacteriol.*, 24(1), 6–20.

Scardovi, V. and Trovatelli, L.D. (1969). New species of bifid bacteria from *Apis mellifica L.* and *Apis indica F.* A contribution to the taxonomy and biochemistry of the genus *Bifidobacterium*. *Zentralbl. Bakteriol. Parasitenkd. Infektionskr. Hyg.*, 123(1), 64–88.

Scardovi, V. and Trovatelli, L.D. (1974). *Bifidobacterium animalis* (Mitsuoka) *comb. nov.* and the "minimum" and "subtile" group of new bifidobacteria found in sewage. *Int. J. Syst. Bacteriol.*, 24(1), 21–28.

Scardovi, V. and Zani, G. (1974). *Bifidobacteriurn magnum sp. nov.*, a large acidophilic *Bifidobacterium* isolated from rabbit feces. *Int. J. Syst. Bacteriol.*, 24(1), 29–34.

Scardovi, V., Trovatelli, L.D., Crociani, F., Sgorbati, B. (1969). Bifid bacteria in bovine rumen. New species of the genus *Bifidobacterium*: *B. globosum n.sp.* and *B. ruminale n.sp. Arch. Microbiol.*, 68(3), 278–294.

Scardovi, V., Trovatelli, L.D., Biavati, B., Zani, G. (1979). *Bifidobacterium cuniculi*, *Bifidobacterium choerinum*, *Bifidobacterium boum*, and *Bifidobacterium pseudocatenulatum*: Four new species and their deoxyribonucleic acid homology relationships. *Int. J. Syst. Bacteriol.*, 29, 291–311.

Scavuzzi, B.M., Henrique, F.C., Miglioranza, L.H.S., Simão, A.N.C., Dichi, I. (2014). Impact of prebiotics, probiotics and synbiotics on components of the metabolic syndrome. *Ann. Nutr. Disord. Ther.*, 1, 1009.

Schell, D. and Beermann, C. (2014). Fluidized bed microencapsulation of *Lactobacillus reuteri* with sweet whey and shellac for improved acid resistance and in-vitro gastro-intestinal survival. *Food Res. Int.*, 62, 308–314.

Schuck, P., Jeantet, R., Dolivet, A. (2012). *Analytical Methods for Food and Dairy Powders*. Wiley, Hoboken.

Schwarz, A., Langmayr, J.J., Ortler, M., Fille, M. (1993). Actinomycosis infection of the cerebellum. *Wien Klin. Wochenschr.*, 105(12), 359–361.

Scuotto, A., Djorie, S., Colavizza, M., Romond, P.-C., Romond, M.-B. (2014). *Bifidobacterium breve* C50 secretes lipoprotein with CHAP domain recognized in aggregated form by TLR2. *Biochimie*, 107, 367–375.

Scuotto, A., Romond, P.-C., Djorie, S., Alric, M., Romond, M.-B. (2016). In silico mining and characterization of bifidobacterial lipoprotein with CHAP domain secreted in an aggregated form. *Int. J. Biol. Macromol.*, 82, 653–662.

Sebald, M., Gasser, F., Werner, H. (1965). GC percentage and classification. Application to the bifidobacteria group and some related genera. *Ann. Inst. Pasteur (Paris)*, 109, 251–269.

Segawa, T., Fukuchi, S., Bodington, D., Tsuchida, S., Mbehang Nguema, P.P., Mori, H., Ushida, K. (2019). Genomic analyses of *Bifidobacterium moukalabense* reveal adaptations to frugivore/folivore feeding behavior. *Microorganisms*, 7(4), 99.

Segers, M.E. and Lebeer, S. (2014). Towards a better understanding of *Lactobacillus rhamnosus* GG-host interactions. *Microb. Cell Factories*, 13(1), 1–16.

Sgorbati, B., Lenaz, G., Casalicchio, F. (1976). Purification and properties of two fructose-6-phosphate phosphoketolases in *Bifidobacterium*. *Antonie van Leeuwenhoek*, 42(1/2), 49–57.

Shakeri, H., Hadaegh, H., Abedi, F., Tajabadi-Ebrahimi, M., Mazroii, N., Ghandi, Y., Asemi, Z. (2014). Consumption of synbiotic bread decreases triacylglycerol and VLDL levels while increasing HDL levels in serum from patients with type-2 diabetes. *Lipids*, 49, 695–701.

Simpson, P.J., Ross, R.P., Fitzgerald, G.F., Stanton, C. (2004). *Bifidobacterium psychraerophilum sp. nov.* and *Aeriscardovia aeriphila gen. nov., sp. nov.*, isolated from a porcine caecum. *Int. J. Syst. Evol. Microbiol.*, 54(2), 401–406.

Skerman, V.B.D., McGowan, V., Sneath, P.H.A. (1980). Approved lists of bacterial names. *Int. J. Syst. Bacteriol.*, 30, 225–420.

Šmid, A., Strniša, L., Bajc, K., Vujić-Podlipec, D., Matijašić, B.B., Rogelj, I. (2016). Randomized clinical trial: The effect of fermented milk with the probiotic cultures *Lactobacillus acidophilus* La-5® and *Bifidobacterium* BB-12® and Beneo dietary fibres on health-related quality of life and the symptoms of irritable bowel syndrome in adults. *J. Funct. Foods*, 24, 549–557.

Splichalova, A., Pechar, R., Killer, J., Splichalova, Z., Bunesova, V.N., Vlkova, E., Salmonova, H.S., Splichal, I. (2020). Colonization of germ-free piglets with mucinolytic and non-mucinolytic *Bifidobacterium boum* strains isolated from the intestine of wild boar and their interference with *Salmonella Typhimurium*. *Microorganisms*, 8(12), 2002.

Stackebrandt, E. and Goebel, B.M. (1994). Taxonomic note: A place for DNA-DNA reassociation and 16S rRNA sequence analysis in the present species definition in bacteriology. *Int. J. Syst. Bacteriol.*, 44(4), 846–849.

Sultana, K., Godward, G., Reynolds, N., Arumugaswamy, R., Peiris, P., Kailasapathy, K. (2000). Encapsulation of probiotic bacteria with alginate–starch and evaluation of survival in simulated gastrointestinal conditions and in yoghurt. *Int. J. Food Microbiol.*, 62(1/2), 47–55.

Suwantarat, N., Romagnoli, M., Wakefield, T., Carroll, K.C. (2014). Ventriculoperitoneal shunt infection caused by *Bifidobacterium breve*. *Anaerobe*, 28, 1–3.

Tajadadi-Ebrahimi, M., Bahmani, F., Shakeri, H., Hadaegh, H., Hijijafari, M., Abedi, F., Asemi, Z. (2014). Effects of daily consumption of synbiotic bread on insulin metabolism and serum high-sensitivity C-reactive protein among diabetic patients: A double-blind, randomized, controlled clinical trial. *Ann. Nutr. Metab.*, 65, 34–41.

Takei, N., Kuda, T., Handa, N., Fujita, S., Takahashi, H., Kimura, B. (2022). Detection and isolation of typical gut indigenous bacteria in ICR mice fed wheat bran and wheat straw fibre. *Food Chem. (Oxf)*, 4, 100071.

Tamang, J.P., Watanabe, K., Holzapfel, W.H. (2016). Review: Diversity of microorganisms in global fermented foods and beverages. *Front. Microbiol.*, 7, 377.

Tanaka, K., Komiyama, A., Sonomoto, K., Ishizaki, A., Hall, S., Stanbury, P. (2002). Two different pathways for D-xylose metabolism and the effect of xylose concentration on the yield coefficient of L-lactate in mixed-acid fermentation by the lactic acid bacterium *Lactococcus lactis* IO-1. *Appl. Microbiol. Biotechnol.*, 60, 160–167.

Tavares, G.M., Croguennec, T., Carvalho, A.F., Bouhallab, S. (2014). Milk proteins as encapsulation devices and delivery vehicles: Applications and trends. *Trends Food Sci. Technol.*, 37(1), 5–20.

Tena, D., Losa, C., Medina, M.J., Sáez-Nieto, J.A. (2015). Peritonitis caused by *Bifidobacterium longum*: Case report and literature review. *Anaerobe*, 27, 27–30.

Tissier, H. (1900). Recherches sur la flore intestinale des nourrissons (état normal et pathologique). Faculty of Medicine Thesis, Université de Paris, Paris.

Touré, R., Kheadr, E., Lacroix, C., Moroni, O., Fliss, I. (2003). Production of antibacterial substances by bifidobacterial isolates from infant stool active against *Listeria monocytogenes*. *J. Applied Microbiol.*, 95, 1058–1069.

Tripathi, M.K. and Giri, S.K. (2014). Probiotic functional foods: Survival of probiotics during processing and storage. *J. Funct. Foods*, 9, 225–241.

Trovatelli, L.D., Crociani, F., Pedinotti, M., Scardovi, V. (1974). *Bifidobacterium pullorum sp. nov.*: A new species isolated from chicken feces and a related group of bifidobacteria isolated from rabbit feces. *Arch. Microbiol.*, 98(3), 187–198.

Truelstrup Hansen, L., Allan-Wojtas, P.M., Jin, Y.-L., Paulson, A.T. (2002). Survival of Ca-alginate microencapsulated *Bifidobacterium* spp. in milk and simulated gastrointestinal conditions. *Food Microbiol.*, 19(1), 35–45.

Tsuchida, S., Takahashi, S., Mbehang Nguema, P.P., Fujita, S., Kitahara, M., Yamagiwa, J., Ngomanda, A., Ohkuma, M., Ushida, K. (2014). *Bifidobacterium moukalabense sp. nov.*, isolated from the faeces of wild west lowland gorilla (*Gorilla gorilla gorilla*). *Int. J. Syst. Evol. Microbiol.*, 64(2), 449–455.

Tsukimi, T., Watabe, T., Tanaka, K., Sato, M.P., Suzuki, H., Tomita, M., Fukuda, S. (2020). Draft genome sequences of *Bifidobacterium animalis* consecutively isolated from healthy Japanese individuals. *J. Genomics*, 8, 37–42.

Tunick, M.H., Mackey, K.L., Shieh, J.J., Smith, P.W., Cooke, P., Malin, E.L. (1993). Rheology and microstructure of low-fat mozzarella cheese. *Int. Dairy J.*, 3, 649–662.

Tuomola, E.M., Ouwehand, A.C., Salminen, S.J. (2000). Chemical, physical and enzymatic pre-treatments of probiotic lactobacilli alter their adhesion to human intestinal mucus glycoproteins. *Int. J. Food Microbiol.*, 60(1), 75–81.

Turroni, F., Milani, C., van Sinderen, D., Ventura, M. (2011). Genetic strategies for mucin metabolism in *Bifidobacterium bifidum* PRL2010: An example of possible human-microbe co-evolution. *Gut Microbes*, 2(3), 183–189.

Tytgat, H.L.P., Douillard, F.P., Reunanen, J., Rasinkangas, P., Hendrickx, A.P.A., Laine, P.K., Paulin, L., Satokari, R., de Vos, W.M. (2016a). *Lactobacillus rhamnosus* GG outcompetes *Enterococcus faecium* via mucus-binding *Pili*: Evidence for a novel and heterospecific probiotic mechanism. *Appl. Environ. Microbiol.*, 82, 5756–5762.

Tytgat, H.L.P., van Teijlingen, N.H., Sullan, R.M.A., Douillard, F.P., Rasinkangas, P., Messing, M., Reunanen, J., Satokari, R., Vanderleyden, J., Dufrêne, Y.F. et al. (2016b). Probiotic gut microbiota isolate interacts with dendritic cells via glycosylated heterotrimeric pili. *PLoS One*, 11, e0151824.

Urashima, T., Katayama, T., Fukuda, K., Hirabayashi, J. (2020). *Human Milk Oligosaccharides and Innate Immunity. Chemistry, Molecular Sciences and Chemical Engineering*. RELX Group/Elsevier, Amsterdam.

Valdez, R.M., Dos Santos, V.R., Caiaffa, K.S., Danelon, M., Arthur, R.A., de Cassia Negrini, T., Delbem, A.C., Duque, C. (2016). Comparative in vitro investigation of the cariogenic potential of bifidobacteria. *Arch. Oral Biol.*, 71, 97–103.

Vandamme, T., Poncelet, D., Subra-Paternault, P. (eds) (2007). *Microencapsulation : des sciences aux technologies*. Lavoisier, Paris.

Ventura, M., Elli, M., Reniero, R., Zink, R. (2001). Molecular microbial analysis of *Bifidobacterium* isolates from different environments by the species-specific amplified ribosomal DNA restriction analysis (ARDRA). *FEMS Microbiol. Ecol.*, 36 (2/3), 113–121.

Vesterlund, S., Salminen, K., Salminen, S. (2012). Water activity in dry foods containing live probiotic bacteria should be carefully considered: A case study with *Lactobacillus rhamnosus* GG in flaxseed. *Int. J. Food Microbiol.*, 157, 319–321.

de Vrese, M. and Schrezenmeir, J. (2008). Probiotics, prebiotics, and synbiotics. In *Food Biotechnology. Advances in Biochemical Engineering Biotechnology*, Stahl, U., Donalies, U.E., Nevoigt, E. (eds). Springer, Berlin/Heidelberg.

de Vries, W. and Stouthamer, A.H. (1967). Pathway of glucose fermentation in relation to the taxonomy of bifidobacteria. *J. Bacteriol.*, 93(2), 574–576.

Wan, L.Y.M., Chen, Z.J., Shah, N.P., El-Nezami, H. (2016). Modulation of intestinal epithelial defense responses by probiotic bacteria. *Crit. Rev. Food Sci. Nutr.*, 56, 2628–2641.

Wasko, N.J., Nichols, F., Clark, R.B. (2020). Multiple sclerosis, the microbiome, TLR2, and the hygiene hypothesis. *Autoimmun. Rev.*, 19(1), 102430.

Watanabe, K., Makino, H., Sasamoto, M., Kudo, Y., Fujimoto, J., Demberel, S. (2009). *Bifidobacterium mongoliense sp. nov.*, from airag, a traditional fermented mare's milk product from Mongolia. *Int. J. Syst. Evol. Microbiol.*, 59, 1535–1540.

Watanabe, M., Kinoshita, H., Huang, I.-N., Eguchi, K., Tsurumi, T., Kawai, Y., Kitazawa, H., Kimura, K., Taketomo, N., Kikuchi, D., et al. (2012). An adhesin-like protein, Lam29, from *Lactobacillus mucosae* ME-340 binds to histone H3 and blood group antigens in human colonic mucus. *Biosci. Biotechnol. Biochem.*, 76(9), 1655–1660.

Wei, Y., Gao, J., Liu, D., Li, Y., Liu, W. (2019). Adaptational changes in physiological and transcriptional responses of *Bifidobacterium longum* involved in acid stress resistance after successive batch cultures. *Microb. Cell. Fact.*, 18(1), 156.

Wickramasinghe, H.K.J.P., Anast, J.M., Schmitz-Esser, S., Serão, N.V.L., Appuhamy, J.A.D.R.N. (2020). Beginning to offer drinking water at birth increases the species richness and the abundance of *Faecalibacterium* and *Bifidobacterium* in the gut of preweaned dairy calves. *J. Dairy Sci.*, 103(5), 4262–4274.

Wilfrid Padonou, S., Nielsen, D.S., Hounhouigan, J.D., Thorsen, L., Nago, M.C., Jakobsen, M. (2009). The microbiota of Lafun, an African traditional cassava food product. *Int. J. Food Microbiol.*, 133(1/2), 22–30.

Won, S.-M., Chen, S., Park, K.W., Yoon, J.-H. (2020). Isolation of lactic acid bacteria from kimchi and screening of *Lactobacillus sakei* ADM14 with anti-adipogenic effect and potential probiotic properties. *LWT*, 126, 109296.

Wong, C.B., Odamaki, T., Xiao, J.-Z. (2019). Beneficial effects of *Bifidobacterium longum subsp. longum* BB536 on human health: Modulation of gut microbiome as the principal action. *J. Funct. Foods*, 54, 506–519.

von Wright, A. and Axelsson, L. (2011). Lactic acid bacteria: An introduction. In *Lactic Acid Bacteria: Microbiological and Functional Aspects*, Vinderola, G., Ouwehand, A., Salminen, S., von Wright, A. (eds). Informa, Zoug, Taylor and Francis, Abingdon-on-Thames.

Wu, S., Tao, N., German, J.B., Grimm, R., Lebrilla, C.B. (2010). Development of an annotated library of neutral human milk oligosaccharides. *J. Proteome Res.*, 9, 4138–4151.

Wu, Y., Li, S., Tao, Y., Li, D., Han, Y., Show, P.L., Wen, G., Zhou, J. (2021). Fermentation of blueberry and blackberry juices using *Lactobacillus plantarum*, *Streptococcus thermophilus* and *Bifidobacterium bifidum*: Growth of probiotics, metabolism of phenolics, antioxidant capacity in vitro and sensory evaluation. *Food Chem.*, 348, 129083.

Xiao, Y., Zhao, J., Zhang, H., Zhai, Q., Chen, W. (2021). Colonized niche, evolution and function signatures of *Bifidobacterium pseudolongum* within bifidobacterial genus. *Foods*, 10(10), 2284.

Yamazaki, S., Machii, K., Tsuyuki, S., Momose, H., Kawashima, T., Ueda, K. (1985). Immunological responses to monoassociated *Bifidobacterium longum* and their relation to prevention of bacterial invasion. *Immunology*, 56(1), 43–50.

Yamazaki, S., Tsuyuki, S., Akashiba, H., Kamimura, H., Kimura, M., Kawashima, T., Ueda, K. (1991). Immune response of bifidobacterium-monoassociated mice. *Bifidobacteria and Microflora*, 10(1), 19–31.

Yang, B., Chen, Y., Stanton, C., Ross, R.P., Lee, Y.-K., Zhao, J., Zhang, H., Chen, W. (2019). *Bifidobacterium* and *Lactobacillus* composition at species level and gut microbiota diversity in infants before 6 weeks. *Int. J. Mol. Sci.*, 20(13), 3306.

Yeung, T.W., Üçok, E.F., Tiani, K.A., McClements, D.J., Sela, D.A. (2016). Microencapsulation in alginate and chitosan microgels to enhance viability of *Bifidobacterium longum* for oral delivery. *Front. Microbiol.*, 7, 494.

Yildirim, Z., Winters, D.K., Johnson, M.G. (1999). Purification, amino acid sequence and mode of action of bifidocin B produced by *Bifidobacterium bifidum* NCFB 1454. *J. Appl. Microbiol.*, 86, 45–54.

Yoshimura, K., Matsui, T., Itoh, K. (2010). Prevention of *Escherichia coli* O157:H7 infection in gnotobiotic mice associated with bifidobacterium strains. *Antonie Van Leeuwenhoek*, 97(2), 107–117.

Yu, L., Liu, X., O'Sullivan, D.J. (2018). Use of *Lactococcus lactis* as a production system for peptides and enzymes encoded by a Lantibiotic gene cluster from *Bifidobacterium longum*. *Microbiology*, 164(12), 1481–1490.

Zbinden, A., Zbinden, R., Berger, C., Arlettaz, R. (2015). Case series of *Bifidobacterium longum* bacteremia in three preterm infants on probiotic therapy. *Neonatology*, 107(1), 56–59.

Zheng, J., Ruan, L., Sun, M., Gänzle, M. (2015). A genomic view of lactobacilli and pediococci demonstrates that phylogeny matches ecology and physiology. *Appl. Environ. Microbiol.*, 81,7233–7243.

Zheng, J., Wittouck, S., Salvetti, E., Franz, C.M., Harris, H.M., Mattarelli, P., O'Toole, P.W., Pot, B., Vandamme, P., Walter, J. (2020). A taxonomic note on the genus *Lactobacillus*: Description of 23 novel genera, emended description of the genus *Lactobacillus* Beijerinck 1901, and union of *Lactobacillaceae* and *Leuconostocaceae*. *Int. J. Syst. Evol. Microbiol.*, 70(4), 2782–2858.

Zhu, L., Li, W., Dong, X. (2003). Species identification of genus *Bifidobacterium* based on partial HSP60 gene sequences and proposal of *Bifidobacterium thermacidophilum Subsp. porcinum Subsp. nov. Int. J. Syst. Evol. Microbiol.*, 53(5), 1619–1623.

Zotta, T., Ricciardi, A., Ianniello, R.G., Storti, L.V., Glibota, N., Parente, E. (2018). Aerobic and respirative growth of heterofermentative lactic acid bacteria: A screening study. *Food Microbiol.*, 76, 117–127.

Zuidam, N.J. and Shimoni, E. (2010). Overview of microencapsulates for use in food products or processes and methods to make them. In *Encapsulation Technologies for Active Food Ingredients and Food Processing*, Zuidam, N.J. and Nedovic, V. (eds). Springer, New York.

Conclusion

Mohamed GHOUL
LRGP, CNRS, Université de Lorraine, Nancy, France

In order to meet the growing consumer demand for healthy, nutritious food products produced by processes with reduced environmental impact, manufacturers in this sector have developed new channels exploiting the specificities of enzymatic biocatalysts and the diversity of microbial systems. Indeed, the use of enzymes and bacteria in raw materials transformation processes makes it possible to obtain products with the desired functionalities under mild temperature and pressure conditions, thus reducing the environmental impact of the waste generated. However, performance of these processes remains a major obstacle to their widespread use. To ensure the competitiveness of these new processes, research and development work has been undertaken in recent years to improve the activity, selectivity and implementation techniques of these biocatalysts.

This work takes advantage of recent advances in molecular biology and process engineering in order to develop enzymes, select microorganisms that are more efficient in terms of functionality and productivity, and to develop intensified processes with higher yields. Major successes have been achieved in several areas, but many obstacles remain to be overcome before processes based on chemical catalysts can be fully replaced. This objective will be the main challenge for research and development in the bioprocessing field over the next few years.

This book presents the progress already made in these fields, as well as the paths for improvement, and illustrates the use of bacteria and enzymes in food processing through the presentation of several industrial applications.

Implementation of Enzymatic Processes and Lactic Bacteria in the Food Industries,
coordinated by Mohamed GHOUL. © ISTE Ltd 2024.

Part 1 is devoted to the presentation of different classes of enzymes and their industrial use in various fields, as well as ways of improving enzymes and processes (new approaches to enzyme design, process intensification, etc.).

Part 2 describes the progress made in the identification and selection of lactic acid bacteria and bifidobacteria, as well as the development of encapsulation processes specific to these bacteria to ensure their viability and industrial exploitation. An illustration of the use of this microbial system for the production of health foods has also been presented and discussed.

List of Authors

Jennifer BURGAIN
LIBio
Université de Lorraine
Nancy
France

Latifa CHEBIL
LRGP, CNRS
Université de Lorraine
Nancy
France

Isabelle CHEVALOT
LRGP, CNRS
Université de Lorraine
Nancy
France

Claire GAIANI
LIBio
Université de Lorraine
Nancy
France

Mohamed GHOUL
LRGP, CNRS
Université de Lorraine
Nancy
France

Catherine HUMEAU
LRGP, CNRS
Université de Lorraine
Nancy
France

Seraphim PAPANIKOLAOU
LFMB
Agricultural University of Athens
Greece

Frank PIVA
Laboratoire de Virologie
Université de Lille
France

Marie-Bénédicte ROMOND
Laboratoire de Virologie
Université de Lille
France

Joël SCHER
LIBio
Université de Lorraine
Nancy
France

Index

A, B, C

D, E, F

G, I, J

K, L, M

O, P, R

S, T

U, W, Y

Printed and bound by CPI Group (UK) Ltd, Croydon, CR0 4YY
13/09/2026
14961068-0003